Collins

CHEMISTRY

AQA A-level
Year 1 and AS
Student Book

Lyn Nicholls
Ken Gadd

CONTENTS

TO THE STUDENT

The aim of this book is to help make your study of advanced chemistry interesting and successful. It includes examples of modern issues, developments and applications that reflect the continual evolution of scientific knowledge and understanding. We hope it will encourage you to study science further when you complete your course.

USING THIS BOOK

Chemistry is fascinating, but complex — underpinned by some demanding ideas and concepts, and by a great deal of experimental data ('facts'). This mass of information can sometimes make its study daunting. So don't try to achieve too much in one reading session and always try to keep the bigger picture in sight.

There are a number of features in the book to help with this:

- Each chapter starts with a brief example of how the chemistry you will learn has been applied somewhere in the world, followed by a short outline of what you should have learned previously and what you will learn through the chapter.

- Important words and phrases are given in bold when used for the first time, with their meaning explained. There is also a glossary at the back of the book. If you are still uncertain, ask your teacher or tutor because it is important that you understand these words before proceeding.

- Throughout each chapter there are many questions, with the answers at the back of the book. These questions enable you to make a quick check on your progress through the chapter.

- Similarly, throughout each chapter there are checklists of key ideas that summarise the main points you need to learn from what you have just read.

- Where appropriate, worked examples are included to show how important calculations are done.

- There are many assignments throughout the book. These are tasks relating to pieces of text and data that show how ideas have been developed or applied. They provide opportunities to apply the science you have learned to new contexts, practise your maths skills and practise answering questions about scientific methods and data analysis.

- Some chapters have information about the 'required practical' activities that you need to carry out during your course. These sections provide the necessary background information about the apparatus, equipment and techniques that you need to be prepared to carry out the required practical work. There are questions that give you practice in answering questions about equipment, techniques, attaining accuracy, and data analysis.

- At the end of each chapter are practice questions. These are examination-style questions which cover all aspects of the chapter.

This book covers the requirements of AS Chemistry and the first year of A-level Chemistry. There are a number of sections, questions, assignments and practice questions that have been labelled 'Stretch and challenge', which you should try to tackle if you are studying for A-level. In places these go beyond what is required for the specification but they will help you build upon the skills and knowledge you acquire and better prepare you for further study beyond advanced level.

Good luck and enjoy your studies. We hope this book will encourage you to study chemistry further when you complete your course.

PRACTICAL WORK IN CHEMISTRY

While they may not all wear white coats or work in a laboratory, chemists and others who use chemistry in their work carry out experiments and investigations to gather evidence. They may be challenging established chemical ideas and models or using their skills, knowledge and understanding to tackle important problems.

Chemistry is a practical subject. Whether in the laboratory or in the field, chemists use their practical skills to find solutions to problems, challenges and questions. Throughout this course you will learn, develop and use these skills.

WRITTEN EXAMINATIONS

Your practical skills will be assessed in the written examinations at the end of the course. Questions on practical skills will account for about 15% of your

marks at AS and 15% at A-level. The practical skills assessed in the written examinations are:

Independent thinking

> solve problems set in practical contexts

> apply scientific knowledge to practical contexts

Figure 1 *Most chemists and others who use chemistry in their work spend time in laboratories. Many also use their practical skills outside of a laboratory.*

Use and application of scientific methods and practices

> comment on experimental design and evaluate scientific methods

> present data in appropriate ways

> evaluate results and draw conclusions with reference to measurement uncertainties and errors

> identify variables including those that must be controlled

Numeracy and the application of mathematical concepts in a practical context

> plot and interpret graphs

> process and analyse data using appropriate mathematical skills

> consider margins of error, accuracy and precision of data

Figure 2 *Chemists record experimental data in laboratory notebooks. They also record, process and present data using computers and tablets.*

Instruments and equipment

> know and understand how to use a wide range of experimental and practical instruments, equipment and techniques appropriate to the knowledge and understanding included in the specification

Figure 3 *You will need to use a variety of equipment correctly and safely.*

Throughout this book there are questions and longer assignments that will give you the opportunity to develop and practise these skills. The contexts of som of the exam questions will be based on the 'required practical activities'.

ASSESSMENT OF PRACTICAL SKILLS

Some practical skills can only be practised when you are doing experiments. For A-level, these **practical competencies** will be assessed by your teacher:

> follow written procedures

> apply investigative approaches and methods when using instruments and equipment

> safely use a range of practical equipment and materials

> make and record observations and measurements

> research, reference and report findings

You must show your teacher that you consistently and routinely demonstrate the competencies listed above during your course. The assessment will not contribut to your A-level grade, but will appear as a 'pass' alongside your grade on the A-level certificate.

These practical competencies must be demonstrated by using a specific range of **apparatus and techniques**. These are:

> use appropriate apparatus to record a range of measurements (to include mass, time, volume of liquids and gases, temperature)

> use a water bath or electric heater or sand bath fo heating

> measure pH using pH charts, or pH meter, or pH probe on a data logger

> use laboratory apparatus for a variety of experimental techniques including:

- titration, using burette and pipette

- distillation and heating under reflux, including setting up glassware using retort stand and clamp

- qualitative tests for ions and organic functional groups

- filtration, including use of fluted filter paper, or filtration under reduced pressure

> use a volumetric flask, including accurate techniqu for making up a standard solution

> use acid–base indicators in titrations of weak/strong acids with weak/strong alkalis

> purify:
  - a solid product by recrystallisation
  - a liquid product, including use of a separating funnel
> use melting point apparatus
> use thin-layer or paper chromatography
> set up electrochemical cells and measuring voltages
> safely and carefully handle solids and liquids, including corrosive, irritant, flammable and toxic substances
> measure rates of reaction by at least two different methods, for example:
  - an initial rate method such as a clock reaction
  - a continuous monitoring method

Figure 4 Many chemists analyse material. They are called analytical chemists. Titration is a commonly used technique.

Figure 5 pH probe

For AS, the above will not be assessed but you will be expected to use these skills and these types of apparatus to develop your manipulative skills and your understanding of the processes of scientific investigation.

REQUIRED PRACTICAL ACTIVITIES

During the A-level course you will need to carry out twelve **required practical** activities. These are the main sources of evidence that your teacher will use to award you a pass for your competency skills. If you are doing the AS, you will need to carry out the first six in this list.

1. Make up a volumetric solution and carry out a simple acid–base titration
2. Measurement of an enthalpy change
3. Investigation of how the rate of a reaction changes with temperature
4. Carry out simple test-tube reactions to identify:
   - cations – Group 2, NH_4^+
   - anions – Group 7 (halide ions), OH^-, CO_3^{2-}, SO_4^{2-}
5. Distillation of a product from a reaction
6. Tests for alcohol, aldehyde, alkene and carboxylic acid
7. Measuring the rate of reaction:
   - by an initial rate method
   - by a continuous monitoring method
8. Measuring the EMF of an electrochemical cell
9. Investigate how pH changes when a weak acid reacts with a strong base and when a strong acid reacts with a weak base
10. Preparation of:
    - a pure organic solid and test of its purity
    - a pure organic liquid
11. Carry out simple test-tube reactions to identify transition metal ions in aqueous solution
12. Separation of species by thin-layer chromatography

Information about the apparatus, techniques and analysis of required practicals 1 to 6 are found in the relevant chapters of this book, and 7 to 12 in Book 2.

You will be asked some questions in your written examinations about these required practicals.

Practical skills are really important. Take time and care to learn, practise and use them.

1 ATOMIC STRUCTURE

PRIOR KNOWLEDGE

You may know that substances are made from atoms and that an element is a substance made from just one sort of atom. You will probably have learnt that an atom consists of a nucleus, made up of protons and neutrons, with electrons moving around it in shells (or energy levels). You may also know about relative electrical charges and masses of protons, neutrons and electrons.

LEARNING OBJECTIVES

In this chapter, you will reinforce and build on these ideas and learn about more sophisticated models of atoms.

(Specification 3.1.1.1, 3.1.1.2, 3.1.1.3)

NASA's Curiosity Rover landed in the Gale Crater on Mars in August 2012. Its main mission was to investigate whether Mars has ever possessed the environmental conditions that could support life, as well as finding out about Martian climate and geology. Curiosity Rover contains an on-board science laboratory, equipped with a sophisticated range of scientific instruments. Many of these instruments have been specially designed for the mission.

The task of the on-board mass spectrometer is to investigate the atoms that are the building blocks of life – carbon, hydrogen, oxygen, phosphorus and sulfur. The spectrometer is making precise measurements of the carbon and oxygen isotopes found in carbon dioxide and methane from the atmosphere and the soil. After one Martian year (687 Earth days) of the mission, scientists have concluded that Mars once exhibited environmental conditions that were favourable for microbial life.

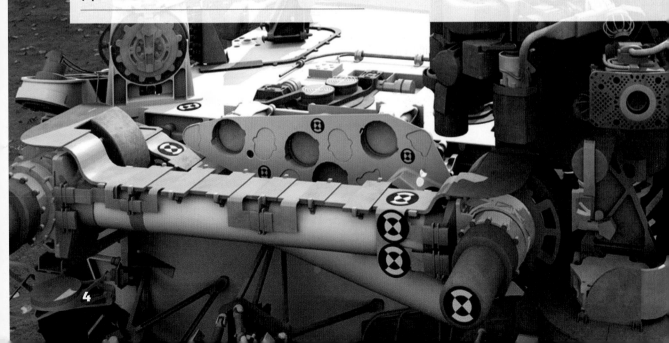

1.1 EARLY IDEAS ABOUT THE COMPOSITION OF MATTER

The nature of matter has interested people since the time of the early Greeks. The ideas that you have learnt about atomic structure have resulted from the work of many people over many centuries. You do not need to remember all of this information but here are some of the major events since 460BCE that led to our understanding of the atom.

Evidence for atomic structure

460–370BCE, Democritus

The Greek philosopher Democritus proposed that matter was made up of particles that cannot be divided further. They became known as atoms from the Greek word *atomos*, meaning 'cannot be divided'.

His ideas were based on reasoning – you cannot keep dividing a lump of matter for ever.

384–322BCE, Aristotle

Aristotle was another ancient Greek philosopher, who proposed that all earthly matter was made from four elements: earth, air, fire and water. These elements have their natural place on Earth and when they are out of place, they move. So, rain falls and bubbles of air rise from water.

A tree grows in the earth, and it needs water and air. So, a tree is made from earth, water and air. Aristotle could analyse most matter in this way.

1627–1691, Robert Boyle

Robert Boyle was a Fellow of the Royal Society of London. His scientific ideas included the notion that matter is made up of tiny identical particles that cannot be subdivided. These tiny particles made up 'mixt bodies' (we now call them compounds). Putting the particles together in different ways made different

compounds. Particles were in fixed positions in solids, but free to move in liquids and gases. Forces between particles made materials solid.

Boyle studied the nature and behaviour of gases, especially the relationship between volume and pressure. His theory of matter supported his experimental observations. He was the first scientist to keep accurate records.

1766–1844, John Dalton

John Dalton was an English chemist and physicist, who named the tiny particles **atoms**. His scientific idea was that atoms are indivisible and indestructible. All atoms of an element are identical and have the same mass and chemical properties. Atoms of different elements have different masses (he called them atomic weights) and different chemical properties. Atoms react together to form 'compound atoms'. These later became known as molecules.

Dalton studied the physical properties of air and gases. This led him to analytical work on ethene (olefiant gas), methane (carburetted hydrogen) and other gases. His atomic theory explained his chemical analyses. He summed up 150 years of ideas with his atomic theories.

1850–1930, Eugen Goldstein

The German physicist Eugen Goldstein's scientific idea was that cathode rays contained negatively charged particles with mass. He assumed that these particles were produced when the gas particles in the cathode ray tube were split. Cathode rays could be deflected by a magnetic field. Goldstein also detected heavier positive particles.

He experimented with electrical discharge tubes – he passed an electric current between a cathode and an anode in a sealed tube containing gas at a very low pressure. He adapted his experiment, inserting a perforated cathode, as in Figure 1.

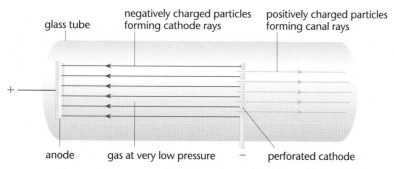

Figure 1 *An electrical discharge tube with a perforated cathode, as used by Goldstein*

1856–1937, Joseph John Thomson

Thomson's idea was that atoms contained **electrons**. He proposed that atoms could be divided into smaller particles. Electrons have very small mass, about one two-thousandths of the mass of a hydrogen atom. They are negatively charged. The negative charge is cancelled out by a sphere of positively charged material, as in Figure 2.

Thomson measured the deflection of the negative particles in cathode rays very accurately and

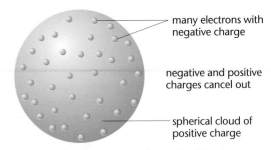

Figure 2 *Thomson's plum pudding model of the atom*

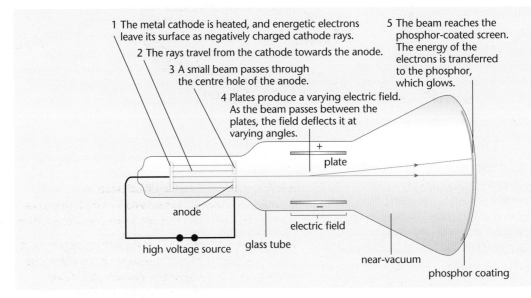

1 The metal cathode is heated, and energetic electrons leave its surface as negatively charged cathode rays.

2 The rays travel from the cathode towards the anode.

3 A small beam passes through the centre hole of the anode.

4 Plates produce a varying electric field. As the beam passes between the plates, the field deflects it at varying angles.

5 The beam reaches the phosphor-coated screen. The energy of the electrons is transferred to the phosphor, which glows.

plate

anode

electric field

high voltage source   glass tube

near-vacuum

phosphor coating

Figure 3 *Cathode ray tubes were used in televisions and computers before flat screens.*

calculated their mass. The cathode ray tubes he used were the forerunners of the cathode ray tubes used in televisions and monitors (Figure 3) before the development of flat screens.

Thomson's model of the atom became known as the 'plum pudding' model.

1871–1937, Ernest Rutherford

From work carried out in Manchester with his research students Hans Geiger and Ernst Marsden, Ernest Rutherford put forward the idea that the mass of the atom is not evenly spread. It is concentrated in a minute central region called the **nucleus**. Rutherford calculated the diameter of the nucleus to be 10^{-14} m.

All the positive charge of the atom is contained in the nucleus.

The electrons circulate in the rest of the atom, being kept apart by the repulsion of their negative charges.

These findings came from his interpretation of the results that are shown in Figure 4 (obtained from the experiment described in Figure 5).

Alpha particles are deflected when they pass close to the nucleus, while the very few that actually hit the nucleus are reflected

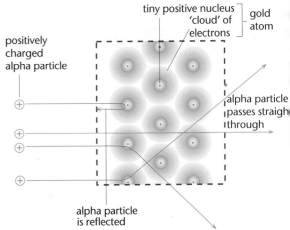

tiny positive nucleus
'cloud' of electrons
gold atom

positively charged alpha particle

alpha particle passes straight through

alpha particle is reflected

Figure 4 *Deflection of alpha particles by gold foil*

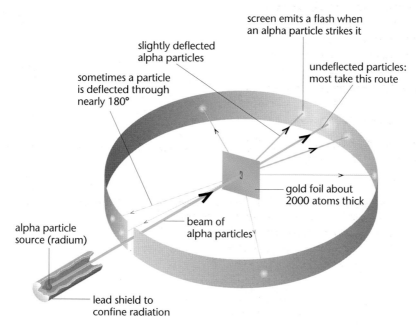

Figure 5 *Rutherford's experiment: the deflection of alpha particles through gold foil*

1888–1915, Henry Moseley and Ernest Rutherford

Rutherford continued the work that he had started, together with Moseley. Their idea was that the nucleus contained positively charged particles called **protons**. The number of protons (the atomic number) corresponds to the element's position in the Periodic Table. Protons make up about half the mass of the nucleus.

Moseley studied X-ray spectra of elements. Mathematically, he related the frequency of the X-rays to a number he called the **atomic number**. This corresponded to the element's position in the Periodic Table. Sadly, Moseley was killed in action at Gallipoli in World War 1. In 1919, Rutherford fired alpha particles at hydrogen gas and produced positive particles, which he called protons. His calculations also showed that the mass of the protons only accounted for half of the mass of the nucleus.

1891–1974, James Chadwick

Chadwick identified the **neutron** in 1932. Neutrons have no charge. They have the same mass as a proton.

He bombarded a beryllium plate with alpha particles and produced uncharged radiation on the other side of the plate. He placed a paraffin wax disc (which contains many hydrogen atoms) in the path of the radiation and showed that the radiation caused protons to be knocked out of the wax (Figure 6).

1885–1962, Niels Bohr

Bohr's scientific idea was that electrons orbit the nucleus in energy levels. Energy levels have fixed energy values – they are **quantised**. Electrons can only occupy these set energy levels.

Bohr studied emission spectra and produced explanations that incorporated the ideas of Einstein and Planck. Electrons orbited the nucleus in energy levels, where each energy level has a fixed energy value.

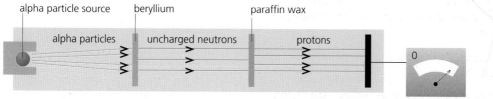

Figure 6 *Chadwick's experiment*

1.2 RELATIVE MASS AND RELATIVE CHARGE OF SUBATOMIC PARTICLES

Further experiments established the masses and charges of protons, electrons and neutrons. These are summarised in Table 1.

Because the values for mass are so small, the idea of **relative mass** is used. The relative mass of a proton is 1 and that of a neutron is 1. The relative mass of the electron is 5.45×10^{-4} or $\frac{1}{1837}$.

Charges on subatomic particles are also given relative to one another. A proton has a relative charge of $+1$ and an electron has a relative charge of -1. A neutron has no charge. The protons and neutrons together are called **nucleons**. Protons in the nucleus do not repel each other because a strong nuclear force acts over the small size of the nucleus and binds all the nucleons together.

Since atoms of any element are neutral, the number of protons (positive charge) must equal the number of electrons (negative charge). The atoms of all elements, except hydrogen, contain these three fundamental particles.

Particle	Mass/kg	Charge/C	Relative mass	Relative charge
Electron	9.109×10^{-31}	1.602×10^{-19}	5.45×10^{-4}	-1
Proton	1.672×10^{-27}	1.602×10^{-19}	1	$+1$
Neutron	1.674×10^{-27}	0	1	0

Note: The mass of the electron is so small compared to the mass of the proton and neutron that chemists often take it to be zero.

Table 1 *The fundamental atomic particles, their mass and charge*

KEY IDEAS

> All matter is composed of atoms.

> The nucleus of an atom contains positive protons, with a relative mass of 1 and relative charge of $+1$, and neutral neutrons (except hydrogen), with a relative mass of 1 and no charge.

> Electrons orbit the nucleus in energy levels (shells). An electron has a very small mass and relative charge of -1.

> The number of electrons in an atom equals the number of protons, to give an uncharged atom.

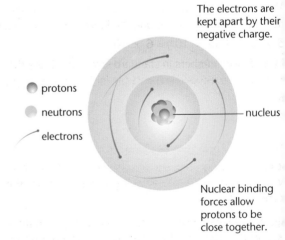

The electrons are kept apart by their negative charge.

protons

neutrons

electrons

nucleus

Nuclear binding forces allow protons to be close together.

Figure 7 *This diagram summarises the model of the atom that scientists often use nowadays.*

1.3 WORKING WITH VERY SMALL AND VERY LARGE NUMBERS

Working with very small numbers can be confusing. To help avoid this, scientists use standard form and standard prefixes when communicating their numerical work.

Standard form

Numbers with many zeros are difficult to follow, so scientists tend to express these in **standard form**. Standard form is a number between one and 10. So, how is the number 769 000 expressed in standard form?

- Locate the decimal point: 769 000.0

- Move the decimal point to give a number between 1 and 10: 7.69000

- Multiply the number by ten raised to the power x, where x is the number of figures the decimal point was moved: 7.69×10^5

Sometimes the decimal point may move the other way. Take the mass of the electron (0.000 545 units) as an example.

- Find the decimal point and move it. This time it goes to the right: 00005.45

- Multiply the number by ten raised to the power x, where x is the number of figures the decimal point was moved. But, this time, the index will be negative: 5.45×10^{-4}

Calculations using standard form

Standard form makes multiplication and division of even the most complex numbers much easier to handle. When you multiply two numbers in standard form, you multiply the numbers and add the indices. For example: $(3 \times 10^2) \times (2 \times 10^3) = 6 \times 10^5$

If you divide numbers in standard form, you divide the standard number and subtract the indices. For example:

$$\frac{8 \times 10^4}{4 \times 10^2} = 2 \times 10^2$$

Units and standard prefixes

Science is based on observations and measurements. When making measurements, it is essential to use the correct units.

Again, to make numbers more manageable, scientists use prefixes that usually have intervals of a thousand. For example, attaching preferred prefixes to the unit metre, you have kilometre, millimetre and nanometre. But other intervals can be used if they are convenient for the task in hand.

A system of prefixes is used to modify units. Prefixes that are commonly used are listed in Table 2.

Prefix	Symbol	Multiplier	Meaning
mega	M	10^6	1 000 000
kilo	k	10^3	1000
deci	d	10^{-1}	0.1
centi	c	10^{-2}	0.01
milli	m	10^{-3}	0.001
micro	μ	10^{-6}	0.000 001
nano	n	10^{-9}	0.000 000 001
pico	p	10^{-12}	0.000 000 000 001

Table 2 Standard prefixes

Significant figures

When carrying out calculations based on measurements made, you must be confident that the answers you give are as precise as the measurements allow. This is done by counting the number of significant figures (sig figs) in the number given for a measurement. So, for example, a measured mass of:

3.4 g (two sig figs) means you are confident to the nearest 0.1 g

3.40 g (three sig figs) means you are confident to the nearest 0.01 g

3.400 g (four sig figs) means you are confident to the nearest 0.001 g.

Worked example 1

Using data from Table 1, calculate how many electrons have the same mass as a nucleus containing one proton and one neutron.

mass of nucleus = $(1.672 \times 10^{-27}) + (1.674 \times 10^{-27})$

number of electrons with the same mass

$$= \frac{(1.672 \times 10^{-27}) + (1.674 \times 10^{-27})}{9.109 \times 10^{-31}}$$

answer given on calculator = 3673.290153

Since the mass of each particle is given to four significant figures, the answer must contain no more than four significant figures. The answer must be rounded up or down. The answer is 3673 electrons.

(Maths Skills 0.0, 0.1, 0.2, 0.4, 1.1)

Remember:

▸ Do not round calculations up or down until you reach the final answer because errors can be carried through.

▸ The answer to a chemical calculation must not have more significant figures than the number used in the calculation with the fewest significant figures.

Fact	Example
all non-zero digits are significant	275 has three sig figs
zero between non-zero digits is significant	205 has three sig figs
zero to the left of the first non-zero digit is not significant	301 has three sig figs, 0.31 has two sig figs
zero to the right of the decimal point is significant	2.9 has two sig figs, 2.90 has three sig figs
numbers ending in zero to the left of the decimal point: the zero may or may not be significant	a mass of 840 g has two sig figs if the balance is accurate to ± 10 g, and three sig figs if the balance is accurate to ± 1 g

Table 3 *Significant figures*

ASSIGNMENT 1: SIZE, SCALE AND SIGNIFICANT FIGURES

(MS 0.0, 0.1, 0.2; PS 1.1, 1.2, 3.2)

A single carbon atom measures about one ten-billionth of a metre across, a dimension so small that it is impossible to imagine. The nucleus is a thousand times smaller again, and the electron a hundred thousand times smaller than that!

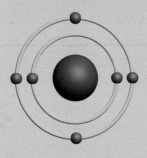

Figure A1 *A single carbon atom measures about one ten-billionth of a metre across.*

Because the numbers are so unimaginably small, scientists do not use grams and metres to describe atoms and subatomic particles. They use a different set of units.

You have already come across the idea of relative masses. Protons and neutrons both have a relative mass of 1. We say these have a mass of 1 relative mass unit. The electron is a mere 0.000 545 relative mass units. Clearly, even with relative masses you have some awkward numbers.

Questions

Give your answers to the appropriate number of significant figures, and in standard form where appropriate.

A1. a. An atom of hydrogen contains only a proton and an electron. Calculate the mass of the hydrogen atom in kilograms.

   b. A molecule of hydrogen contains two atoms. Calculate the mass of a hydrogen molecule in grams.

   c. How many electrons have the same mass as a single neutron?

A2. Convert these quantities into measurements in grams, expressed in standard form:

   a. The mass of a neutron.

   b. 200 million electrons.

   c. 10 gold coins weighing a total of 0.311 kg.

A3. A uranium atom contains 92 electrons. Calculate the mass, in kilograms, of protons in the atom.

A4. How many times heavier is the nucleus of a helium atom (two protons and two neutrons) than its electrons?

1.4 ATOMIC NUMBER, MASS NUMBER AND ISOTOPES

Different elements have different numbers of electrons, protons and neutrons in their atoms. It is the number of protons in the nucleus of an atom that identifies the element. Remember that, if an atom forms an ion by gain or transfer of electrons, it is still an ion of the same element. An atom can also have one or two more or fewer neutrons and still remain the same element. Using this information, you can define an element using two numbers: the atomic number and the mass number (Figure 8).

Atomic (proton) number (Z). The atomic number of an element is the number of protons in the nucleus of the atom. It has the symbol Z and is also known as the proton number. Its value is placed in front of the element's symbol, below its mass number. Since atoms are neutral, the number of protons equals the number of electrons orbiting the nucleus. All atoms of the same element have the same atomic number.

Mass number (A). The mass number of an element is the total number of protons and neutrons in the nucleus of an atom. It is a measure of its mass compared with other types of atom. Even in heavy atoms, the electron's mass is so small that it makes little difference to the overall mass of the atom. Protons and neutrons both have a mass of 1, so:

mass number (A) = number of protons (Z) + number of neutrons (n)

$$A = Z + n$$

The symbol for the mass number is A, and its value goes above the atomic number in front of the element's symbol.

You can calculate the number of neutrons in the nucleus using:

number of neutrons (n) = mass number (A) − atomic number (Z).

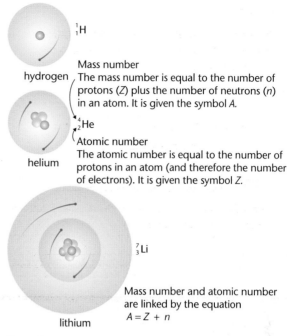

$^{1}_{1}H$

hydrogen

Mass number
The mass number is equal to the number of protons (Z) plus the number of neutrons (n) in an atom. It is given the symbol A.

$^{4}_{2}He$

Atomic number
The atomic number is equal to the number of protons in an atom (and therefore the number of electrons). It is given the symbol Z.

helium

$^{7}_{3}Li$

Mass number and atomic number are linked by the equation
$$A = Z + n$$

lithium

Figure 8 *Mass number and atomic number*

QUESTIONS

5. How many protons, neutrons and electrons do the following atoms and ions have?

   a. An element with mass number 19 and atomic number 9.

   b. An element with mass number 210 and atomic number 85.

   c. An ion with one positive charge, a mass number of 23 and atomic number 11.

   d. An ion with three negative charges, a mass number of 31 and atomic number 15.

   e. An ion with three positive charges, a mass number 52 and atomic number 24.

Isotopes

All atoms of the same element have the same number of protons and the same atomic number, Z. However, they may have a different number of neutrons and so a different mass number, A. Atoms of the same element with different mass numbers are called **isotopes**. Nitrogen has two isotopes, $^{14}_{7}N$ and $^{15}_{7}N$. Both isotopes have seven protons, but $^{14}_{7}N$ has seven neutrons and $^{15}_{7}N$ has eight neutrons. The notation for an isotope shows the mass number and the atomic number:

Carbon   mass number ——— 12

atomic number ——— 6

This is also written as carbon-12

The isotopes of carbon and their sub atomic particles are summarised in Table 4.

Name		No. of protons	No. of neutrons	No. of electrons	Relative abundance %
carbon-12		6	6	6	98.93
carbon-13		6	7	6	1.07
carbon-14		6	8	6	10^{-10}

Table 4 *Isotopes of carbon*

Properties of isotopes

The chemical properties of an element depend on the number and arrangement of the electrons in its atoms. Since all the isotopes of an element have the same number and arrangement of electrons, they also all have the same chemical properties. However, because of the difference in mass, isotopes differ slightly in their physical properties, such as in the rate of diffusion (which depends on mass), and their nuclear properties, such as radioactivity.

Isotopes that are not radioactive, such as chlorine-35 and chlorine-37, are called **stable isotopes**. Data books give you the relative abundance of each isotope present in such stable, naturally occurring elements.

Relative abundance of isotopes

Most elements have isotopes. The percentage of each isotope that naturally occurs on Earth is referred to as its **relative isotopic abundance**. Chlorine has two isotopes, $^{35}_{17}Cl$ and $^{35}_{17}Cl$. Any sample of naturally occurring chlorine will contain 75.53% of chlorine-35 and 24.47% of chlorine-37.

Hydrogen has three isotopes: $^{1}_{1}H$, $^{2}_{1}H$ (called deuterium) and $^{3}_{1}H$ (called tritium). Elements that occur in space may contain different percentages of isotopes. These percentages are known as its **isotope signature**.

Mass spectrometry

Performance enhancing drugs are illegal in most sports and most organisations use drug tests to

ensure that the competition is fair. An athlete may be asked to produce a urine sample and, sometimes, a blood sample. This is sent to a testing facility. The drug tests detect the presence of compounds that are produced by chemical reactions in the body as it processes the drug. **Mass spectrometers** are used to help analyse the sample. Drug analysis is just one of a vast range of applications of mass spectrometry i analytical chemistry.

There are several different types of mass spectrometer. They can be used to identify the mass of an element, an isotope or a molecule. Knowing the mass of a particle helps scientists to identify the particle. One type of spectrometer is called a **time-of-flight mass spectrometer** (Figure 11).

> **KEY IDEAS**
>
> ‣ The atomic (proton) number, Z is equal to the number of protons in the nucleus.
>
> ‣ The mass number, A is equal to the number of protons plus the number of neutrons in the nucleus.
>
> ‣ Isotopes of an element have the same number of protons but different numbers of neutrons.

ASSIGNMENT 2: ISOTOPE DETECTIVE

(MS 0.0, 0.1; PS 1.1, 1.2, 3.2)

Human beings have long been obsessed with the idea that life might exist or have existed on Mars, one of the closest planets to Earth. In 2003 the European Mars Express detected methane gas, CH_4. We use methane to heat our homes and for cooking food. On Earth, 90% of all methane comes from living things, such as the decay of organic material. This is how the gas in our homes originated. The remaining 10% was produced from geological activity.

The big question is: 'What is the origin of the methane gas on Mars?'. Perhaps it was formed by the decay of organic material billions of years ago. Or maybe it is being given off by present-day microbes that exist under the surface in areas heated by volcanic activity. Alternatively, was the methane gas a result of geological processes? The answer may be found in isotopes.

Carbon has three isotopes: carbon-12, carbon-13 and carbon-14. Their abundances on Earth, as shown by the isotopic signature of methane, are given in Table A1.

Hydrogen has two naturally occurring isotopes: hydrogen-1, which has a 99.9885% abundance on Earth, and hydrogen-2, which has a 0.01115% abundance. (The other hydrogen isotope, hydrogen-3, is not naturally occuring. It is produced in nuclear reactors.)

Figure A1 *Nili Fossae, one of the regions on Mars emitting methane. Are the plumes of methane gas evidence for life on Mars?*

Chemical reactions involve electrons; the presence of extra neutrons in isotopes does not affect these reactions. So, methane molecules may contain a mixture of these different isotopes. The three most common methane molecules are formed when one carbon-12 combines with four hydrogen-1 atoms, one atom of carbon-13 combines with four hydrogen-1 atoms, and when one atom of carbon-12 combines with three hydrogen-1 atoms and one hydrogen-2 atom. These can be written as:

$^{12}CH_4$, $^{13}CH_4$ and $^{12}CH_3D$ respectively. (D stands for deuterium, which is the name given to hydrogen-2.)

Methane formula	Natural percentage abundance on Earth
$^{12}CH_4$	0.99827
$^{13}CH_4$	0.01110
$^{12}CH_3D$	0.00062

Table A1 *The isotopic signature of naturally occurring methane on Earth*

When scientists determine the isotopic signature of Martian methane, they can compare it with that on Earth. They may then be a step nearer to deciding its origin.

Questions

A1. What are the atomic numbers and mass numbers of:

    a. the isotopes of carbon

    b. the isotopes of hydrogen?

A2. Why is carbon-14 ignored in possible methane formulae?

A3. What is the difference in mass between $^{12}CH_4$ and $^{13}CH_4$?

A4. What is the difference in mass between $^{12}CH_4$ and $^{12}CH_3D$?

A5. Which form of methane is most common and why?

A6. Suggest a formula for an extremely rare type of methane.

A7. When scientists compare the isotopic signature of Martian methane with that of methane on Earth, what assumptions are they making?

Figure 9 *Athletes undergo drugs tests during training and competition. Any found using performance enhancing drugs face bans from the sport.*

Figure 10 *Inside the flight tube of a triple quadrupole time-of-flight mass spectrometer*

Electrospray ionisation. The sample being tested is vaporised and injected into the spectrometer. A beam of electrons is fired at the sample and knocks out electrons to produce ions. A technique called electrospray ionisation is used because this reduces the number of molecules that break up or fragment. Most atoms or molecules lose just one electron, but a few lose two. Positively charged ions are produced. Only the most energetic electrons can knock out two electrons, so most ions have a single positive charge. If M is a molecule of the sample, then:

$$M(g) \rightarrow M^+(g) + e^- \text{ or, } M(g) \rightarrow M^{2+}(g) + 2e^-.$$

Acceleration. The electric field has a fixed strength – the potential difference is constant. It accelerates the ions so that all the ions with the same charge have the same kinetic energy – they are travelling at the same speed.

Ion detector. The ions are distinguished by different flight times at the ion detector. The electronic signal is used by computer software to produce a mass spectrum. The position of each peak on the mass spectrum is related to the *m/z* charge of the ions. Since most ions have a +1 charge, this will be the same as the mass of the ion. The size of each peak is proportional to the abundance of the ion in the sample.

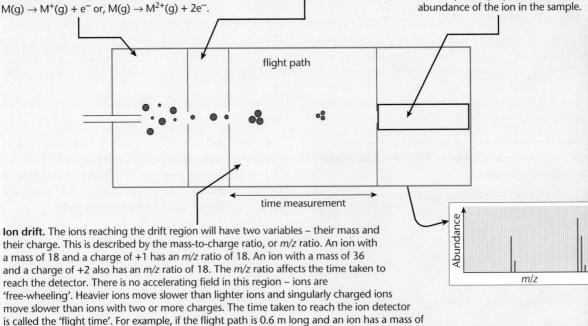

Ion drift. The ions reaching the drift region will have two variables – their mass and their charge. This is described by the mass-to-charge ratio, or *m/z* ratio. An ion with a mass of 18 and a charge of +1 has an *m/z* ratio of 18. An ion with a mass of 36 and a charge of +2 also has an *m/z* ratio of 18. The *m/z* ratio affects the time taken to reach the detector. There is no accelerating field in this region – ions are 'free-wheeling'. Heavier ions move slower than lighter ions and singularly charged ions move slower than ions with two or more charges. The time taken to reach the ion detector is called the 'flight time'. For example, if the flight path is 0.6 m long and an ion has a mass of 26 atomic units, the flight time will be 6×10^{-6} seconds.

Figure 11 *The basic principles of a time-of-flight mass spectrometer*

The mass spectrum of magnesium

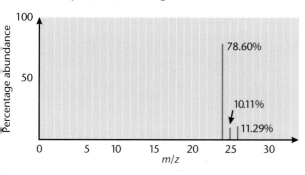

Figure 12 *Mass spectrum of magnesium*

The chart produced by a mass spectrometer is called a mass spectrum (plural: mass spectra).

When a sample of magnesium vapour is fed into the mass spectrometer, it is bombarded by high energy electrons in the **ionisation region**. These fast-moving, energetic electrons knock electrons off the magnesium atoms to produce positively charged magnesium ions, Mg^+.

$$Mg(g) \rightarrow Mg^+(g) + e^-$$

If bombarded with very high energy electrons, some magnesium ions lose a second electron:

$$Mg^+(g) \rightarrow Mg^{2+}(g) + e^-$$

These positively charged magnesium ions pass through the spectrometer and are accelerated by an electric field to give all ions with the same charge the same kinetic energy. The sample then passes into the drift region. Magnesium has three isotopes: magnesium-24, magnesium-25 and magnesium-26. If ions of all isotopes have the same charge, then the lighter magnesium-24 will take a shorter time to reach the ion detector than the heavier ions. The flight time will be less.

A mass spectrum is produced. The *y*-axis is the percentage abundance. The *x*-axis is the mass/charge (or *m/z*) ratio (Figure 12).

The spectrum of magnesium has three lines. These correspond to the three isotopes of magnesium. The heights of the lines are proportional to the amounts of each isotope present. The sample in Figure 12 contains 78.60% of magnesium-24, 10.11% of magnesium-25 and 11.29% of magnesium-26.

The mass spectrum of an element shows:

> the mass number of each isotope present (since mass numbers are masses compared with carbon-12, this number is called the relative isotopic mass)

> the relative abundance of each isotope.

Calculating the relative atomic mass from isotopic abundance

You can use information from mass spectra to calculate the **relative atomic mass**, A_r, of an element (see Section 1.5).

For magnesium, the isotopes ^{24}Mg, ^{25}Mg and ^{26}Mg are present in the ratio 78.60 : 10.11 : 11.29. This means that the mass of 100 magnesium atoms will be:

$$(78.60 \times 24) + (10.11 \times 25) + (11.29 \times 26)$$

The average mass of one magnesium atom will be:

$$\frac{(78.60 \times 24) + (10.11 \times 25) + (11.29 \times 26)}{100}$$

= 24.3 (to one decimal place)

This is the relative atomic mass of magnesium. The A_r value for magnesium in your data book is 24.3.

Note that atoms of magnesium with this actual mass do not exist. This is the *average* mass of all the naturally occurring isotopes of magnesium, taking abundance into account.

The mass spectrum of lead

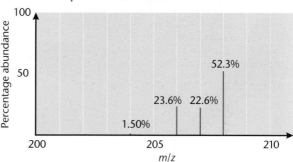

Figure 13 *Mass spectrum of lead*

The mass spectrum of lead (Figure 13) shows that lead has four isotopes with mass numbers of 204, 206, 207 and 208. The heights of the lines show that these are in the ratio of 1.50 : 23.6 : 22.6 : 52.3. These are percentage abundances and add up to 100.

To calculate the relative atomic mass of lead:

mass of 100 atoms

$$= (1.540 \times 204) + (23.6 \times 206) + (22.6 \times 207) + (52.3 \times 208)$$

relative atomic mass

$$= \frac{(1.540 \times 204) + (23.6 \times 206) + (22.6 \times 207) + (52.3 \times 208)}{100}$$

= 207.2

15

QUESTIONS

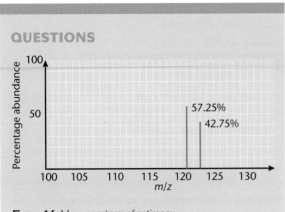

Figure 14 *Mass spectrum of antimony*

6. Look at the mass spectrum of a sample of antimony (Figure 14).

a. How many isotopes does antimony have?

b. What are their mass numbers?

c. What is the percentage abundance of each isotope?

d. Calculate the relative atomic mass of antimony from this spectrum.

7. Find the relative atomic mass of naturally occurring uranium that contains 0.006% uranium-234, 0.72% uranium-235 and 99.2% uranium-238.

8. Silver has two isotopes, silver-107 and silver-109. These are present in the ratio of 51.35 : 48.65 in naturally occurring silver. Calculate the relative atomic mass of silver.

ASSIGNMENT 3: ANALYSING HAIR

(MS 1.1, 1.2; PS 1.1, 1.2, 3.2)

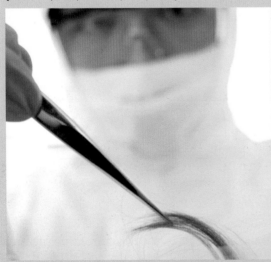

Figure A3 *Scientists analyse hair to provide forensic evidence.*

Hair grows at a fairly uniform rate. Its composition depends partly on diet and the water you drink. The ratios of different isotopes in the water supply vary with your location and the rocks the water percolates through. For example, the isotopes of strontium (^{87}Sr and ^{88}Sr) and the isotopes of oxygen (^{16}O, ^{17}O and ^{18}O) vary all over the world. As your hair grows, the isotope ratios from your environment are captured in your hair.

For the forensic scientist, analysing the isotopes in a sample of your hair can tell where you are from and where you have been. This is a very useful tool in cases such as deciding whether a terrorist suspect has been to a particular location.

Questions

A1. Strontium has an atomic number of 38. How many protons, neutrons and electrons are in one atom of strontium-87 and in one atom of strontium-88?

A2. Naturally occurring strontium has four isotopes with these percentage abundances:

Strontium isotope	Percentage abundance
strontium-84	0.56
strontium-86	9.86
strontium-87	7.00
strontium-88	82.58

Table A2

a. Sketch strontium's mass spectrum.

b. Calculate the relative atomic mass of strontium.

A3. The data given in Table A2 for strontium isotopes are average figures for all naturally occurring strontium isotopes on Earth. What information do forensic scientists need when using hair analysis to track people?

1.5 RELATIVE ATOMIC MASS, A_r

Atoms have very small masses, from 10^{-24} g to 10^{-22} g. Instead of using these masses, scientists use **relative atomic mass** (symbol A_r). Relative here means the mass of one atom compared with another. Originally, the mass of each atom was compared with the mass of a hydrogen atom, where hydrogen had a mass of one. As mass spectroscopy developed and gave more accurate values for the masses of atoms, it was discovered that hydrogen's mass is slightly more than one.

Relative atomic mass is now defined as the average mass of an atom compared with $\frac{1}{12}$ the mass of a carbon-12 atom.

relative atomic mass, $A_r =$

$$\frac{\text{average mass of one atom of an element}}{\frac{1}{12}\text{ the mass of one carbon-12 atom}}$$

Relative atomic masses have no units because they show how many times heavier one atom is compared with another. Books give different numbers of decimal places for these values. The A_r for magnesium is given as 24 in most GCSE Periodic Tables. You will now need to use the more precise value of 24.3 for most calculations.

Remember, relative atomic mass is the average mass of all isotopes of an element, taking relative abundance into consideration. These values can be found using mass spectroscopy and the calculations you did earlier.

1.6 RELATIVE MOLECULAR MASS, M_r

In chemistry, you also need to know the mass of molecules. The same relative atomic mass scale is used. The **relative molecular mass** is the mass of a molecule compared with $\frac{1}{12}$ the mass of a carbon-12 atom.

relative molecular mass, $M_r =$

$$\frac{\text{average mass of one molecule}}{\frac{1}{12}\text{ the mass of one carbon-12 atom}}$$

Finding M_r values

The mass spectrometer can also be used to find relative molecular mass values. This is dealt with in more detail in Chapter 16. If a sample of vaporised molecules is introduced into the mass spectrometer, the bombarding electrons can knock an electron off a molecule in the same way as they did with a sample of atoms. This produces a positively charged ion called the **molecular ion, M$^+$**.

$$M(g) \rightarrow M^+(g) + e^-$$

Most of these molecular ions are now split into fragments by the bombarding electrons, but some remain intact.

The line produced by these molecular ions on the mass spectrum represents the relative molecular mass of the sample – if the atoms in the molecule have isotopes, there will be more than one molecular ion peak. The mass spectrum of methane (Figure 15) shows the molecular ion peak at $m/z = 16$ and another at $m/z = 17$. The one at 16 is due to $^{12}CH_4{}^+$ and the much smaller one at 17 is due to $^{13}CH_4{}^+$. These can be used to calculate the relative molecular mass of methane.

You will find more information about relative atomic mass and relative molecular mass in Chapter 2.

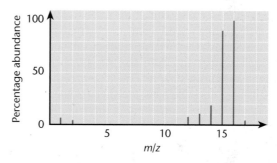

Figure 15 *The mass spectrum of methane, CH_4*

KEY IDEAS

› In a time-of-flight mass spectrometer, samples are ionised, accelerated to constant kinetic energy, allowed to drift and detected.

› A mass spectrum can be used to find the relative isotopic mass and abundance of isotopes of an element.

› The mass spectrum of a compound can be used to find its M_r and provide clues about its structure.

› Relative atomic mass is the average mass of an atom compared with $\frac{1}{12}$ the mass of a carbon-12 atom.

› Relative molecular mass is the average mass of a molecule compared with $\frac{1}{12}$ the mass of a carbon-12 atom.

1.7 DESCRIBING ELECTRONS

Main shell	Sub-shell	Max no. electron pairs in sub-shell	Max no. electrons in sub-shell	Max. no. electrons in main shell
1	s	1	2	2
2	s	1	2	8
	p	3	6	
3	s	1	2	18
	p	3	6	
	d	5	10	
4	s	1	2	32
	p	3	6	
	d	5	10	
	f	7	14	

Figure 16 *Shells, sub-shells and number of electrons*

Electrons are arranged in **electron shells** around the nucleus. Each electron shell has a particular energy value. Electrons can be described as being in a particular shell. Within each shell, there are sub-shells (or orbitals). The number of sub-shells in each shell is shown in Figure 16. The sub-shells are given the letters s, p, d and f. The letters come from words used to describe emission spectral lines (this is discussed further a little later in this chapter). Figure 16 shows that the first shell has a maximum of two electrons and that they are both in sub-shell s. The second electron shell has a maximum of eight electrons, two of which are in sub-shell s and six in sub-shell p. This sequence of sub-shells corresponds to an increase in energy (Figure 17). Each additional electron goes into the sub-shell with the next lowest energy. The order of filling is the same as the order of the elements in the Periodic Table.

Electron orbitals

Electrons are constantly moving, and it is impossible to know the exact position of an electron at any given time. However, measurements of the density of electrons as they move round the nucleus show that there are regions where it is highly probable to find an electron. These regions of high probability are called **orbitals**. Each s, p, d and f sub-shell corresponds to a differently shaped orbital.

The shapes of s and p orbitals are shown in Figure 18. Each orbital can hold two electrons, which spin in opposite directions. Table 5 shows the numbers of electrons and orbitals in the sub-shells.

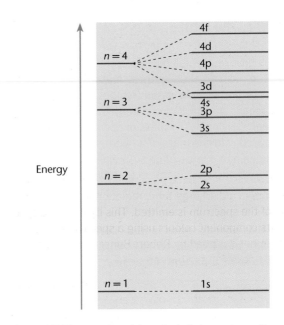

Figure 17 *The energies of the sub-shells in an atom with many electrons*

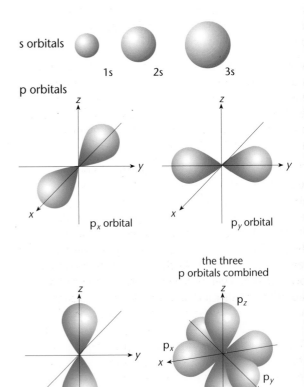

Figure 18 *The three-dimensional (3D) shape of the s and p orbitals*

Sub-level	s	p	d	f
Number of orbitals in sub-shell	1	3	5	7
Maximum number of electrons	2	6	10	14

Table 5 *Number of orbitals and maximum number of electrons per sub-shell*

Emission spectra and electrons

When an electrical voltage is applied to a gas at low pressure in a discharge tube, radiation in the visible part of the spectrum is emitted. This light can be split into its component colours using a spectroscope, an instrument designed by Robert Bunsen (the same Bunsen as in Bunsen burner).

One theory of light considers it to consist of particles, called photons, that move in a wave-like motion. Each photon has its own amount of energy, depending on the wavelength of the photon. The shorter the wavelength, the more energetic the photon and the higher the energy. Ultraviolet (UV) radiation has a shorter wavelength than infrared (IR) radiation, so a photon of UV radiation has more energy than a photon of IR radiation.

Many advertising signs are gas-discharge tubes, which are often filled with neon. When neon absorbs electrical energy, electrons become excited to higher energy levels. As the electrons fall back, radiation in the yellow part of the visible spectrum is emitted. Other colours are usually obtained by using tinted glass.

The electrons in the atoms of gas in the discharge tube absorb electrical energy. This excites the electrons and they move into a higher energy level. This is not a stable arrangement and the electrons fall back to their original position, called the ground state, in one or more steps. You can see this in Figure 19. As the electron returns to a lower energy level (or shell), energy is emitted as radiation. If this radiation is in the visible range, you see it as coloured light. A spectroscope will split this radiation into lines of a particular colour. The energy gaps between the energy levels in the atom determine the wavelength of the radiation emitted. All the lines for an element make up its emission spectrum.

If the energies of electrons were not fixed, the **emission spectra** would be continuous, with no lines.

Niels Bohr suggested that electrons could only exist at fixed energies. He gave each energy level (shell) the symbol n and numbered them 1, 2, 3, and so on, so that $n = 1$ is the first energy level and the ground state for hydrogen's 1 electron (see Figure 19).

Each line in hydrogen's emission spectrum represents the difference between the energy of the level to which the electron becomes excited and the level to which it falls back.

This is the basis of the **quantum theory**. Whereas Rutherford thought that the electron moved smoothly, Bohr showed that it moved in small jumps, or quanta.

Emission spectra provide evidence for electrons in shells. You will read about other evidence later in this chapter.

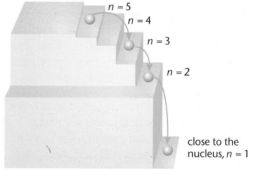

far from the nucleus

$n = 5$
$n = 4$
$n = 3$
$n = 2$

close to the nucleus, $n = 1$

Figure 19 *The staircase model for the levels of the energies emitted by the hydrogen electron. As shown, energy jumps can be down one step or more than one step.*

QUESTIONS

Stretch and challenge

9. Why did Bohr suggest that electrons have fixed amounts of energy?

10. Draw a hydrogen atom with seven energy levels (shells). Show hydrogen's electron in the first shell. Annotate your diagram to show what happens when the electron absorbs energy and moves to $n = 3$ before falling back to the ground state. Label your diagram to show the outcome.

Electron configuration of atoms

The arrangement of electrons in an atom can be written as symbols in an **electron configuration**. The electron configuration includes sub-shells as well as shells, and shows the number of electrons in each.

> Hydrogen has one electron in shell 1, sub-shell s. Its electron configuration is 1s.

> Helium has two electrons with opposite spin. Its electron configuration is $1s^2$.

> Lithium has two electrons in 1s and one in 2s. Its electron configuration is $1s^2\ 2s^1$.

You can also draw a **spin diagram** for each sub-shell that shows the direction of spin of all the electrons. So, you can represent the 12 electrons in the shells/ sub-shells of magnesium in two ways, electron configuration or a spin diagram.

Shells/sub-shells

	1s	2s	2p	3s
Electron configuration	$1s^2$	$2s^2$	$2p^6$	$3s^2$
Spin diagram	↑↓	↑↓	↑↓ ↑↓ ↑↓	↑↓

Between hydrogen and argon, electrons of increasing energy are added, one per element, in sub-shell order 1s, 2s, 2p, 3s, 3p. Then, for potassium, the next electron skips sub-shell 3d and goes into 4s. Though shell three energies are lower overall than shell four energies, the 3d sub-shell has a higher energy than the 4s sub-shell as shown in Table 6 (and Figure 17). The order of filling is the order of elements in the Periodic Table and 4s is filled before 3d. Later, you will see that the chemical properties of elements reflect the energy levels of electrons.

Filling orbitals

You have seen that the arrows in electron spin diagrams indicate their direction of spin and whether there are one or two electrons per orbital. The electrons fill the orbitals in a set order.

Electrons organise themselves so that they remain unpaired and fill the maximum number of sub-shells possible.

As you have seen, for the p sub-shells, this means that electrons first occupy empty orbitals and are parallel spinned. When these orbitals each have one electron, additional electrons are spin-paired; the second electron in an orbital will spin in the opposite direction.

Electron $2p^1$ in boron is:

↑ ☐ ☐

Electrons $2p^2$ in carbon are:

↑ ↑ ☐

Electrons $2p^3$ in nitrogen are:

↑ ↑ ↑

There is now one electron in each orbital. The next electron goes into the first orbital and spins in the opposite direction, so that:

Electrons $2p^4$ in oxygen are:

↑↓ ↑ ↑

Electrons $2p^5$ in fluorine are:

↑↓ ↑↓ ↑

Electrons $2p^6$ in neon are:

↑↓ ↑↓ ↑↓

In Table 6, shell one in helium is filled. The next element with a filled level is neon, which has the electron configuration $1s^2\ 2s^2\ 2p^6$.

Since the outermost shell is complete, these elements are very stable and are known as the noble gases. Noble gas configurations are used to write abbreviated electron configurations. For example, the full electron configuration for potassium is $1s^2\ 2s^2\ 2p^6\ 3s^2\ 3p^6\ 4s^1$. The abbreviated form is $[Ar]\ 4s^1$.

Similarly, the abbreviated electron configuration for phosphorus is $[Ne]\ 3s^2\ 3p^3$.

QUESTIONS

11. Use Table 6 to help you write abbreviated electron configurations for:

   a. sulfur

   b. aluminium

   c. calcium

   d. scandium

   e. silicon

   f. iron

   g. krypton

   h. copper.

Z	Element	Electron configuration	Electron spin diagram (1s 2s 2p 3s 3p 3d 4s 4p)
1	H	$1s^1$	
2	He	$1s^2$	
3	Li	$1s^22s^1$	
4	Be	$1s^22s^2$	
5	B	$1s^22s^22p^1$	
6	C	$1s^22s^22p^2$	
7	N	$1s^22s^22p^3$	
8	O	$1s^22s^22p^4$	
9	F	$1s^22s^22p^5$	
10	Ne	$1s^22s^22p^6$	
11	Na	$1s^22s^22p^63s^1$	
12	Mg	$1s^22s^22p^63s^2$	
13	Al	$1s^22s^22p^63s^23p^1$	
14	Si	$1s^22s^22p^63s^23p^2$	
15	P	$1s^22s^22p^63s^23p^3$	
16	S	$1s^22s^22p^63s^23p^4$	
17	Cl	$1s^22s^22p^63s^23p^5$	
18	Ar	$1s^22s^22p^63s^23p^6$	
19	K	$1s^22s^22p^63s^23p^64s^1$	
20	Ca	$1s^22s^22p^63s^23p^64s^2$	
21	Sc	$1s^22s^22p^63s^23p^63d^14s^2$	
22	Ti	$1s^22s^22p^63s^23p^63d^24s^2$	
23	V	$1s^22s^22p^63s^23p^63d^34s^2$	
24	Cr	$1s^22s^22p^63s^23p^63d^54s^1$	
25	Mn	$1s^22s^22p^63s^23p^63d^54s^2$	
26	Fe	$1s^22s^22p^63s^23p^63d^64s^2$	
27	Co	$1s^22s^22p^63s^23p^63d^74s^2$	
28	Ni	$1s^22s^22p^63s^23p^63d^84s^2$	
29	Cu	$1s^22s^22p^63s^23p^63d^{10}4s^1$	
30	Zn	$1s^22s^22p^63s^23p^63d^{10}4s^2$	
31	Ga	$1s^22s^22p^63s^23p^63d^{10}4s^24p^1$	
32	Ge	$1s^22s^22p^63s^23p^63d^{10}4s^24p^2$	
33	As	$1s^22s^22p^63s^23p^63d^{10}4s^24p^3$	
34	Se	$1s^22s^22p^63s^23p^63d^{10}4s^24p^4$	
35	Br	$1s^22s^22p^63s^23p^63d^{10}4s^24p^5$	
36	Kr	$1s^22s^22p^63s^23p^63d^{10}4s^24p^6$	

Table 6 Electron configurations and spin diagrams for the first 30 elements

Electron configuration of ions

An ion is an atom in which either:

> one or more electrons have been removed, producing a positively charged ion, or

> one or more electrons have been added, producing a negatively charged ion.

Worked example

What is the electron configuration of the sodium ion, Na^+?

The electron configuration of the sodium atom is $1s^2\ 2s^2\ 2p^6\ 3s^1$.

In Na^+ the outermost electron, $3s^1$, has been removed.

This is the electron of highest energy in sodium, and so takes the least energy to remove. The electron configuration of Na^+ is $1s^2\ 2s^2\ 2p^6$. Inner electron shells have the effect of shielding outermost electrons from the positive charge of the nucleus. A full shell has a strong shielding effect on a single outermost electron, which is then easy to remove, as in the case of Na^+.

QUESTIONS

12. Explain the meaning of 2, p and 6 in $2p^6$.

13. Write the electron configuration for each of:
    a. Ca^{2+}
    b. Cl^-
    c. Al^{3+}
    d. Br^-
    e. N^{3-}

KEY IDEAS

> Electrons in an atom are arranged in shells, with the first shell closest to the nucleus and with least energy.

> Each shell consists of one or more sub-shells, also called orbitals, of which there are four types: s, p, d and f.

> An orbital contains a maximum of two electrons spinning in opposite directions.

> The electron configuration of an atom specifies the number of electrons in each shell and sub-shell.

1.8 IONISATION ENERGIES

The energy required to remove an electron from an atom in its gaseous state is called the **ionisation energy**. The energy required to remove the first electron is called the **first ionisation energy** and can be written as:

$$M(g) \rightarrow M^+(g) + e^-$$

The energy required to remove the second electron from an atom is called the second ionisation energy and can be written as:

$$M^+(g) \rightarrow M^{2+}(g) + e^-$$

Ionisation energy values for removing the second and subsequent electrons are called **successive ionisation energies**.

The ionisation energy for one atom is so small that, for convenience, ionisation energies are measured per mole of atoms, in $kJ\ mol^{-1}$.

QUESTIONS

14. Write equations, using M, to show the third and fourth ionisation energies.

The first ionisation energy is the enthalpy change (energy change) when one mole of gaseous atoms forms one mole of gaseous ions with a single positive charge.

Ionisation energies have been calculated for all but a few of the very heavy elements in the Periodic Table. Figure 20 shows the first ionisation energies for the elements from hydrogen to caesium.

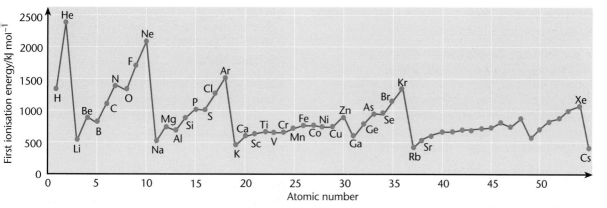

Figure 20 *First ionisation energies of the elements, from hydrogen to caesium*

1.9 EVIDENCE FOR SHELLS AND SUB-SHELLS

Patterns in first ionisation energies provide evidence for the existence of electron shells and sub-shells. You can see this if you look at the first ionisation energies down Group 2 and across Period 3. Successive ionisation energies of an element provide further evidence.

First ionisation energies of Group 2 elements

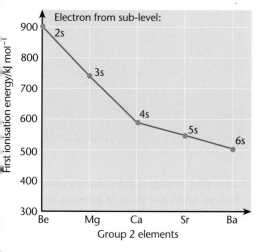

Figure 21 *First ionisation energies of Group 2 elements, from beryllium to barium*

The Group 2 elements, beryllium to barium, are reactive metals. They are also known as the alkaline earth metals because they react with water to form an alkaline solution. The outer sub-shells of these elements contain a pair of electrons in an s orbital. The first ionisation energy measures how much energy is needed to remove one mole of these electrons from a mole of atoms.

Figure 21 shows how the first ionisation energies decrease down Group 2. That means that the first electron becomes easier to remove. This is because:

❯ the number of electron shells between the outer electron and the nucleus is increasing; the electron shells shield the outer electron from the attraction of the nucleus, and

❯ the radius of each atom is increasing as you go down Group 2; the distance between the outer electron and the nucleus is increasing.

So, the outer electrons are easier to remove and the first ionisation energies decease. This is evidence for the existence of electron shells.

First ionisation energies of Period 3 elements

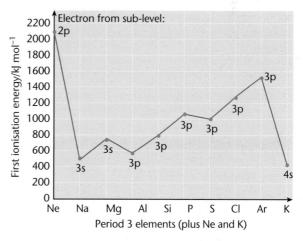

Figure 22 *First ionisation energies of Period 3 elements, from sodium to argon*

As you move across Period 3, each element has one more electron than the last. This electron fills the first available empty orbital. The electron for sodium fills the 3s orbital. The electrons in the first and second

23

shells shield the 3s electron from the positive charge of the nucleus and it is relatively easy to remove.

Magnesium has one more electron than sodium, and this completes the 3s orbital and spins in the opposite direction. Magnesium also has an extra proton, so the positive charge on the nucleus has increased. More energy is needed to remove magnesium's first electron. Magnesium's first ionisation energy is higher than sodium's.

The extra electron that aluminium has compared with magnesium is the first to fill a 3p orbital. p orbitals have higher energy than s orbitals. Aluminium's first electron is easier to remove than the 3s electron of magnesium. The first ionisation energy drops.

The extra electron that silicon has compared with aluminium, and that phosphorus has compared with silicon, fill the remaining empty 3p orbitals. At the same time, the positive charge on the nucleus is increasing and more energy is needed to remove these electrons. The first ionisation energies increase from aluminium to phosphorus.

Sulfur's first electron enters a 3p orbital already containing one electron. These spin in opposite directions and repel each other. It takes less energy to remove the first electron from sulfur than to remove the first electron from phosphorus. Its first ionisation energy is lower.

The electrons for chlorine and argon fill the remaining 3p orbitals. The positive charge on the nucleus continues to increase and the first ionisation energy increases as more energy is needed to remove an electron.

The general trends for first ionisation energy are:

> A sharp fall in ionisation energy between neon and sodium and between argon and potassium as electrons enter a new shell. This is evidence that the outer electron is on its own in a new shell and is shielded from the charge on the nucleus by electrons in the inner shells.

> An overall increase in the first ionisation energy across Period 3 as the positive charge on the nucleus increases and electrons are attracted more strongly.

> An increase in ionisation energy for each sub-shell as the charge on the nucleus increases and electrons are attracted more strongly.

> A fall in ionisation energy between magnesium and aluminium as electrons start to fill a new sub-shell, 3p. This is evidence that a new sub-shell is being filled.

> A fall in ionisation energy between phosphorus and sulfur as electrons start to pair up in the 3p sub-shells. This is evidence that electrons are pairing up in sub-shells.

Successive ionisation energies

Magnesium has the electron configuration 2,8,2. Its first three successive ionisation energies are:

$Mg(g) \rightarrow Mg^+(g) + e^-$
first ionisation energy = $+738$ kJ mol^{-1}

$Mg^+(g) \rightarrow Mg^{2+}(g) + e^-$
second ionisation energy = $+1451$ kJ mol^{-1}

$Mg^{2+}(g) \rightarrow Mg^{3+}(g) + e^-$
third ionisation energy = $+7733$ kJ mol^{-1}

Figure 23 shows a graph of the $\log_{10}$ successive (ionisation energy) against the number of the electron removed (ionisation number) for magnesium. We use $\log_{10}$ to make the numbers easier to handle.

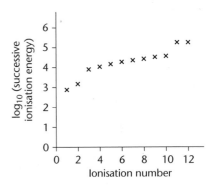

Figure 23 *The trend in the successive ionisation energies of magnesium*

The first two electrons are removed from the third or outer shell. The increase between the second and third electron is because the third electron is taken from the second shell. The gradual increase from the third to the tenth electron shows electrons being removed from the second shell. The large increase between the tenth electron and eleventh electron is because the eleventh electron is taken from the first shell.

QUESTIONS

Stretch and challenge

15. Use Table 7 (which gives the successive ionisation energies for magnesium) to plot a graph of successive ionisation energy divided by the charge on the remaining ion against the number of electrons removed.

16. Using your knowledge of shells and sub-shells, explain the shape of the graph you obtained.

17. These are the first five successive ionisation energies for elements X, Y and Z:

    X   578, 1817, 2745, 11578, 14831

    Y   496, 4563, 6913, 9544, 13352

    Z   738, 1451, 7733, 10541, 13629

    In which group of the Periodic Table are these elements found?

Ionisation	Ionisation energy/kJ mol^{-1}
1	738
2	1451
3	7733
4	10 543
5	13 630
6	18 020
7	21 711
8	25 661
9	31 653
10	35 458
11	169 988
12	189 368

Table 7 *The successive ionisation energies of magnesium*

ASSIGNMENT 4: WHY DO SCIENTISTS THINK ELECTRONS ARE ARRANGED IN SHELLS AND SUB-SHELLS?

(MS 3.1, 3.2)

One piece of evidence for this theory came from patterns from ionisation energy plots. These are the first ionisation energies for Period 2:

Element	Li	Be	B	C	N	O	F	Ne
First ionisation energy/kJ mol^{-1}	520	899	801	1087	1402	1313	1681	2080

Questions

A1. Plot a graph of first ionisation energy against atomic number, Z.

A2. Why is there an overall increase across Period 2 from lithium to neon?

A3. Why is the first ionisation energy of beryllium higher than that of lithium?

A4. Why are there dips in the pattern at boron and oxygen?

A5. Why is there an increase in the first ionisation energy between:

   a. boron and nitrogen

   b. oxygen and neon?

A6. If there was a regular increase in the first ionisation energy across Period 2, what might scientists conclude about the existence of sub-shells?

A7. What is the evidence for the existence of:

   a. electron shells

   b. electron sub-shells?

A8. How does lithium's first ionisation energy help to predict its reactivity?

A9. How does neon's first ionisation energy help to predict its stability and/or lack of reactivity?

KEY IDEAS

> The first ionisation energies decrease down Group 2 because the outermost electrons are increasingly shielded from the attraction of the nucleus.

> There is an overall increase in the first ionisation energy across a period because of the increasing nuclear charge.

> The first ionisation energies provide evidence for the existence of shells and sub-shells.

PRACTICE QUESTIONS

1. The element rubidium exists as the isotopes ^{85}Rb and ^{87}Rb.

   a. State the number of protons and the number of neutrons in an atom of the isotope ^{85}Rb.

   b. i. Explain how the gaseous atoms of rubidium are ionised in a mass spectrometer.

      ii. Write an equation, including state symbols, to show the process that occurs when the first ionisation energy of rubidium is measured.

   c. Table Q1 shows the first ionisation energies of rubidium and some other elements in the same group.

Element	Sodium	Potassium	Rubidium
First ionisation energy/kJ mol^{-1}	494	418	402

   Table Q1

   State one reason why the first ionisation energy of rubidium is lower than the first ionisation energy of sodium.

   d. i. State the block of elements in the Periodic Table that contains rubidium.

      ii. Deduce the full electron configuration of a rubidium atom.

   e. A sample of rubidium contains the isotopes ^{85}Rb and ^{87}Rb only. The isotope ^{85}Rb has an abundance 2.5 times greater than that of ^{87}Rb. Calculate the relative atomic mass of rubidium in this sample. Give your answer to one decimal place.

   f. By reference to the relevant part of the mass spectrometer, explain how the abundance of an isotope in a sample of rubidium is determined.

   AQA June 2012 Unit 1 Question 1

2. The element nitrogen forms compounds with metals and non-metals.

   a. Nitrogen forms a nitride ion with the electron configuration $1s^2\ 2s^2\ 2p^6$. Write the formula of the nitride ion.

   b. An element forms an ion Q with a single negative charge that has the same electron configuration as the nitride ion. Identify the ion Q.

   c. Use the Periodic Table and your knowledge of electron arrangement to write the formula of lithium nitride.

   AQA Jan 2012 Unit 1 Question 5a, b

3. Mass spectrometry can be used to identify isotopes of elements.

   a. i. In terms of fundamental particles, state the difference between isotopes of an element.

      ii. State why isotopes of an element have the same chemical properties.

   b. Give the meaning of the term relative atomic mass.

(Continued

c. The mass spectrum of element X has four peaks. Table Q2 gives the relative abundance of each isotope in a sample of element X.

m/z	64	66	67	68
Relative abundance	12	8	1	6

Table Q2

i. Calculate the relative atomic mass of element X. Give your answer to one decimal place.

ii. Use the Periodic Table to identify the species responsible for the peak at m/z = 64.

d. Explain how the detector in a mass spectrometer enables the abundance of an isotope to be measured.

AQA June 2011 Unit 1 Question 1

4. Indium is in Group 3(13) in the Periodic Table and exists as a mixture of the isotopes ^{113}In and ^{115}In.

a. Use your understanding of the Periodic Table to complete the electron configuration of indium.
$1s^2\ 2s^2\ 2p^6\ 3s^2\ 3p^6\ 4s^2\ 3d^{10}\ 4p^6$ _____

b. A sample of indium must be ionised before it can be analysed in a mass spectrometer.

i. State what is used to ionise a sample of indium in a mass spectrometer.

ii. Write an equation, including state symbols, for the ionisation of indium that requires the minimum energy.

iii. State why more than the minimum energy is not used to ionise the sample of indium.

iv. Give two reasons why the sample of indium must be ionised.

c. A mass spectrum of a sample of indium showed two peaks at m/z = 113 and m/z = 115. The relative atomic mass of this sample of indium is 114.5

i. Give the meaning of the term *relative atomic mass*.

ii. Use these data to calculate the ratio of the relative abundances of the two isotopes.

d. State and explain the difference, if any, between the chemical properties of the isotopes ^{113}In and ^{115}In.

AQA Jan 2011 Unit 1 Question 2

5. a. Copy and complete Table Q3.

	Relative mass	Relative charge
Proton		
Electron		

Table Q3

b. An atom has twice as many protons and twice as many neutrons as an atom of ^{19}F. Deduce the symbol, including the mass number, of this atom.

c. The Al^{3+} ion and the Na^+ ion have the same electron arrangement.

i. Give the electron arrangement of these ions in terms of s and p electrons.

ii. Explain why more energy is needed to remove an electron from the Al^{3+} ion than from the Na^+ ion.

d. The first ionisation energies of a group of elements provides evidence for the existence of electron shells.

i. Describe the trend in first ionisation energies down Group 2.

ii. Explain how the trend you have described in d.i. provides evidence for the existence of electron shells.

e. First ionisation energies across a period provide evidence for the existence of electrons in sub-shells.

i. Describe the trend in first ionisation energies in Period 3.

ii. Explain how the trend you have described in e.i. provides evidence for the existence of electron sub-shells.

AQA January 2007 2 Unit 1 Question 1

6. a. Copy and complete Table Q4.

	Relative mass	Relative charge
Proton		
Electron		

Table Q4

(Continued)

b. An atom of element Q contains the same number of neutrons as are found in an atom of ^{27}Al. An atom of Q also contains 14 protons.

   i. Give the number of protons in an atom of ^{27}Al.

   ii. Deduce the symbol, including mass number and atomic number, for this atom of element Q.

c. Define the term *relative atomic mass* of an element.

d. Table Q5 gives the relative abundance of each isotope in a mass spectrum of a sample of magnesium.

m/z	24	25	26
Relative abundance	73.5	10.1	16.4

Table Q5

   Calculate the relative atomic mass of this sample of magnesium, using the data in Table Q5. Give your answer to one decimal place.

e. State how the relative molecular mass of a covalent compound is obtained from its mass spectrum.

AQA June 2004 Unit 1 Question 1

7. A sample of iron from a meteorite was found to contain the isotopes ^{54}Fe, ^{56}Fe and ^{57}Fe.

a. The relative abundances of these isotopes can be determined using a time-of-flight (TOF) mass spectrometer. In the mass spectrometer, the sample is first vaporised and then ionised.

   i. State what is meant by the term *isotopes*.

   ii. Give an equation to show a gaseous iron atom producing one electron and an iron ion in the ionisation area.

   iii. State the two variables that determine the time taken for an ion to move across the drift area.

   iv. Explain why it is difficult to distinguish between an ^{56}Fe$^+$ ion and a ^{112}Cd^{2+} ion in a mass spectrometer.

b. **i.** Define the term *relative atomic mass of an element*.

   ii. The relative abundances of the isotopes in this sample of iron are shown in Table Q6.

m/z	54	56	57
Relative abundance	5.8	91.6	2.6

Table Q6

   Calculate the relative atomic mass of iron in this sample, using the data in Table Q6. Give your answer to one decimal place.

AQA June 2005 Unit 1 Question 1

8. **a.** Titanium is a d block element in Period 4.

   i. State what is meant by a d block element.

   ii. Write the full electron configuration for titanium in terms of s, p and d electrons.

b. Titanium has five stable isotopes, with ^{48}Ti being the most abundant.

   i. State one difference and two similarities between the stable isotopes of titanium.

   ii. Explain why stable isotopes of titanium have the same chemical properties.

c. Mass spectroscopy can be used to determine the relative abundance of titanium isotopes. Why is it difficult to distinguish between ^{48}Ti^{2+} and ^{24}Mg$^+$ ions on a mass spectrum?

Stretch and challenge

9. Scandium is a d block element and is used in alloys to make sporting equipment such as golf clubs and fishing rods.

a. Give the electron configuration for scandium.

(Continued)

b. When d block elements form ions, the s electrons are lost first, then d electrons. Most d block elements form ions with more than one charge, but the scandium ions has a +3 charge in most of its compounds.

   i. Give the electron configuration for the Sc^{3+} ion.

   ii. Suggest why most d block elements have ions with a +2 charge.

c. Like scandium, zinc only forms one type of ion. It has a +2 charge. Give the electron configuration for a Zn^{2+} ion.

d. Iron forms ions with a +2 charge and with a +3 charge. Give the electron configuration for the Fe^{3+} ion.

Multiple choice

10. Which element has an isotope with an atomic number of 35 and a mass number of 79?

   A. Chlorine

   B. Gold

   C. Bromine

   D. Selenium

11. How many neutrons are present in an atom of $^{27}_{13}Al$?

   A. 13

   B. 27

   C. 14

   D. 40

12. What is 0.00859 in standard form?

   A. 8.59×10^{-1}

   B. 8.59×10^{-2}

   C. 8.59×10^{-3}

   D. 8.59×10^{-4}

13. What determines the flight time of ions in the drift region of a time-of-flight spectrometer?

   A. Mass only

   B. Charge only

   C. m/z ratio

   D. The number of electrons removed by electrospray ionisation only

2 AMOUNT OF SUBSTANCE

PRIOR KNOWLEDGE

You read in Chapter 1 that we use relative atomic mass and relative molecular mass to compare the masses of atoms and molecules. You may already know how to use a chemical equation to describe a chemical reaction. You may have also learned how chemists use moles to count particles and you have possibly carried out some calculations using moles.

LEARNING OBJECTIVES

In this chapter, will build on these ideas and learn how amounts of chemical substances in solids, liquids, gases and solutions can be measured in moles.

(Specification 3.1.2.1, 3.1.2.2, 3.1.2.3, 3.1.2.4, 3.1.2.5)

Ibuprofen is an anti-inflammatory medicine that was first patented in the 1960s. It is now available without prescription (over-the-counter), and is sold under several different brand names, such as *Nurofen* and *Ibuleve*. The early manufacture of ibuprofen involved a series of six different chemical reactions, but each stage generated unwanted products. This meant that there was a lot of waste, which was associated with expense and the potential for environmental harm. The process also resulted in only 40.1% of all the atoms in the reactants ending up in the ibuprofen molecules. During the 1990s, Boots™ developed an alternative process for the manufacture of ibuprofen that involved only three stages. Less waste was produced and 77.4% of atoms in the reactants ended up in the ibuprofen molecules. Scientists use the idea of atom economy to calculate percentages of waste products. You will find out more about atom economy in this chapter.

2.1 RELATIVE MASSES

Relative atomic mass, A_r

Relative atomic mass values for elements can be found on your Periodic Table.

The masses of atoms vary from 1×10^{-24} g for hydrogen to 1×10^{-22} g for the heavier elements. These are very small masses that are impossible to imagine. Small numbers like these also complicate calculations.

As described in Chapter 1, you can use the idea of relative atomic mass, A_r, with the mass of a carbon-12 atom as the standard, to deal with the mass of atoms.

Relative atomic mass is the average mass of an atom of an element compared with one-twelfth of the mass of an atom of carbon-12.

For an element,

$$A_r = \frac{\text{average mass of an atom of an element}}{\frac{1}{12} \text{ the mass of one atom of carbon-12}}$$

You use the average mass of an element because most elements have more than one isotope. Therefore, not all the element's atoms have the same mass.

For example, chlorine has two naturally occurring isotopes: chlorine-35 and chlorine-37. Any sample of chlorine contains 75.53% of $^{35}_{17}Cl$ and 24.47% of $^{37}_{17}Cl$. This gives a relative atomic mass of 35.453, which is usually rounded up to 35.5.

Relative molecular mass, M_r

For molecules rather than atoms, you use relative molecular mass, symbol M_r. As with A_r, M_r has no units.

Relative molecular mass (M_r) is the sum of the relative atomic masses of all the atoms in a molecule.

You can also define relative molecular mass as:

$$M_r = \frac{\text{average mass of one molecule of an element or compound}}{\frac{1}{12} \text{ the mass of one atom of carbon-12}}$$

Worked example 1

To calculate the relative molecular mass of ammonia, NH_3:

Step 1 Look up the relative atomic mass values for each atom in the formula: nitrogen $= 14$; hydrogen $= 1$.

Step 2 Add up the masses of all the atoms in the molecule:

nitrogen: $1 \times 14 = 14$; hydrogen: $3 \times 1 = 3$;
M_r of ammonia $= 17$.

QUESTIONS

1. Calculate the relative molecular mass of:
   a. sulfur dioxide, SO_2
   b. ethane, C_2H_6
   c. ethanol, C_2H_5OH
   d. phosphorus(V) chloride, PCl_5
   e. glucose, $C_6H_{12}O_6$

   (A_r: H 1; C 12; O 16; P 31; S 32; Cl 35.5)

Relative formula mass, M_f

Relative molecular mass applies to molecules, which are covalently bonded. Many of the formulae that you meet in this course have giant structures with ionic or covalent bonding. Sodium chloride has ionic bonding and consists of a large number of sodium ions and an equally large number of chloride ions held together in a lattice by electrostatic charges. The formula NaCl is called the **formula unit** and shows the ratio of each type of atom in the lattice. Silicon dioxide, SiO_2, has covalent bonding and a macromolecular structure (a giant covalent structure). The formula SiO_2 is the formula unit. The mass of the formula unit is called the relative formula mass. It has the symbol M_f, though you may find that M_r is still used. It is calculated in the same way as the relative molecular mass. Many ionic and molecular compounds have brackets in their formulae.

You can also define relative formula mass as:

$$M_f = \frac{\text{average mass of one formula unit of an element or compound}}{\frac{1}{12} \text{ the mass of one atom of carbon-12}}$$

Worked example 2

To find the relative formula mass of ammonium sulfate, $(NH_4)_2SO_4$:

Step 1 Look up the relative atomic mass value for each atom in the formula: nitrogen $= 14$; hydrogen $= 1$; sulfur $= 32$; oxygen $= 16$

Step 2 Add up the masses of the atoms in the formula: nitrogen: $2 \times 14 = 28$; hydrogen: $8 \times 1 = 8$; sulfur: $1 \times 32 = 32$; oxygen: $4 \times 16 = 64$. Total: 132

The relative formula mass of ammonium sulfate is 132.

Remember, atoms (or ions) inside a bracket are multiplied by the subscript number after the bracket.

QUESTIONS

2. Calculate M_f for the following:

   a. $MgBr_2$

   b. $Ca(OH)_2$

   c. $Al(NO_3)_3$

   d. $Al_2(SO_4)_3$

   e. $(CH_3COO)_2Ca$

   (A_r: H 1; C 12; N 14; O 16; Mg 24; Al 27; S 32; Ca 40; Br 80)

KEY IDEAS

 › Relative atomic mass is the average mass of an atom of an element compared with $\frac{1}{12}$ the mass of one atom of carbon-12.

 › Relative molecular mass is the sum of the relative atomic masses of all the atoms in a molecule.

 › Relative formula mass is the sum of the atoms that make a formula unit of a compound with a giant structure.

2.2 THE MOLE AND THE AVOGADRO CONSTANT

When ammonia is manufactured, the amount of product that can be generated is calculated from the amounts of reactants (nitrogen and hydrogen) that are used.

From the equation for the reaction you know that three hydrogen molecules react with one nitrogen molecule:

$$3H_2(g) + N_2(g) \rightarrow 2NH_3(g)$$

However, you cannot count out molecules to get reactants in the right proportion. Chemists count the number of particles in moles, for which we use the symbol mol. A mole of particles contains 6.023×10^{23} particles. For ammonia manufacturers, a mole of hydrogen refers to 6.023×10^{23} molecules of hydrogen and a mole of nitrogen refers to 6.023×10^{23} molecules of nitrogen.

But a mole can be 6.023×10^{23} particles of anything – atoms, molecules, ions, electrons. It is important to state the type of particles, for example, a mole of chlorine atoms, Cl, or a mole of chlorine molecules, Cl_2. A mole of chlorine molecules has twice the mass of a mole of chlorine atoms.

Since atoms have different masses, moles of atoms will have different masses. The mass of a mole of atoms of an element, in grams, is the relative atomic mass in grams.

The standard for relative atomic masses is the carbon-12 isotope of carbon. It is assigned the value 12.000 and all other atoms are measured relative to this.

One mole of carbon-12 atoms $(6.023 \times 10^{23} \text{ atoms})$ has a mass of 12.000 g.

Two moles of carbon-12 atoms $(1.205 \times 10^{24} \text{ atoms})$ have a mass of 24.000 g.

And so on.

The number 6.023×10^{23} is the Avogadro constant (symbol L). It was named after Amedeo Avogadro, a 19th-century Italian lawyer who was interested in mathematics and physics. He hypothesised that, at the same temperature and pressure, equal volumes of different gases contain the same numbers of particles. However, he did not calculate the number of particles in a mole. The first person to do this was an Austrian school teacher called Josef Loschmidt, in 1825.

Moles and relative mass

The relative atomic mass of carbon-12 is 12 and the relative atomic mass of magnesium-24 is 24. Therefore:

 › one magnesium-24 atom has twice the mass of one carbon-12 atom

 › 6.023×10^{23} magnesium atoms have twice the mass of 6.023×10^{23} carbon atoms

 › 1 mole of magnesium-24 atoms has twice the mass of 1 mole of carbon-12 atoms

 › 1 mole of carbon-12 atoms weighs 12 grams, 1 mole of magnesium-24 atoms weighs 24 grams.

The mass of a mole of molecules, in grams, is the relative molecular mass in grams, and the mass of a mole of formula units in grams is the relative formula mass in grams.

Earlier you learned that relative atomic mass is the average mass of an atom of an element compared with one-twelfth of the mass of an atom of carbon-12. Chlorine, for example, has two isotopes: chlorine-35 and chlorine-37. Taking into account their relative abundances, the relative atomic mass for chlorine is 35.5. It is these average values that we use in calculations. So, we can say that:

- 1 mole of carbon dioxide, CO_2, molecules has a mass of $12 + (2 \times 16) = 44$ g
- 0.1 mole of carbon dioxide molecules has a mass of 4.4 g
- 1 mole of sodium chloride, NaCl, has a mass of $23 + 35.5 = 58.5$ g
- 5 moles of sodium chloride has a mass of $58.5 \times 5 = 292.5$ g

Calculations involving the Avogadro constant

We can use the Avogadro constant to find the number of atoms or molecules in an amount of moles.

Worked example 3

(MS 0.1)

Helium is a monatomic gas, He. How many atoms are in 0.200 mol helium gas?

Step 1 1 mol helium contains 6.023×10^{23} atoms

Step 2 0.200 mol helium contains $0.2 \times 6.023 \times 10^{23} = 1.205 \times 10^{23}$ atoms

Carbon dioxide exists as molecules, CO_2. How many molecules are in 0.125 mol carbon dioxide?

Step 1 1 mol carbon dioxide contains 6.023×10^{23} molecules

Step 2 0.125 mol carbon dioxide contains $0.125 \times 6.023 \times 10^{23} = 7.529 \times 10^{22}$ molecules

QUESTIONS

3. How many atoms are there in:

a. 3.000 mol aluminium

b. 0.0100 mol argon

c. 0.2750 mol chromium

d. 1.750 mol lithium

e. 0.007 mol calcium?

4. How many molecules are there in:

a. 5.00 mol water

b. 0.725 mol sulfur dioxide

c. 20 mol oxygen

d. 0.025 mol ammonia

e. 0.001 mol hydrogen chloride?

Converting mass to moles

You will need to be able to convert the mass of a substance into the number of moles and vice versa.

To convert the mass of an element (consisting of atoms) to moles, you can use the formula:

$$\text{mass} = A_r \times \text{number of moles}$$

To convert the mass of a substance consisting of molecules to moles, simply substitute M_r for A_r.

You may find the triangle in Figure 1 useful, where mass = moles × M_r, moles = mass ÷ M_r and M_r = mass ÷ moles.

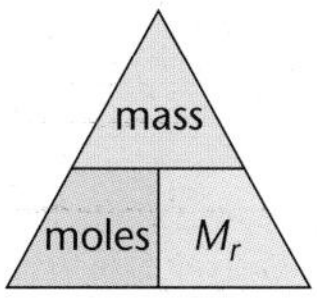

Figure 1 *This relationship triangle can help you to calculate the mass and the number of moles of substances.*

Worked example 4

(MS 2.1 and 2.2)

How many moles are there in 414 g of lead?

Step 1 A_r lead = 207

Step 2 $\text{moles} = \frac{414}{207} = 2 \text{ mol}$

How many moles are there in 1 kg of glucose, $C_6H_{12}O_6$?

Step 1
M_r glucose $= (6 \times 12) + (12 \times 1) + (6 \times 16) = 180$

1 kg = 1000 g

Step 2 $\text{moles} = \frac{1000}{180} = 5.56 \text{ mol}$

Converting moles to mass

For elements consisting of atoms:

$$\text{mass (g)} = \text{moles} \times \text{relative atomic mass } (A_r)$$

For substances consisting of molecules or formula units:

$$\text{mass (g)} =$$

$$\text{moles} \times \text{relative molecular or formula mass } (M_r)$$

Worked example 5

(MS 2.1 and 2.2)

What is the mass in grams of 2 moles of argon gas? (A_r: Ar 40)

$$\text{mass (g)} = 2 \times 40$$

$$= 80 \text{ g}$$

What is the mass in grams of 2.5 moles of ethanol, C_2H_5OH?

Step 1 M_r ethanol $= (2 \times 12) + (5 \times 1) + 16 + 1 = 46$

Step 2 mass (g) $= 2.5 \times 46$

$$= 115 \text{ g}$$

Figure 2 *The photograph shows 1 mole of different compounds. They have different masses, but they each contain 6.023×10^{23} formula units.*

QUESTIONS

5. Calculate the number of moles of:

   a. atoms in 6.00 g magnesium

   b. formula units in 60.0 g calcium carbonate, $CaCO_3$

   c. molecules in 109.5 g hydrogen chloride, HCl

   d. formula units in 303 g potassium nitrate, KNO_3

   e. formula units in 26.5 g anhydrous sodium carbonate, Na_2CO_3

   f. molecules in 4.00 g hydrogen gas

   g. atoms in 4.00 g hydrogen gas

   h. atoms in 336 g iron

   i. atoms in 27.0 kg aluminium

   j. molecules in 9.80 g sulfuric acid, H_2SO_4

6. Calculate the mass in grams of:

   a. 0.500 mol chromium

   b. 0.200 mol bromine atoms

   c. 10.0 mol lead

   d. 0.100 mol zinc(II) chloride, $ZnCl_2$

   e. 0.500 mol potassium hydroxide, KOH

   f. 0.010 mol ethanol, C_2H_5OH

   g. 0.001 mol sulfuric acid

   h. 5.00 mol nitrogen atoms

   i. 5.00 mol nitrogen molecules

   j. 0.025 mol sodium hydroxide, NaOH

More calculations using the Avogadro constant

We can use the Avogadro constant to work out the number of atoms or molecules in the mass of a substance (Figure 2).

Worked example 6

(MS 0.1, 2.1 and 2.2)

How many atoms are in 19.70 g gold (Au)?

Step 1: convert the mass to moles. (A_r: Au 197)

$$19.70 \text{ g gold} = \frac{19.70}{197} \text{mol gold}$$

$$= 0.10 \text{ mol}$$

Step 2: 0.10 mol contains $0.1 \times 6.023 \times 10^{23}$ atoms $= 6.023 \times 10^{22}$ atoms

How many molecules are in 3.910 g ammonia?

Step 1: convert mass to moles. (A_r: H 1; N 14)

$$M_r \text{ NH}_3 = 14 + (3 \times 1) = 17$$

$$3.910 \text{g NH}_3 = \frac{3.910}{17} \text{mol NH}_3$$

$$= 0.230 \text{ mol}$$

Step 2: 0.23 mol contains $0.23 \times (6.023 \times 10^{23})$

molecules $= 1.385 \times 10^{23}$ molecules

You may find the diagram in Figure 3 useful when doing calculations involving the Avogadro constant.

Remember, if you are given a mass in kilograms, you will need to convert it to grams first.

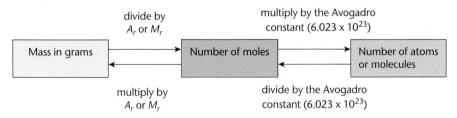

Figure 3 *This diagram shows the relationship between mass, moles and the number of atoms or molecules in an element or compound.*

QUESTIONS

7. How many atoms are there in:
   a. 0.040 g helium
   b. 2.30 g sodium
   c. 0.054 g aluminium
   d. 2.10 g lithium
   e. 0.238 g uranium?

8. How many molecules are there in:
   a. 4.00 g oxygen gas
   b. 1.80 g water
   c. 0.014 g carbon monoxide
   d. 0.335 g chlorine gas
   e. 2.00 kg water?

   (A_r: H 1; He 4; Li 6.9; C 12; O 16; Na 23.0; Al 27.0; Cl 35.5; U 238.0)

ASSIGNMENT 1: HOW MANY ATOMS ARE THERE IN THE WORLD?

(MS 0.0, 0.1, 0.2, 0.4; PS 1.1. 1.2, 3.2)

The world is made up of atoms too numerous to count, but you can estimate the number of atoms if you know the mass of the Earth. This is estimated to be 5 980 000 000 000 000 000 000 000 000 grams or, put in standard form, 5.98×10^{27} g.

All atoms are tiny (see Figure A1), but atoms of different elements have different masses. Table A1 gives the percentage abundance of the most common types of atom that make up the Earth. Most percentage abundance figures for elements in the Earth only include the amount of element found in the Earth's crust. So, as we do not know exactly what is in the centre of the Earth, a bit of estimating based on some good scientific theories is needed.

Table A1 also includes the mass of a mole of atoms of each element. This is the relative atomic mass of the element in grams. Moles are used to count particles. One mole contains 6.023×10^{23} particles.

Figure A1 *An image of gold atoms obtained using scanning tunnelling microscopy. Gold atoms are about 0.1441 nm in diameter. That is 1.441×10^{-10} m.*

Element	Mass of 1 mole/g	Percentage of the Earth/%	Mass in the Earth/g	Number of atoms
iron	55.8	35	2.09×10^{27}	2.26×10^{49}
oxygen	16.0	30	1.79×10^{27}	6.75×10^{49}
silicon	16.0	15	8.97×10^{26}	1.92×10^{49}
magnesium	24.3	13	7.77×10^{26}	1.93×10^{49}
sulfur	32.1	2	1.20×10^{26}	2.24×10^{48}
calcium	40.1	1	5.98×10^{25}	8.98×10^{47}
aluminium	27.0	1	5.98×10^{25}	1.33×10^{48}

Table A1 *The abundance and estimated mass of the most common elements that make up the Earth*

Questions

A1. If the Earth has a mass of 5.98×10^{27} g, what is the mass of iron in the Earth?

A2. How many moles of iron are in the Earth? [A_r: Fe 55.8]

A3. If one mole of particles contains 6.023×10^{23} particles, how many iron atoms are there in the Earth?

Stretch and challenge

A4. Repeat the steps in questions A1, A2 and A3 for the other elements in the table and calculate the total number of atoms in the Earth.

2.3 THE IDEAL GAS EQUATION

Measuring the mass of a gas is not easy. Gases are usually measured by volume, but the volume of a gas depends upon its temperature and pressure. So, to find the mass of a gas after measuring its volume, you need to understand the way that gases change in volume when temperature and pressure change. Each gas behaves very slightly differently compared with other gases but, for most practical purposes, the differences are small enough to assume that gases all behave like an imaginary 'ideal gas'.

An **ideal gas** has a number of assumed properties:

> It is made up of identical particles in continuous random motion.

> The particles can be thought of as point-like; with position but with zero volume (which means that the volume of the gas particles is taken to be zero).

> The particles do not react when they collide.

> Collisions between particles are perfectly elastic – the total kinetic energy (energy of motion) of the

particles after a collision is the same as that before the collision.

> The particles have no intermolecular forces, meaning that they do not attract or repel each other. (You will find out more about intermolecular forces in Chapter 3.)

As shown in Figure 4, no real gas follows this model exactly. In an ideal gas there are no intermolecular forces, so the particles are not attracted to each other. In a real gas, attractive intermolecular forces divert the paths of particles. These forces explain why, under the right conditions, gases can be liquefied.

Some gases do behave like an ideal gas over a limited range of temperatures and pressures. Hydrogen, nitrogen, oxygen and the inert (noble) gases behave most like ideal gases.

The effect of pressure

Robert Boyle was one of the first chemists to study the behaviour of gases under different conditions. He noticed that when he kept the number of moles of gas and the temperature constant, but increased the

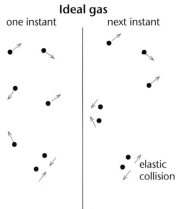

Ideal gas

one instant | next instant

elastic collision

There are no intermolecular forces, so the particles are not attracted to each other

Real gas

one instant | next instant

close particles interact

collision not elastic

Attractive intermolecular forces divert the paths of particles

Figure 4 *Movement of particles in an ideal gas and a real gas*

pressure on the gas, then its volume became smaller. In 1662, based on this evidence, he stated:

> At constant temperature T, the volume V of a fixed mass of gas is inversely proportional to the pressure p applied to it.

When this theory proved generally to be true, it became known as Boyle's law. You can write it mathematically as:

$$p \, \alpha \frac{1}{V}$$

when T and the mass of gas are constant,

which is the same as:

$$p \times V = \text{constant}$$

when T and the mass of gas are constant.

The effect of temperature

About one hundred years after Robert Boyle made his conclusions, Frenchman Jacques Charles was studying during a period when ballooning was all the rage in France. Balloons were initially filled with hydrogen. Since the density of hydrogen was less than that of air, balloons filled with enough hydrogen floated up from the ground. But disastrous fires, exemplified by the Hindenburg disaster (Figure 5), put an end to the use of hydrogen. Ballons were subsequently filled with air that was heated to reduce its density. As soon as the total mass of the balloon, passengers and air in the balloon was less than that of the air displaced, the balloon could rise into the air. This is the method that is still used today (Figure 6).

Charles was looking at the effect of changing the temperature of gas and measuring the resulting

Figure 5 *The Hindenburg was the largest airship that used a hydrogen-filled balloon. In 1937, it burst into flames when trying to dock at the end of its first North American transatlantic journey. This marked the end of the airship era.*

changes in volume. As temperature was increased, the gases expanded because their molecules moved faster and were further apart, so their density decreased. As temperature was decreased, the gases became more dense as they contracted. The effect was the same for a wide range of gases and, in 1787, Charles published his law:

> At constant pressure, the volume of a fixed mass of gas is proportional to its temperature.

$$V = \text{constant} \times T$$

when p and the mass of gas are constant.

The equation for Charles's law implies that, at constant pressure, as the temperature goes down, the volume of any sample of gas decreases until, at

37

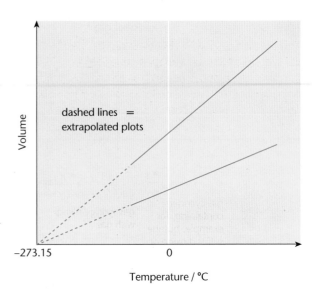

Figure 6 *Balloon rides today use hot air and work on the principle that air expands and becomes less dense.*

Figure 7 *The relationship between volume of a gas and temperature*

a certain very low temperature, the volume becomes zero. Plotting the volumes of most gases against temperature and extrapolating (Figure 7) produces a surprising result – they would all reach zero volume at the same temperature, −273.15 °C, which is known as the **absolute zero** temperature.

At this temperature, the atoms or molecules are assumed to have no kinetic energy, to have ceased moving and colliding, and to be so close together as to occupy a negligible volume. Of course, at normal pressures, the gases would be solids at −273.15 °C, which is why you have to extrapolate to obtain the value (Figure 7). On the Kelvin scale, this temperature is 0 kelvin (0 K), at which an ideal gas is assumed to occupy zero volume. In your calculations using gas laws, always remember to state temperatures in kelvin, converting degrees celsius to kelvin by adding 273.

Combining Boyle's law and Charles's law

Boyle's law and Charles's law can be combined into a single ideal gas equation, but one more piece of information is needed, which was discovered by the Italian Amedeo Avogadro. He found that:

> Equal volumes of all gases at the same pressure and temperature contain the same number of particles.

This means that the molar volume, which is the volume occupied by one mole of a gas, is the same for all gases at the same temperature and pressure. For example, Figure 8 shows that the volume of a mole of different gases at 298 K (25 °C) is 24 dm^3.

As a consequence, the volume of a gas is proportional to the number of moles of the gas present when the pressure and temperature are constant. Thus:

$$V = \text{constant} \times n$$

when p and T are constant and n is the number of moles of the gas.

The **ideal gas equation** combines all the equations above for gases into one equation:

$$pV = nRT$$

where:

p = pressure of the gas in pascals (Pa)

V = volume of the gas in m^3

n = number of moles of the gaseous particles

R = molar gas constant $\left(8.31\,\text{J}\,\text{K}^{-1}\text{mol}^{-1}\right)$

T = temperature in kelvin (K)

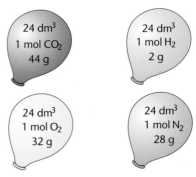

Figure 8 *The molar volume of gases at 298 K and 100 kPa*

Determining the molar gas constant

The apparatus shown in Figure 9 can be used to calculate the values for all the quantities in the ideal gas equation that are needed to calculate R. It includes a gas syringe with a friction-free plunger, which allows gas in the syringe to reach atmospheric pressure.

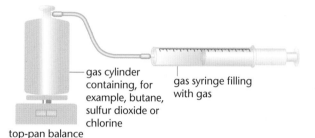

gas cylinder containing, for example, butane, sulfur dioxide or chlorine

gas syringe filling with gas

top-pan balance

Figure 9 *This apparatus is used to calculate the molar gas constant, R, used in the ideal gas equation. (Drawing not to scale.)*

The syringe is filled several times from the cylinder. This gives a sum of volumes $= V$

The total mass of gas lost from the cylinder is recorded and the number of moles calculated $= n$

The temperature is recorded $= T$

The atmospheric pressure is recorded $= p$

Then the gas constant is calculated from:

$$R = \frac{pV}{nT}$$

You need to be able to use the ideal gas equation to carry out calculations made using this apparatus, and other examples.

Finding the M_r of a volatile liquid

The apparatus used to find M_r of a volatile liquid this is shown in Figure 10. A small syringe is used to inject a known mass of the volatile liquid into a large gas syringe. At the recorded temperature, the liquid changes to a gas and its volume in the large syringe is measured.

Some sample results from an experiment to find the M_r of hexane are given. You can use these data to calculate the M_r of hexane.

volume of air $= 10$ cm^3

volume of vapour $+$ air $= 100$ cm^3

mass of sample $= 2.59$ g

temperature $= 87$ °C

pressure $= 100$ kPa [this is a pressure of 1 atmosphere (atm)]

First, convert the measurements to SI units (as in the gas equation):

volume of sample $= 100 - 10$ cm^3

$= 90$ cm^3

$= 90 \times 10^{-6}$ m^3

temperature $= 87 + 273$ K

$= 360$ K

pressure $= 100\ 000$ Pa

Rearrange the ideal gas equation to give M_r as the subject:

$$\text{number of moles} = \frac{\text{mass (g)}}{M_r}$$

$$pV = \frac{mRT}{M_r}$$

$$M_r = \frac{mRT}{pV}$$

Insert values into the ideal gas equation:

$$M_r = \frac{2.59 \times 8.31 \times 360}{100\ 000 \times 90 \times 10^{-6}}$$

$$M_r = 86$$

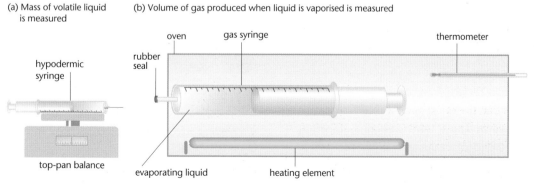

(a) Mass of volatile liquid is measured

(b) Volume of gas produced when liquid is vaporised is measured

hypodermic syringe

top-pan balance

oven

rubber seal

gas syringe

evaporating liquid

heating element

thermometer

Figure 10 *Apparatus to find the M_r of a volatile liquid*

Remember that the answer can only contain two significant figures because the volume of the sample is given as 90 cm^3. Do not round up or down until the end of the calculation because you may lose accuracy.

Calculating moles from gas volume

Worked example 7

(MS 2.1, 2.2, 2.3, and 2.4)

How many moles of oxygen are there in 500 cm^3 of gas at 25 °C and 100 KPa?

Step 1 Convert the units to the units of the ideal gas equation:

$p = 100$ kPa $= 100\,000$ Pa

$V = 500$ cm$^3 = 500 \times 10^{-6}$m^3

$T = 25$ °C $= 298$ K

Step 2 Rewrite the ideal gas equation to make n the subject:

$$n = \frac{pV}{RT}$$

Step 3 Insert the values:

$$n = \frac{105 \times 500 \times 10^{-6}}{8.31 \times 298}$$

moles of oxygen $= 0.0200$ mol

Calculating the volume of a reactant gas and the mass of a gas product

Worked example 8

(MS 2.1, 2.2, 2.3, and 2.4)

Methane reacts with oxygen to produce carbon dioxide and water. Calculate the volume of oxygen needed to react with 20 dm^3 methane, and the mass of carbon dioxide produced at 120 kPa and 30 °C.

The equation for the reaction is:

$CH_4\,(g) + 2O_2\,(g) \rightarrow CO_2\,(g) + 2H_2O\,(l)$

Step 1 The equation shows that one mole of methane reacts with two moles of oxygen to give one mole of carbon dioxide.

From Avogadro's rule, at constant temperature and pressure, 20 dm^3 of methane therefore requires 40 dm^3 oxygen for the reaction.

The volume of carbon dioxide produced is 20 dm^3 .

Step 2 To find the mass of 20 dm^3 CO_2 , first use the ideal gas equation to find the number of moles.

Using $n = \frac{pV}{RT}$, insert the values converted to the correct units:

$$n = \frac{120 \times 100 \times 20 \times 10^{-3}}{8.31 \times 303}$$

$= 0.953$ mol

The reaction produces 0.95 mol of carbon dioxide.

Step 3 Convert mass (g) into moles:

mass (g) $=$ moles $\times M_r$

$= 0.953 \times 44$

$= 41.9$ g

The mass of carbon dioxide produced in the reaction $= 41.9$ g.

Calculating the volume of gas produced in a reaction involving a non-gas reactant

Worked example 9

(MS 2.1, 2.2, 2.3, and 2.4)

When potassium nitrate is heated, it gives off oxygen and becomes potassium nitrite. Calculate the volume, in dm^3, of oxygen produced from 345 g potassium nitrate at 38 °C and 100 kPa.

The equation for the reaction is:

$2KNO_3\,(s) \rightarrow 2KNO_2\,(s) + O_2\,(g)$

Step 1 Work out the number of moles in 345 g of potassium nitrate:

$M_r\,KNO_3 = 101 \qquad (A_r:$ K 39; N 14; O 16$)$

moles $KNO_3 = \frac{mass}{M_r}$

$= \frac{345}{101}$

$= 3.42$ moles

Step 2 From the reaction equation, calculate the moles of O_2:

3.42 moles KNO_3 produces 0.5×3.42 moles $O_2 = 0.171$ moles

Step 3 Work out the temperature in kelvin:

$T = 38 + 273 = 311 \, \text{K}$

Step 4 Write the ideal gas equation, making V the subject, and insert the values:

$$V = \frac{nRT}{p}$$

$$V = \frac{1.71 \times 8.31 \times 31}{100 \times 10^3}$$

$$= 0.0442 \, \text{m}^3 \, O_2$$

The volume of oxygen produced is $44 \, \text{dm}^3$.

QUESTIONS

9. a. Convert these temperatures into kelvin:
   i. 25 °C
   ii. 250 °C
   iii. −78 °C

   b. What is the volume, in m^3, of 5 mol of oxygen gas at 25 °C and 100 kPa?

   c. What volume, in dm^3, of hydrogen gas is produced when 19.5 g zinc metal dissolves in excess hydrochloric acid at 30 °C and 100 kPa?

   d. A fairground balloon is filled with $1000 \, \text{cm}^3$ of helium gas. The temperature is 25 °C and the pressure is 100 kPa. How many moles of gas does the balloon contain?

KEY IDEAS

> Theories and equations about the behaviour of gases assume that all gases are ideal gases.

> Ideal gases have point-like particles in random motion. The particles do not react on collision, undergo elastic collisions and there are no intermolecular forces between them.

> The ideal gas equation is $pV = nRT$, where p = pressure (Pa), V = volume (m^3), n = amount (mol), R = molar gas constant, T = temperature (K).

2.4 EMPIRICAL AND MOLECULAR FORMULAE

Remember:

> The empirical formula is the simplest whole number ratio of atoms of each element that are in a compound.

> The chemical formula (molecular formula or formula unit) is the actual number of atoms of each element used to make a molecule of formula unit.

When chemists need to find the composition of a compound, they measure the mass of each element in that compound. They use this information to work out the empirical formula of the compound. The empirical formula gives the simplest ratio of atoms of each element present in the compound. For example, the molecular formula of ethane is C_2H_6, but it has the empirical formula CH_3, because the simplest whole number ratio of carbon to hydrogen is 1:3.

As an example of calculating an empirical formula, consider a compound that was found to contain 40% by mass of calcium, 12% by mass of oxygen and 48% by mass of oxygen. From these figures, you can calculate that 100 g of the compound would contain 40 g calcium, 12 g carbon and 48 g oxygen. If you convert these masses to amounts in moles, you will know the ratio of each element (Table 1).

Element	Calcium	Carbon	Oxygen
mass/g	40	12	48
amount/mol	$\frac{40}{40}$	$\frac{12}{12}$	$\frac{48}{16}$
ratio of elements	1 :	1 :	3

Table 1

If the substance has a lattice structure (either bonded covalently or ionically), the **formula unit** shows the ratio of atoms of each element in the substance. For example, calcium carbonate, $CaCO_3$, consists of ions held in a giant structure, as shown in Figure 11. The formula unit is $CaCO_3$, showing that it is made from one calcium atom, one carbon atom and three oxygen atoms.

The ratio of atoms in a chemical formula (molecular formula or formula unit) stays the same.

Therefore, one molecule of CO_2 is made from one atom of carbon and two atoms of oxygen, and 100 molecules

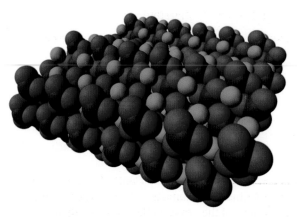

Figure 11 *Calcium carbonate giant lattice. The green spheres represent calcium ions, Ca^{2+}, and the grey spheres each with three red spheres attached represent carbonate ions, CO_3^{2-} (grey for carbon and red for oxygen).*

Element	Carbon	Hydrogen	Oxygen
mass/g	0.34	0.057	0.46
amount/mol	$\dfrac{0.34}{12}$	$\dfrac{0.057}{1}$	$\dfrac{0.46}{16}$
=	0.028	0.057	0.029
ratio of elements (divide by the lowest)	$\dfrac{0.028}{0.028}$	$\dfrac{0.057}{0.028}$	$\dfrac{0.029}{0.029}$
=	1.0 :	2.0 :	1.0

Table 2

Step 3 Since the M_r of the carbohydrate is 180, if you divide 30 into 180, the answer will show how many times you need to multiply the empirical formula to give the molecular formula:

$$\frac{180}{30} = 6$$

The molecular formula is $6 \times$ (empirical formula)

$= 6 \times (CH_2O)$

$= C_6H_{12}O_6$

of CO_2 are made from 100 atoms of carbon and 200 atoms of oxygen.

Since you can count the number of particles in moles:

one mole of CO_2 is made from one mole of carbon atoms and two moles of oxygen atoms.

Calculating a molecular formula

Worked example 10

(MS 0.2)

An investigation to identify a carbohydrate found that it contained 0.34 g carbon, 0.057 g hydrogen and 0.46 g oxygen. Find its empirical formula.

Step 1 Note that the mass of each element is given in the question, rather than the percentage composition. Treat this in exactly the same way as described earlier. You may need some common sense to sort out the whole-number ratio (Table 2).

The empirical formula is CH_2O.

Step 2 To find the molecular formula, you need to know the relative molecular mass.

If the M_r of this carbohydrate is 180, then to calculate the molecular formula you first find the M_r of the empirical formula:

$M_r\ CH_2O = 12 + (2 \times 1) + 16$

$= 30$

QUESTIONS

10. **a.** Find the empirical formula of a compound containing 84 g magnesium and 56 g oxygen.

 b. 3.36 g of iron combine with 1.44 g of oxygen to form an oxide of iron. What is its empirical formula?

 c. A compound contains, by mass, 20.14% iron, 11.51% sulfur, 63.31% oxygen and 5.04% hydrogen. Find its empirical formula.

 d. A hydrocarbon contains 82.8% by mass of carbon. Find its empirical formula. (Hint: what types of atoms are hydrocarbons made from?)

 e. Another hydrocarbon has a relative molecular mass of 28 and contains 85.7% by mass of carbon and 14.3% by mass of hydrogen. Calculate its empirical formula and its molecular formula.

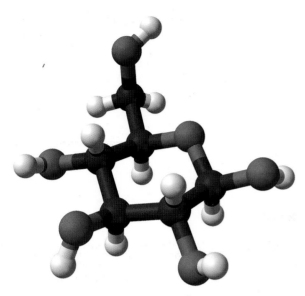

Figure 12 *Model of a glucose molecule, which has the empirical formula CH_2O. The molecular formula is $C_6H_{12}O_6$. The grey spheres represent carbon, the white are hydrogen and the red are oxygen atoms.*

KEY IDEAS

 ❭ A mole contains 6.023×10^{23} particles.

 ❭ The empirical formula is the simplest whole number ratio of atoms of each element used to make a compound.

 ❭ The molecular formula is the actual number of atoms of each element used to make a molecule.

 ❭ The chemical formula of an element or a compound with a giant structure is the number of atoms of each element used to make a formula unit.

2.5 CHEMICAL EQUATIONS

Balancing chemical equations

A **chemical equation** is a shorthand way of writing a chemical reaction. The equation uses **chemical formulae** for the reactants and the products. Formulae use letter symbols to represent the elements and numbers to show how many atoms of each element are involved.

A chemical equation can also include the **state** of reactants and products:

(s) means the solid state

(l) means the liquid state

(g) means the gaseous state

(aq) means aqueous, that is, in aqueous (dissolved in water) solution

As an example, sulfur burns in oxygen to produce sulfur dioxide. This is how you write the reaction using chemical formulae:

reactants   product

$$S(s) + O_2(g) \rightarrow SO_2(g)$$

A chemical equation must balance. This means there must be the same number of each type of atom on both sides of the arrow. In this example, there is one sulfur atom on the left and one sulfur atom on the right. There are two oxygen atoms on the left and two oxygen atoms on the right. So, the equation is balanced. We call it a balanced equation.

Sulfur dioxide reacts with oxygen to produce sulfur trioxide. You could use chemical formulae to write the reaction like this:

$$SO_2(g) + O_2(g) \rightarrow SO_3(g)$$

Is this equation balanced? There is one sulfur atom on the left and one on the right. But there are four oxygen atoms on the left and only three on the right. So, the equation needs to be altered so that it is balanced.

An SO_2 molecule is made from one S atom and two O atoms. An SO_3 molecule is made from one S atom and three O atoms. So, one extra O atom is needed to make SO_3 rather than SO_2. One oxygen molecule is made from enough oxygen atoms to react with two molecules of sulfur dioxide and produce two molecules of sulfur trioxide. So, we can now write a balanced equation:

$$2SO_2(g) + O_2(g) \rightarrow 2SO_3(g)$$

Figure 13 *The manufacture of sulfuric acid includes two oxidation reactions: $S(s) + O_2(g) \rightarrow SO_2(g)$ and $2SO_2(g) + O_2(g) \rightarrow 2SO_3(g)$*

This is a very simple equation. Others are more complicated, but the basic procedure is the same. Take it step by step and you should have no problems. Balancing chemical equations is trial and error. With each step you take, you need to check whether you need more steps to rebalance the equation. Worked example 11 shows how to balance the equation for ammonia burning in oxygen.

Worked example 11

$$NH_3 + O_2 \rightarrow NO_2 + H_2O$$

Look at nitrogen. There is one atom on either side, so the nitrogen balances.

Look at hydrogen. There are three atoms on the left and two on the right. Use two ammonia molecules and three water molecules to give six hydrogen atoms on each side:

$$2NH_3 + O_2 \rightarrow NO_2 + 3H_2O$$

Now the nitrogen is unbalanced, so double the nitrogen dioxide [nitrogen(IV) oxide] on the right:

$$2NH_3 + O_2 \rightarrow 2NO_2 + 3H_2O$$

Now check on the oxygen. There are two atoms on the left and seven on the right. First, double both sides to get even numbers:

$$4NH_3 + 2O_2 \rightarrow 4NO_2 + 6H_2O$$

The nitrogen and hydrogen are still balanced. You have 14 oxygen atoms on the right, so you need seven oxygen molecules on the left:

$$4NH_3 + 7O_2 \rightarrow 4NO_2 + 6H_2O$$

Add state symbols:

$$4NH_3(g) + 7O_2(g) \rightarrow 4NO_2(g) + 6H_2O(g)$$

QUESTIONS

11. Balance these equations:
    a. $C_2H_5OH(l) + O_2(g) \rightarrow CO_2(g) + H_2O(g)$
    b. $Al(s) + NaOH(aq) \rightarrow Na_3AlO_3(aq) + H_2(g)$
    c. $CO_2(g) + H_2O(l) \rightarrow C_6H_{12}O_6(aq) + O_2(g)$
    d. $C_3H_8(g) + O_2(g) \rightarrow CO_2(g) + H_2O(l)$
    e. $Al(s) + H_2SO_4(l) \rightarrow Al_2(SO_4)_3(aq) + SO_2(g) + H_2O(l)$

12. Write balanced symbol equations for these reactions, and include state symbols.
    a. iron(III) chloride + ammonia in water $\rightarrow$ iron(III) hydroxide + ammonium chloride
    b. copper(II) carbonate + hydrochloric acid $\rightarrow$ copper(II) chloride + carbon dioxide + water
    c. sodium hydroxide + phosphoric acid $(H_3PO_4) \rightarrow$ sodium phosphate + water (phosphate is PO_4^{3-})
    d. iron + chlorine $\rightarrow$ iron(III) chloride
    e. copper(II) oxide + sulfuric acid $\rightarrow$ copper(II) sulfate + water

Using chemical equations and moles

Chemical equations enable you to calculate the masses of reactants and the masses of products formed in a chemical reaction.

Look at this reaction:

$$CaCO_3(s) + 2HCl(aq) \rightarrow CaCl_2(s) + CO_2(g) + H_2O(l)$$

To carry out the reaction without leaving a surplus of either reactant, one mole of $CaCO_3$ is needed for every two moles of HCl. So, you can write the moles in the reaction as:

1 mol $CaCO_3$ reacts with 2 mol HCl to produce 1 mol $CaCl_2$ + 1 mol CO_2 + 1 mol H_2O

But since you can calculate the mass of a mole, you can take a given mass of calcium carbonate and work out exactly what mass of hydrochloric acid in solution reacts with it. You can also calculate the mass of each product that is made.

Calcium carbonate decomposes when it is heated in a furnace to make calcium oxide, sometimes called quicklime. The simple equation is:

$$CaCO_3(s) \rightleftharpoons CaO(s) + CO_2(g)$$

This is a reversible reaction, but if the carbon dioxide is allowed to escape, then all the calcium carbonate decomposes.

So, one mole of calcium carbonate breaks down to give one mole of calcium oxide and one mole of carbon dioxide.

Calculating the mass of product formed

Worked example 12

(MS 2.1, 2.2, 2.3 and 2.4)

How much calcium oxide could be obtained from 800 kg of calcium carbonate?

Step 1 Write a balanced equation:

$$CaCO_3 (s) \rightleftharpoons CaO (s) + CO_2 (g)$$

Step 2 Calculate the M_f for the calcium carbonate and calcium oxide:

(A_r: Ca 40; C 12; O 16)

M_f $CaCO_3$ = 40 + 12 + 48 = 100

M_f CaO = 40 + 16 = 56

Step 3 Calculate the moles of $CaCO_3$ used:

$$moles = \frac{mass\ of\ CaCO_3 (s)}{M_f}$$

$$= \frac{800 \times 10^3}{100}$$

$$= 8000\ mol$$

Step 4 From the equation, find the moles of calcium oxide produced:

1 mol $CaCO_3$ produces 1 mol CaO

8000 mol $CaCO_3$ produces 8000 mol CaO

Step 5 Convert the moles of calcium oxide into mass (g)

$$Mass\ (g) = moles \times M_f$$

$$= 8000 \times 56$$

$$= 448\ 000\ g$$

$$Mass = 448\ kg$$

Calculating reacting masses

Worked example 13

(MS 2.1, 2.2, 2.3 and 2.4)

Copper carbonate reacts with hydrochloric acid to produce copper chloride. What mass of copper chloride is made when 24.7 g copper carbonate react with excess acid?

Step 1 Write a balanced equation:

$$CuCO_3 (s) + 2HCl (aq) \rightarrow CuCl_2 (aq) + H_2O (l) + CO_2 (g)$$

Step 2 Calculate the M_r for the substances involved in the question.

(A_r: Cu 63.5; C 12; O 16; Cl 35.5)

M_f $CuCO_3$ = 63.5 + 12 + 48

$$= 123.5$$

M_f $CuCl_2$ = 63.5 + (2 × 35.5)

$$= 134.5$$

Step 3 Calculate the moles of $CuCO_3$ used:

$$moles\ CuCO_3 = \frac{mass (g)}{M_f}$$

$$= \frac{24.7}{123.5}$$

$$= 0.2\ mol$$

Step 4 Use the equation to find the moles of $CuCl_2$ produced:

1 mol $CuCO_3 \rightarrow$ 1 mol $CuCl_2$;
0.2 mol $CuCO_3 \rightarrow$ 0.2 mol $CuCl_2$

M_f $CuCl_2$ = 63.5 + (2 × 35.5) = 134.5

Step 5 Convert the moles of $CuCl_2$ into mass (g)

$$mass\ (g) = moles \times M_f$$

$$= 0.2 \times 134.5$$

$$= 26.8\ g$$

Percentage yield

Worked example 12 showed that 800 kg of calcium carbonate could produce 448 kg of calcium oxide. This is called the theoretical yield. The reality is that, although industries strive to achieve the maximum yield possible and make the maximum profit, it is never possible to achieve the theoretical yield. Calcium carbonate only decomposes when the temperature is greater than 1000 °C. The reaction is also reversible, so complete decomposition depends on the carbon dioxide escaping. Some calcium oxide may be lost when transferring it from the kiln to the next stage.

The mass of product actually obtained is called the actual yield. It can only be found by actually doing the reaction. Percentage yield is used to express how close the actual yield is to the theoretical yield.

$$\text{percentage yield} = \frac{\text{actual yield}}{\text{theoretical yield}} \times 100$$

Worked example 14

(MS 0.2, 2.1, 2.2, 2.3 and 2.4)

Calcium oxide is used to make cement (Figure 14). If 800 kg of calcium carbonate produced 314 kg of calcium oxide in an actual process, calculate the percentage yield.

Figure 14 *In the cement kiln, calcium oxide is heated with silicon dioxide and aluminium oxide at 1400 °C. Calcium sulfate is added to the produce from the kiln and the mixture is then ground to the fine grey powder you probably know as Portland cement. Cement manufacturers are actively researching ways to increase their percentage atom ecomony.*

Step 1 Calculate the theoretical yield from the equation:

$$CaCO_3(s) \rightleftharpoons CaO(s) + CO_2(g)$$

800 kg calcium carbonate can produce a theoretical yield of 448 kg calcium oxide (from the calculation in Worked example 12).

Step 2 $\text{percentage yield} = \dfrac{\text{actual yield}}{\text{theoretical yield}} \times 100$

$= \dfrac{314}{448} \times 100\%$

$= 70.0\%$

Percentage atom economy

Chemists have traditionally measured the efficiency of a reaction by its percentage yield. However, this is only half the story because many atoms in the reactants are not needed in the product. They make up the waste or side products.

Percentage atom economy is a way to compare the maximum mass of a product that can be obtained with the mass of the reactants. It is calculated using the equation:

percentage atom economy =

$$\frac{\text{molecular mass of desired product}}{\text{sum of molecular masses for all reactants}} \times 100$$

Developing chemical processes with high atom economies can have economic, ethical and environmental benefits for industry and the society it serves.

Copper can be extracted by heating copper(II) oxide with carbon, as in the equation:

$$2CuO(s) + C(s) \rightarrow 2Cu(s) + CO_2(g)$$

Two moles (159 g) of copper(II) oxide react with one mole (12 g) of carbon. The reactants need to be added in these reacting quantities. Theoretically, the reaction can produce 127 g of copper from 159 g of copper oxide and 12 g of carbon. So:

$$\text{percentage atom economy} = \frac{127}{171} \times 100$$

$$= 74\%$$

The reaction uses 74% of the mass of the reactants to give the product required. The remaining 26% ends up as waste or side products, assuming the percentage yield is 100%. The reality is that the yield will be lower because the percentage yield is unlikely to be 100%.

QUESTIONS

13. **a.** What theoretical yield of iron can be obtained from reacting 320 tonnes of iron(III) oxide with carbon monoxide in the blast furnace? The equation for the reaction is:

    $$Fe_2O_3(s) + 3CO(g) \rightarrow 2Fe(s) + 3CO_2(g)$$

    b. If the actual yield in a. is 200 tonnes, what is the percentage yield?

    c. What is the percentage atom economy of the reaction in a?

    d. Sulfuric acid is reacted with calcium carbonate to produce calcium sulfate for use in making plaster. Calculate the theoretical mass of calcium sulfate that can be made from 490 tonnes of sulfuric acid.

e. If the actual yield is 550 tonnes of calcium sulfate, calculate the percentage yield.

f. Calculate the percentage atom economy for the reaction in d.

g. A student is extracting copper metal by displacing it from copper(II) sulfate using zinc metal. She uses 23.93 g of copper sulfate and obtains 4.76 g of copper. What is the percentage yield?

Stretch and challenge

14. Cracking can be used to produce alkenes from alkanes. Decane can be cracked to give two products:

$$C_{10}H_{22} \rightarrow C_2H_4 + C_8H_{18}$$

a. If only the alkene can be sold, what is the percentage atom ecomony?

b. If both products can be sold, what is the percentage atom ecomony?

ASSIGNMENT 2: GREEN CHEMISTRY

(MS 0.2; PS 1.1, 1.2, 3.2)

The ethos of green chemistry is that the chemical industry must not adversely affect the environment so that the environment is protected now, and for future generations. A key aim is to reduce the production of waste. Taking into account the percentage atom economy of industrial reactions represents an important step towards reducing waste from industrial processes. In fact, atom economy is the key idea behind green chemistry.

Phenol is the starting point for making many products. Its formula is C_6H_5OH. The traditional method for manufacturing phenol used benzene, C_6H_6, sulfuric acid and sodium hydroxide. The overall equation for the reaction is:

$$C_6H_6\,(l) + H_2SO_4\,(aq) + 2NaOH(aq) \rightarrow$$
$$C_6H_5OH(l) + Na_2SO_3\,(aq) + 2H_2O\,(l)$$

So, 1 mol of benzene (78 g) should yield 1 mol of phenol (94 g). In practice, 1 mol of benzene yields about 77 g of phenol, which is a good percentage yield.

However, the reaction also produces 1 mol of sodium sulfite (Na_2SO_3) for every mole of phenol produced. At present, we do not have any uses for sodium sulfite. It creates serious problems for waste disposal, which adds to the costs.

A better method is to manufacture phenol from benzene and propene, CH_3CHCH_2. The reaction also uses oxygen. The overall equation is:

$$C_6H_6\,(l) + CH_3CHCH_2\,(l) + O_2(g) \rightarrow$$
$$C_6H_5OH(l) + CH_3COCH_3\,(l)$$

CH_3COCH_3 is propanone, which is commonly called acetone. It has many uses, including as a component

Figure A1 *Phenol is an important chemical. It is used to make plastics, detergents and many other products.*

of nail polish removers. So, by manufacturing phenol using an alternative reaction it is possible to generate another product that is useful, which means there is no waste.

Questions

A1. What is the percentage yield of phenol when it is manufactured from benzene, sulfuric acid and sodium hydroxide?

A2. Calculate the percentage atom economy for the production of phenol from benzene, sulfuric acid and sodium hydroxide.

A3. Calculate the percentage atom economy when phenol is manufactured using propene. Consider both products as desired.

A4. Explain why manufacturing phenol using benzene and propene may be a better method than using benzene, sulfuric acid and sodum hydroxide.

A5. How might percentage yield affect your answer to question A4?

2.6 IONIC EQUATIONS

Ionic equations provide a shorthand way to show the essential chemistry involving the ions in a reaction. They enable you to make generalisations about a reaction and to pick out the species that have lost or gained electrons.

Working from a full equation to an ionic equation

Consider how to work out the ionic equation for the reaction between hydrochloric acid and sodium hydroxide that produces sodium chloride and water.

First write out the full equation:

$$HCl(aq) + NaOH(aq) \rightarrow NaCl(aq) + H_2O(l)$$

Now write the equation as ions and cancel the ions that appear on both sides:

$$H^+(aq) + Cl^-(aq) + Na^+(aq) + OH^-(aq) \rightarrow$$
$$Na^+(aq) + Cl^-(aq) + H_2O(l)$$

Water molecules are covalently bonded, so they form no ions.

The ionic equation is:

$$H^+(aq) + OH^-(aq) \rightarrow H_2O(l)$$

The ions that appear unchanged on both sides of the ionic equation are cancelled out and are known as **spectator ions**. They are not written in the final ionic equation.

Worked example 15

Write the ionic equation for the reaction between zinc and hydrochloric acid.

Step 1 First write out the full equation:

$$Zn(s) + 2HCl(aq) \rightarrow ZnCl_2(aq) + H_2(g)$$

Step 2 Then write the equation as ions:

$$Zn(s) + 2H^+(aq) + 2Cl^-(aq) \rightarrow$$
$$Zn^{2+}(aq) + 2Cl^-(aq) + H_2(g)$$

Hydrogen molecules are covalently bonded, so they form no ions.

Step 3 Finally, cancel out ions that appear on both sides to give the ionic equation:

$$Zn(s) + 2H^+(aq) \rightarrow Zn^{2+}(aq) + H_2(g)$$

In this reaction the zinc has been **oxidised** (it has lost electrons) and the hydrochloric acid has been **reduced** (the hydrogen has gained electrons).

QUESTIONS

15. Iron displaces silver from silver nitrate solution. The ionic equation is:

    $$Fe(s) + 2Ag^+(aq) \rightarrow Fe^{2+}(aq) + 2Ag(s)$$

    What is the maximum mass of silver that can be displaced using 9.52 g iron and excess silver nitrate solution? (A_r: Fe 55.8; Ag 107.9)

Calculations from ionic equations

You can also calculate amounts from ionic equations.

One mole of sodium chloride has a mass of 58.5 g. This is calculated as the mass of a mole of sodium atoms plus the mass of a mole of chlorine atoms $(23 + 35.5)$ g.

But sodium chloride consists of sodium ions, Na^+, and chloride ions, Cl^-.

Since the mass of the electrons lost or gained is negligible, the mass of 1 mol Na^+ is taken as the same as the mass of 1 mol sodium atoms. Similarly, the mass of 1 mol Cl^- is taken as the same as the mass of 1 mol of chlorine atoms.

Worked example 16

(MS 0.2)

You can use an ionic equation to calculate an amount. Zinc metal reacts with copper sulfate solution to deposit copper metal. Calculate the mass of copper that can be obtained from 130 g of zinc, using excess copper sulfate.

Step 1 The chemical equation is:

$$Zn(s) + CuSO_4(aq) \rightarrow ZnSO_4(aq) + Cu(s)$$

Step 2 Write the equation as separate ions:

$$Zn(s) + Cu^{2+}(aq) + SO_4^{2-}(aq) \rightarrow$$
$$Zn^{2+}(aq) + SO_4^{2-}(aq) + Cu(s)$$

Step 3 Write the ionic equation:

$$Zn(s) + Cu^{2+}(aq) \rightarrow Zn^{2+}(aq) + Cu(s)$$

Step 4 From the ionic equation:

1 mol zinc atoms $\rightarrow$ 1 mol copper atoms

so, 65 g Zn $\rightarrow$ 63.5 g Cu

and, 130 g Zn $\rightarrow$ 128 g Cu

You will meet many ionic equations in this course.

2.7 REACTIONS IN SOLUTIONS

A solution contains a solute dissolved in a solvent. Water and ethanol are common solvents. The solute dissolved in the solvent can be solid, liquid or gas. Solutions in the home are commonplace. In fizzy drinks, for example, the solutes are carbon dioxide, which makes them fizzy, and other ingredients such as flavouring, colouring and sweeteners. The solvent is water. In shampoo, the solutes are detergent, perfume, preservatives and other ingredients, which you can read on the label. The solvent is, again, water.

You will use solutions in most of the practical work that you do. You need to be able to do calculations that involve concentrations. Concentration is measured in mol dm^{-3} (see Figure 15). Occasionally, you may be asked to calculate concentration in g dm^{-3}.

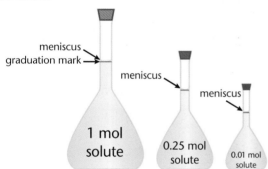

Volumetric flask volume	1 dm³ 1000 cm³	250 cm³	100 cm³
	1 mol solute	0.25 mol solute	0.01 mol solute
Concentration	1 mol dm⁻³	1 mol dm⁻³	0.1 mol dm⁻³

Figure 15 *Concentrations of solutions*

To calculate concentration, you need to know:

› the mass of solute

› the volume of the solution.

Then you can use this equation:

$$\text{concentration (mol dm}^{-3}) = \frac{\text{moles of solute}}{\text{volume of solution (dm}^3)}$$

You may find it useful to use Figure 16, where:

n = moles of solute

c = concentration (mol dm^{-3})

v = volume (dm^3)

Then:

$$n = c \times v$$

$$c = \frac{n}{V} \quad \text{and} \quad v = \frac{n}{c}$$

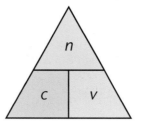

Figure 16 *This relationship can help to calculate the number of moles of a solute, the concentration of solution and its volume.*

Worked example 17

(MS 2.1 and 2.2)

4 g of sodium hydroxide is dissolved in 1 dm³ of solution. Calculate its concentration in mol dm^{-3}.

$$M_r \, \text{NaOH} = 23 + 16 + 1 = 40$$

$$\text{moles of NaOH in 4 g} = \frac{4}{40}$$

$$= 0.1 \, \text{mol}$$

Since $c = \dfrac{n}{V}$

$$\text{concentration (mol dm}^{-3}) = \frac{0.1}{1}$$

$$= 0.1 \, \text{mol dm}^{-3}$$

Worked example 18

(MS 2.1 and 2.2)

4 g of sodium hydroxide is dissolved in 250 cm³ of solution. Find its concentration.

4 g NaOH = 0.1 mol (from Worked example 17)

250 cm³ = 0.25 dm³

$$\text{concentration (mol dm}^{-3}) = \frac{0.1}{0.25}$$

$$= 0.025 \, \text{mol dm}^{-3}$$

QUESTIONS

16. Calculate the concentration of:

   a. 0.98 g sulfuric acid dissolved in 1 dm³ solution

   b. 1.00 g sodium hydroxide dissolved in 1 dm³ solution

   c. 1.00 g sodium hydroxide dissolved in 2 dm³ solution

d. 49.0 g sulfuric acid dissolved in 2 dm^3 solution

e. 49.0 g sulfuric acid dissolved in 250 cm^3 solution

17. Calculate the number of moles of solute in:

a. 500 cm^3 of 0.50 mol dm^{-3} sodium hydroxide solution

b. 25 cm^3 of 0.10 mol dm^{-3} sodium hydroxide solution

c. 100 cm^3 of 0.25 mol dm^{-3} hydrochloric acid solution

d. 10 cm^3 of 2.00 mol dm^{-3} sulfuric acid solution

e. 2.00 cm^3 of 0.50 mol dm^{-3} potassium manganate(VII) solution

Making a volumetric solution

When chemists carry out practical work to determine unknown concentrations, they usually carry out a **titration**, sometimes called a **volumetric analysis**. They need to make a volumetric (or standard) solution. A volumetric solution is one in which the precise concentration is known. It involves dissolving a known mass of solute in a solvent and making the solution up to a known volume.

To make 1 dm^3 of a standard solution of sodium carbonate with a concentration of 0.1 mol dm^{-3}, you first need to calculate the relative formula mass, M_f, for Na$_2$CO$_3$

$$M_f \ Na_2CO_3 = (23 \times 2) + 12 + (16 \times 3) = 106$$

106 g is the mass of sodium carbonate we would need to dissolve in 1 dm^3 of solution to produce a concentration of 1.0 mol dm^{-3}.

To make 1 dm^3 of 0.1 mol dm^{-3} solution, you need 0.1 × 106 g of Na$_2$Co$_3$ or 10.6 g.

QUESTIONS

18. a. What mass of sodium carbonate would need to be dissolved in 250 cm^3 of solution to give a standard solution of concentration 0.100 mol dm^{-3}?

b. If the mass of sodium carbonate in a. was dissolved in 100 cm^3 of solution, what would be its new concentration?

c. What mass of sodium carbonate would need to be dissolved in 1 dm^3 solution to produce a concentration of 0.01 mol dm^{-3}?

Email: djcshh8mhjggz20@marketplace.amazon.co.uk

Items:

Qty	Item	Locator	Condition	Price
1	Collins AQA A-level Science - AQA A-level Chemistry Year 1 and AS Student Book Gadd, Ken, Nicholls, Lyn SKU: mon00046555287 ISBN: 9780007590216 - Books	W6 -1-01-024-001-1369	Very Good	£3.80

Notes:

Thanks for your order!

Subtotal:	£3.80
Shipping:	£2.80
Total:	£6.60

If you have any questions or concerns regarding this order, please contact us at Books@Revivalbooks.co.uk or call us on **01706 227207**

Margin scheme second-hand goods VAT Number GB 901 5786 27

Revival Books Ltd.

Unit 11 Hugh Business Park, Bacup Road
Waterfoot, Lancashire, BB4 7BT
United Kingdom
Books@RevivalBooks.co.uk
01706 227207
www.revivalbooks.co.uk

Order Date: 13/04/2016

Order Number: 4975763

Marketplace: Amazon UK

Marketplace Order #: 204-7500492-6057153

Ship Method: Standard

REQUIRED PRACTICAL ACTIVITY 1: APPARATUS AND TECHNIQUES (PART 1)

(PS 4.1, AT a, d, e, k)

Make a volumetric solution

This is the first part of the required practical activity 'Make up a volumetric solution and carry out a simple acid–base titration'. It gives you the opportunity to show that you can:

> use appropriate apparatus to record (a) mass, (b) volume of liquids

> use a volumetric flask, including an accurate technique for making up a standard solution

> safely and carefully handle solids and liquids, including corrosive, irritant, flammable and toxic substances.

Apparatus

A volumetric flask, such as the one shown in Figure P2, is used to make solutions of known concentration accurately. These solutions are called volumetric (or standard) solutions.

Technique

To prepare a volumetric solution, a known number of moles of solute must be dissolved in deionised water and the solution made up to a known volume. Occasionally other solvents are used.

Preparation involves weighing a small container such as a weighing bottle or boat, and measuring the required mass of compound into it. The number of moles can be calculated from the mass and the relative molecular or formula mass of the compound.

The compound needs to be transferred quantitatively, meaning that the compound weighed out is all transferred. This is generally done by putting a small funnel in the neck of a volumetric flask. The compound is tipped or poured in and any remaining in the container is washed in using a wash bottle containing deionised water.

It is important to dissolve the compound completely before filling the flask to its graduation mark with water. Typically, the flask is filled to about one-third and the contents swirled until the compound dissolves. More deionised water is added, swirling the flask regularly to ensure thorough mixing, until the solution is about 1 cm below the flask's graduation mark. Deionised water is then added drop-by-drop until the bottom of the meniscus is just level with the graduation mark, as shown in Figure P3. With the stopper held in place, the flask is repeatedly turned up and down until the solution is thoroughly mixed.

Sodium Hydroxide

White, odorless, hydroscopic flakes, lumps, or pellets. Highly corrosive! Causes severe eye, skin, and respiratory tract burns. Repeated skin contact can cause dermatitis. Reacts with water producing excessive heat.

CAS No. 1310-73-2

Figure P1 *Hazard cards give the information you need to complete a risk assessment.*

Figure P2 *Volumetric flasks come in a range of sizes. The graduation mark is on the neck of the flask. The temperature for which it is graduated is also shown on the flask.*

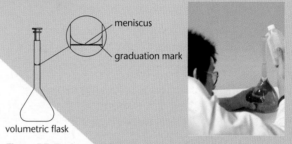

meniscus

graduation mark

volumetric flask

Figure P3 *The bottom of the meniscus must be level with the graduation mark.*

An alternative method is to dissolve the compound in water in a beaker then transfer the solution to the flask. This is better if the compound is difficult to dissolve and may need warming. If it is warmed to dissolve the compound, the solution must be allowed to cool to room temperature before transferring to the volumetric flask.

QUESTIONS

P1. Why can 100 cm^3 of a liquid be measured more accurately in a 100 cm^3 volumetric flask than in a 100 cm^3 measuring cylinder?

P2. Look at the design of a volumetric flask. Why is the graduation mark in the neck of the flask?

P3. A solid is added to a beaker containing water. The mixture is heated to dissolve the solid. Why must the solution be allowed to cool to room temperature before transferring it to a volumetric flask and making it up to volume?

Finding concentrations and volumes of solutions

Sodium hydroxide reacts with hydrochloric acid in a neutralisation reaction:

$$NaOH(aq) + HCl(aq) \rightarrow NaCl(aq) + H_2O(l)$$

If you add the precise amount of hydrochloric acid required to neutralise the sodium hydroxide, the resulting solution will contain only sodium chloride and water and have pH 7. This procedure is called a **titration**.

If you are given 0.100 mol dm^{-3} hydrochloric acid (the standard solution) you can carry out a titration to determine the concentration of a sample of sodium hydroxide solution.

A burette is filled with 0.100 mol dm^{-3} hydrochloric acid. 25.0 cm^3 aliquots of the sodium hydroxide solution are titrated against the acid. The volume of acid required to neutralise each aliquot is called a **titre**. To improve reliability, a minimum of three titres within 0.1 cm^3 are usually obtained and an average titre calculated.

From the sample results in Table 3, you can now calculate the concentration of the sodium hydroxide.

Here is the calculation:

25 cm^3 of sodium hydroxide solution is neutralised by 19.80 cm^3 0.100 mol dm^3 hydrochloric acid. We now need to calculate the concentration of the alkali.

19.80 cm^3 0.100 mol dm^{-3} hydrochloric acid contains

$$\frac{19.80}{1000} \times 0.100 \text{ mol HCl}$$

$$= 1.98 \times 10^3 \text{ mol HCl}$$

From the equation

$$HCl(aq) + NaOH(aq) \rightarrow NaCl(aq) + H_2O(l)$$

Titration			
	Rough	1	2	3
Final burette reading/cm^3	23.80	43.60	21.40	41.15
Initial burette reading/cm^3	3.60	23.80	1.55	21.40
Titre /cm^3	20.20	19.80	19.85	19.75
Used in mean	No	Yes	Yes	Yes
	Mean titre = 19.80 cm^3			

Table 3 *Sample results for the titration of 25 cm^3 sodium hydroxide solution against 0.100 mol dm^{-3} hydrochloric acid*

1 mol HCl reacts with 1 mol NaOH

Therefore, there must be 1.98×10^{-3} mol NaOH in 25.0 cm^3 of the sodium hydroxide solution.

If there are 1.98×10^{-3} mol NaOH in 25.0 cm^3, then the moles in 1 dm^3 are:

$$1.98 \times 10^{-3} \times \frac{1000}{25}$$

$$= 0.0792 \text{ mol NaOH}$$

Therefore, the concentration of sodium hydroxide solution $= 0.0792$ mol dm^{-3}.

Worked example 19

(MS 0.1, 0.1, 2.1, 2.2, 2.3 and 2.4)

Sodium carbonate is a soluble compound used as washing soda. Calculate the volume of 0.100 mol dm^{-3} hydrochloric acid needed to react exactly with 25 cm^3 of 0.220 mol dm^{-3} sodium carbonate solution.

Step 1 First work out how many moles of sodium carbonate is contained in the solution:

25 cm^3 of 0.220 mol dm^{-3} sodium carbonate solution contains

$$0.220 \times \frac{25}{1000} \text{ mol Na}_2CO_3$$

$$= 5.5 \times 10^{-3} \text{ mol Na}_2CO_3$$

Step 2 From the equation

$$2HCl(aq) + Na_2CO_3(aq) \rightarrow 2NaCl(aq) + CO_2(g) + H_2O$$

1 mol sodium carbonate reacts with 2 mol hydrochloric acid.

Therefore, 5.5×10^{-3} mol Na$_2$CO$_3$ reacts with $2 \times 5.5 \times 10^{-3}$ mol HCl $= 0.0110$ mol HCl.

Step 3 To find the volume of 0.100 mol dm^{-3} hydrochloric acid containing 0.0110 mol HCl :

0.100 mol HCl in 1 dm^3 0.100 mol dm^{-3} hydrochloric acid

1 mol HCl in 10 dm^3 of 0.100 mol dm^{-3} hydrochloric acid

0.0110 mol HCl in (10×0.0110) dm^3 of 0.100 mol dm^{-3} hydrochloric acid

$$= 0.110 \text{ dm}^3 = 110 \text{ cm}^3$$

Therefore, 110 cm^3 of 0.100 mol dm^{-3} hydrochloric acid is needed to react with 25 cm^3 of 0.220 mol dm^{-3} sodium carbonate solution.

REQUIRED PRACTICAL ACTIVITY 1: APPARATUS AND TECHNIQUES (PART 2)

(PS 4.1, AT a, d, e, k)

Carry out a simple acid–base titration

This is the second part of the required practical activity 'Make up a volumetric solution and carry out a simple acid–base titration'. It gives you the opportunity to show that you can:

> use appropriate apparatus to record the volume of liquids

> use laboratory apparatus for a titration, including a burette and pipette

> safely and carefully handle solids and liquids, including corrosive, irritant, flammable and toxic substances.

Apparatus

A pipette (Figure P1) is used to transfer an accurately measured volume of liquid. A burette (Figure P2) is used to measure accurately any volume within its capacity.

Technique

Using a pipette:

> The pipette must be clean and dry and some of the solution to be analysed should be poured into a clean, dry beaker. This avoids possible contamination of the rest of the solution.

> A safety filler must be used to draw the solution into the pipette. Never suck it up by mouth. The tip of the pipette must be below the surface of the solution being pipetted at all times.

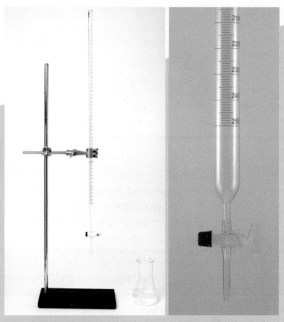

Figure P2 *A 25 cm³ burette can measure between 0 and 25 cm³ of liquid to an accuracy of about 0.02 cm³. However, the reading is usually taken to the nearest 0.05 cm³ by an experienced user or to the nearest 0.1 cm³ by people who are less experienced.*

> The solution is drawn up to just above the graduation line, the pipette removed from the solution and the outside of the pipette wiped with a tissue.

> Controlling the release with the safety filler, the solution is drained out until the bottom of the meniscus is level with the graduation line on the pipette. The tip of the pipette is touched on a clean glass surface.

> To empty the solution and run it freely into the conical flask ready for titration, the filler is removed. When the solution has emptied, touch the tip of the pipette on the surface of the solution in the flask. The pipette is designed to deliver a known volume of solution and leave a small amount in its tip, so it is important not to blow out any solution remaining in the pipette.

Using a burette:

> The burette must be clean and dry and clamped vertically. Before use it should be rinsed with three lots of about 10 cm³ to 15 cm³ of the volumetric solution.

> The volumetric solution is poured in through a funnel until it is 3 cm to 4 cm above the zero graduation.

Figure P1 *Like a volumetric flask, a pipette is calibrated at a fixed temperature (usually 20 °C) and has a tolerance, for example 25 ± 0.06 cm³.*

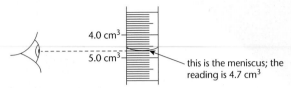

Figure P3 *A burette has graduation lines for each 1 cm³ quantity. Each 1 cm³ is subdivided into 0.1 cm³ sections. Read the burette with your eye level with the bottom of the meniscus.*

> The solution is run out, making sure that there is no trapped air bubble below the tap, until the bottom of the meniscus is just level with the 0.00 cm³ mark. If it falls below this mark it does not matter, but the burette reading must be recorded.

> To record a burette reading your eye should be level with the meniscus and the reading taken at the bottom of the meniscus (Figure P3). A piece of white card held behind it can make the meniscus easier to see.

Carrying out a titration:

> The volumetric solution is added from a burette in 1–2 cm³ amounts, swirling the flask between additions. As the end point gets closer, the colour takes longer to change.

> The burette reading is recorded when the colour change is permanent (does not disappear when the flask is swirled). It does not matter if a little too much is added. This is a 'rough' titration.

> The titration is repeated, but this time the volumetric solution is added 1–2 cm³ at a time, swirling the contents of the flask thoroughly after each addition, until it is within 2–3 cm³ of the 'rough' titration. At this point the solution is added drop by drop until the end point is reached. This should be repeated until three concordant results are obtained.

Figure P4 *Carrying out a titration accurately requires care and practice.*

QUESTIONS

P1. Which apparatus will measure 20 cm³ with greater precision: a 20 cm³ pipette or a 50 cm³ burette?

P2. Explain why a 'rough titration' is carried out and why its value is usually ignored when taking an average of the repeated titration values.

Stretch and challenge

P3. You will often find that a 50 cm³ burette is used for titrations and the concentration of the volumetric solution for routine analyses of similar samples is adjusted so that each titre is between 20 cm³ and 25 cm³. Why do you think this is?

ASSIGNMENT 3: HOW ACCURATE ARE YOUR TITRATION RESULTS?

(MS 1.1, 1.2, 1.3; PS 2.1, 2.3, 4.1)

There are two sources of error in a titration experiment.

> **Procedural errors:** errors due to the way the experiment was carried out. These are errors that can be avoided with care and practice.

> **Apparatus errors:** errors due to the limitations of precision of the apparatus used. These limitations are due to the manufacturing of the apparatus and are usually printed on the side of glassware.

A top-pan balance may measure mass in grams to two decimal places (to the nearest 0.01 g). The second decimal place is an approximation of the third decimal place. So, a reading of 2.56 g may be anything between 2.555 g and 2.564 g. This means that the second decimal place is uncertain and has an error of ±0.005 g.

A 250 cm³ volumetric flask has an error of ±0.25 cm³ (Figure A1).

A 25 cm³ pipette has an error of ±0.06 cm³.

One drop from a burette has a volume of 0.05 cm³. This is why all burette readings should include two decimal places and the second place should be 0 or 5.

Since measuring a volume from a burette involves two readings, the error on a volume measured from a burette is $±2 \times 0.05 = ±0.10$ cm³.

Errors are often written as percentage errors:

percentage error =

$$\frac{\text{error due to limitation of the apparatus's precision}}{\text{reading taken using the apparatus}} \times 100\%$$

For example, to calculate the percentage error on a balance reading of 3.69 g:

% error $= \dfrac{0.05}{3.69} \times 100$

$= 1.36\%$

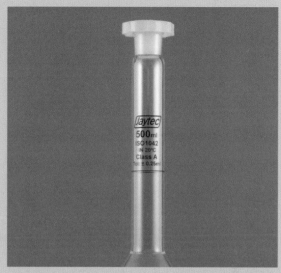

Figure A1 *This 250 cm³ volumetric flask has an error of ±0.25 cm³.*

Questions

A1. Calculate the percentage error:

    a. on a 250 cm³ volumetric flask

    b. on a 25 cm³ pipette

    c. on a burette reading of 24.00 cm³

    d. on a balance reading of 55.60 g

    e. on a balance reading of 5.56 g

A2. Which has the bigger percentage error, a larger or smaller balance reading?

A3. The overall percentage error for an experiment can be found by adding together all the percentage errors for the apparatus used. What is the overall percentage error for using a burette and a 25 cm³ pipette to carry out a titration?

ASSIGNMENT 4: HOW MUCH ASPIRIN IS IN AN ASPIRIN TABLET?

(MS 1.1, 1.2, 1.3; PS 2.1, 2.3, 4.1)

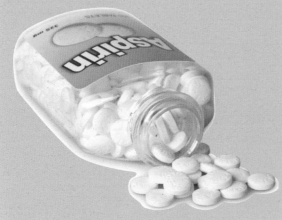

Figure A1 *Extracts from willow and similar plants have been used to reduce fever for hundreds of years. In the mid-nineteenth century, scientists started experimenting with the active ingredient in willow to find a related substitute. Today we use acetylsalicylic acid in aspirin tablets.*

Not all of an aspirin tablet is aspirin (Figure A1). It contains other ingredients. The ingredients list for a medicine tablet may include, amongst other substances, corn starch, cellulose and binders.

The active ingredient in an aspirin tablet is acetylsalicylic acid. The fact that it is an acid allows us to carry out an acid–base titration to find out how much is in the tablet.

The formula for acetylsalicylic acid is $CH_3COOC_6H_4COOH$.

The acid–base titration reaction is:

$$CH_3COOC_6H_4COOH + NaOH \rightarrow$$
$$CH_3COOC_6H_4COONa + H_2O$$

A student carried out the following practical.

The tablet was ground up and added to a conical flask. It was then dissolved in 10.0 cm³ ethanol (this is because aspirin does not dissolve in water easily).

The aspirin solution was titrated against 0.100 mol dm⁻³ sodium hydroxide, using phenolphthalein as the indicator.

The rough titre is an outlier (a measurement that lies outside the range of the others).

Questions

A1. From the results in Table A2, calculate the average titre.

	Titration			
	Rough	1	2	3
Final burette reading/cm³	16.35	31.05	45.70	40.80
Initial burette reading/cm³	0.00	16.35	31.05	26.05
Titre/cm³	16.35	14.70	14.65	14.75

Table A2 *Results of an acid–base titration to find the mass of aspirin in a tablet*

A2. How many moles of sodium hydroxide were used to neutralise the aspirin?

A3. From the reaction equation, how many moles of sodium hydroxide react with one mole of acetylsalicylic acid?

A4. How many moles of acetylsalicylic acid reacted with the sodium hydroxide in the titration?

A5. Calculate the relative molecuar mass, M_r, of acetylsalicylic acid.

A6. What is the mass of acetylsalicylic acid in the aspirin tablet? Note: decide on a suitable number of significant figures for your answer.

A7. Calculate the percentage errors for each apparatus used and an overall percentage error.

KEY IDEAS

> Concentration is measured in mol dm⁻³.

> Concentration $(mol\ dm^{-3}) = \dfrac{moles\ of\ solute}{volume\ of\ solution\ (dm^3)}$

> A volumetric solution is one in which the concentration is known.

> A titration is used to find an unknown volume or concentration of a reactant.

PRACTICE QUESTIONS

1. The metal lead reacts with warm dilute nitric acid to produce lead(II) nitrate, nitrogen monoxide and water, according to the following equation:

$$3Pb(s) + 8HNO_3(aq) \rightarrow$$
$$3Pb(NO_3)_2(aq) + 2NO(g) + 4H_2O(l)$$

   a. In an experiment, an 8.14 g sample of lead reacted completely with a 2.00 mol dm^{-3} solution of nitric acid. Calculate the volume, in dm^3, of nitric acid required for the complete reaction. Give your answer to 3 significant figures.

   b. In a second experiment, the nitrogen monoxide gas produced in the reaction occupied 638 cm^3 at 101 kPa and 298 K. Calculate the amount, in moles, of NO gas produced. (The gas constant $R = 8.31$ J K^{-1} mol^{-1})

   c. When lead(II) nitrate is heated it decomposes to form lead(II) oxide, nitrogen dioxide and oxygen.

      i. Balance the following equation that shows this thermal decomposition.

      $$........Pb(NO_3)_2(s) \rightarrow ......PbO(s) +$$
      $$.......NO_2(g) + ...O_2(g)$$

      ii. Suggest one reason why the yield of nitrogen dioxide formed during this reaction is often less than expected.

      iii. Suggest one reason why it is difficult to obtain a pure sample of nitrogen dioxide from this reaction.

      AQA Jan 2012 Unit 1 Q6

2. Norgessaltpeter was the first nitrogen fertiliser to be manufactured in Norway. It has the formula $Ca(NO_3)_2$.

   a. Norgessaltpeter can be made by the reaction of calcium carbonate with dilute nitric acid as shown by the following equation:

   $$CaCO_3(s) + 2HNO_3(aq) \rightarrow$$
   $$Ca(NO_3)_2(aq) + CO_2(g) + H_2O(l)$$

   In an experiment, an excess of powdered calcium carbonate was added to 36.2 cm^3 of 0.586 mol dm^{-3} nitric acid.

   i. Calculate the amount, in moles, of HNO$_3$ in 36.2 cm^3 of 0.586 mol dm^{-3} nitric acid. Give your answer to 3 significant figures.

   ii. Calculate the amount, in moles, of CaCO$_3$ that reacted with the nitric acid. Give your answer to 3 significant figures.

   iii. Calculate the minimum mass of powdered CaCO$_3$ that should be added to react with all of the nitric acid. Give your answer to 3 significant figures.

   iv. State the type of reaction that occurs when calcium carbonate reacts with nitric acid.

   b. Norgessaltpeter decomposes on heating as shown by the following equation:
   $$2Ca(NO_3)_2(s) \rightarrow 2CaO(s) + 4NO_2(g) + O_2(g)$$
   A sample of Norgessaltpeter was decomposed completely. The gases produced occupied a volume of 3.50×10^{-3} m^3 at a pressure of 100 kPa and a temperature of 31°C. (The gas constant $R = 8.31$ J K^{-1} mol^{-1})

   i. Calculate the total amount, in moles, of gases produced.

   ii. Hence calculate the amount, in moles, of oxygen produced.

   c. Hydrated calcium nitrate can be represented by the formula $Ca(NO_3)_2.xH_2O$ where x is an integer. A 6.04 g sample of $Ca(NO_3)_2.xH_2O$ contains 1.84 g of water of crystallisation. Use this information to calculate a value for x. Show your working.

   AQA Jun 2011 Unit 1 Q2

3. a. An unknown metal carbonate, M_2CO_3, reacts with hydrochloric acid according to the following equation:

   $$M_2CO_3(aq) + 2HCl(aq) \rightarrow$$
   $$2MCl(aq) + CO_2(g) + H_2O(l)$$

   A 3.44 g sample of M_2CO_3 was dissolved in distilled water to make 250 cm^3 of solution. A 25.0 cm^3 portion of this solution required 33.2 cm^3 of 0.150 mol dm^{-3} hydrochloric acid for complete reaction.

(Continued)

i. Calculate the amount, in moles, of HCl in 33.2 cm^3 of 0.150 mol dm^{-3} hydrochloric acid. Give your answer to 3 significant figures.

ii. Calculate the amount, in moles, of M_2CO_3 that reacted with this amount of HCl. Give your answer to 3 significant figures.

iii. Calculate the amount, in moles, of M_2CO_3 in the 3.4 g sample. Give your answer to 3 significant figures.

iv. Calculate the relative formula mass, M_r, of M_2CO_3. Give your answer to 1 decimal place.

v. Hence determine the relative atomic mass, A_r, of the metal M and deduce its identity.

b. In another experiment, 0.658 mol of CO_2 was produced. This gas occupied a volume of 0.0220 m^3 at a pressure of 100 kPa. Calculate the temperature of this CO_2 and state the units. (The gas constant $R = 8.31$ J K^{-1} mol^{-1})

c. Suggest one possible danger when a metal carbonate is reacted with an acid in a sealed flask.

d. In a different experiment, 6.27 g of magnesium carbonate was added to an excess of sulfuric acid. The following reaction occurred:

$MgCO_3 + H_2SO_4 \rightarrow MgSO_4 + CO_2 + H_2O$

i. Calculate the amount, in moles, of $MgCO_3$ in 6.27 g of magnesium carbonate.

ii. Calculate the mass of $MgSO_4$ produced in this reaction assuming a 95% yield.

AQA Jan 2011 Unit 1 Q3

4. In this question give all your answers to 3 significant figures.

Magnesium nitrate decomposes on heating to form magnesium oxide, nitrogen dioxide and oxygen, as shown in the following equation:

$2Mg(NO_3)_2(s) \rightarrow 2MgO(s) + 4NO_2(g) + O_2(g)$

a. Thermal decomposition of a sample of magnesium nitrate produced 0.741 g of magnesium oxide.

i. Calculate the amount, in moles, of MgO in 0.741 g of magnesium oxide.

ii. Calculate the total amount, in moles, of gas produced from this sample of magnesium nitrate.

b. In another experiment, a different sample of magnesium nitrate decomposed to produce 0.402 mol of gas. Calculate the volume, in dm^3, that this gas would occupy at 333 K and 1.00×10^5 Pa. (The gas constant $R = 8.31$ J K^{-1} mol^{-1})

c. A 0.0152 mol sample of magnesium oxide, produced from the decomposition of magnesium nitrate, was reacted with hydrochloric acid. The equation for the reaction is:

$MgO + 2HCl \rightarrow MgCl_2 + H_2O$

i. Calculate the amount, in moles, of HCl needed to react completely with the 0.0152 mol sample of magnesium oxide.

ii. This 0.0152 mol sample of magnesium oxide required 32.4 cm^3 of hydrochloric acid for complete reaction. Use this information and your answer to part c) i) to calculate the concentration, in mol dm^{-3}, of the hydrochloric acid.

AQA Jun 2010 Unit 1 Q3

5. a. An acid, H_2X, reacts with sodium hydroxide as shown in the following equation:

$H_2X(aq) + 2NaOH(aq) \rightarrow$
$\quad\quad 2Na^+(aq) + X^{2-}(aq) + 2H_2O(l)$

A solution of this acid was prepared by dissolving 1.92 g of H_2X in water and making the volume up to 250 cm^3 in a volumetric flask.

A 25.0 cm^3 sample of this solution required 21.70 cm^3 of 0.150 mol dm^{-3} aqueous NaOH for complete reaction.

i. Calculate the number of moles of NaOH in 21.70 cm^3 of 0.150 mol dm^{-3} aqueous NaOH.

ii. Calculate the number of moles of H_2X that reacted with this amount of NaOH. Hence, deduce the number of moles of H_2X in the 1.92 g sample.

iii. Calculate the relative molecular mass, M_r, of H_2X.

(Continued)

b. Analysis of a compound, Y, showed that it contained 49.31% of carbon, 6.85% of hydrogen and 43.84% of oxygen by mass. (The M_r of Y is 146.0)

  i. State what is meant by the term *empirical formula*.

  ii. Use the above data to calculate the empirical formula and the molecular formula of Y.

c. Sodium hydrogencarbonate decomposes on heating as shown in the equation below:

$$2NaHCO_3(s) \rightarrow Na_2CO_3(s) + CO_2(g) + H_2O(g)$$

A sample of $NaHCO_3$ was heated until completely decomposed. The CO_2 formed in the reaction occupied a volume of 352 cm^3 at 1.00×10^5 Pa and 298 K.

  i. State the ideal gas equation and use it to calculate the number of moles of CO_2 formed in this decomposition. (The gas constant $R = 8.31$ J K^{-1} mol^{-1})

  ii. Use your answer from part c) i) to calculate the mass of the $NaHCO_3$ that has decomposed. (If you have been unable to calculate the number of moles of CO_2 in part c) i), you should assume this to be 0.0230 mol. This is not the correct value.)

6. Nitroglycerine, $C_3H_5N_3O_9$, is an explosive that, on detonation, decomposes rapidly to form a large number of gaseous molecules. The equation for this decomposition is:

$$4C_3H_5N_3O_9(l) \rightarrow 12CO_2(g) + 10H_2O(g) + 6N_2(g) + O_2(g)$$

a. A sample of nitroglycerine was detonated and produced 0.350 g of oxygen gas.

  i. State what is meant by the term *one mole* of molecules.

  ii. Calculate the number of moles of oxygen gas produced in this reaction, and hence deduce the total number of moles of gas formed.

  iii. Calculate the number of moles, and the mass, of nitroglycerine detonated.

b. A second sample of nitroglycerine was placed in a strong sealed container and detonated. The volume of this container was 1.00×10^{-3} m^3. The resulting decomposition produced a total of 0.873 mol of gaseous products at a temperature of 1100 K. State the ideal gas equation and use it to calculate the pressure in the container after detonation. (The gas constant $R = 8.31$ J K^{-1} mol^{-1})

7. Magnesium chloride is also know as E511 and is used in the preparation of tofu. Magnesium chloride can be prepared in the lab using the reaction:.

$$MgCO_3(s) + 2HCl(aq) \rightarrow MgCl_2(aq) + H_2O(l) + CO_2(g)$$

a. What is the maximum yield of magnesium chloride that can be obtained from 8.4 g magnesium carbonate?

b. If the actual yield is 6.3 g, what is the percentage yield?

c. State one reason why the percentage yield is less than 100%.

d. What is the percentage atom ecomony for the reaction?

e. Give one advantage of using a reaction with a high atom ecomony in the chemical industry?

8. Students were determining the concentration of a sample of hydrochloric acid by titrating it against 0.100 mol dm^{-3} sodium hydroxide solution.

a. A pipette was used to transfer 25 cm^3 sodium hydroxide solution into a conical flask. Explain how the students could use phenolphthalein indicator and a burette to carry out the titration.

b. State the changes that they would need to make in the procedure if they used a pH probe instead of an indicator.

Table Q1 shows their results using phenolphthalein indicator.

c. Calculate the titres for each titration and select the three best results to calculate an average titre.

d. Calculate the following:

  i. the number of moles of 0.100 mol dm^{-3} sodium hydroxide in 25 cm^3 solution

(Continued)

ii. the number of moles of hydrochloric acid that this reacted with.

iii. concentration of the hydrochloric acid to the appropriate number of significant figures.

e. The error on a burette reading is $\pm 0.05 \text{ cm}^3$. What is the error on a titre?

	Titration 1	Titration 2	Titration 3	Titration 4
Final burette reading/cm³	24.00	46.50	24.75	47.15
Initial burette reading/cm³	0.00	24.00	2.15	24.75

Table Q1

10. Which of **A–D** is the number of atoms in one mole of atoms?

   A 6.023×10

   B 6.023×10^{-23}

   C 6.023×10^{10}

   D 6.023×10^{23}

11. Which of **A–D** is the correct ideal gas equation?

   A $nR = pVT$

   B $pV = nRT$

   C $nV = pRT$

   D $pT = nRV$

12. Which of **A–D** represents the empirical formula of glucose, $C_6H_{12}O_6$?

   A CH_2O

   B $C_6H_{12}O_6$

   C CHO

   D CHO_2

Stretch and challenge

13. The acid in vinegar is ethanoic acid (its old-fashioned name was acetic acid). The concentration of ethanoic acid in vinegar can be determined by titration. Ethanoic acid reacts with sodium hydroxide to produce sodium ethanoate and water:

$$CH_3COOH(aq) + NaOH(aq) \rightarrow$$
$$CH_3COONa(aq) + H_2O(l)$$

The graph in Figure Q1 shows the changes in pH when vinegar was titrated against sodium hydroxide solution.

Multiple Choice

9. For which of the compounds **A–D** is relative formula mass used?

   A sodium chloride

   B water

   C carbon dioxide

   D glucose

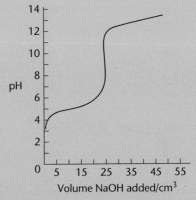

Figure Q1 pH titration curve of weak acid (CH_3COOH) and strong base ($NaOH$)

a. Assuming the pH of sodium ethanoate solution is 7, what volume of sodium hydroxide is needed to neutralise the vinegar?

b. What is present in the reaction mixture when:

   i. 10 cm^3 sodium hydroxide solution has been added

   ii. 35 cm^3 sodium hydroxide solution has been added?

c. Explain the changes in pH as the titration proceeds.

d. A different brand of vinegar, X, was similarly tested. This brand was found to contain a higher concentration of ethanoic acid. Copy the pH titration curve in Figure Q1 and add a second curve to show the changes in pH when vinegar X was titrated against the same concentration of sodium hydroxide solution. Label the second curve 'Vinegar X'.

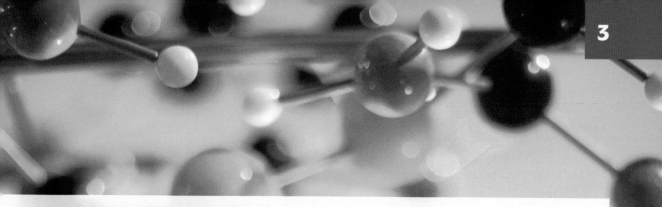

3 BONDING

PRIOR KNOWLEDGE

You may have learnt about covalent, ionic and metallic bonding in your GCSE course and the properties of these compounds. You may have heard of intermolecular forces, which are the forces of attraction between one molecule and another.

LEARNING OBJECTIVES

In this chapter, you will reinforce and build on these ideas, learning about polarisation of covalent bonds and the relationship between covalent and ionic bonds. You will learn how bonding determines the shapes of molecules and about the different types of bond between one molecule and another (intermolecular bonds).

(Specification 3.1.3.1, 3.1.3.2, 3.1.3.3, 3.1.3.4, 3.1.3.5, 3.1.3.6, 3.1.3.7)

It started with Scotch Tape and a pencil. Scientists had long predicted that a layer of graphite one atom thick might have amazing properties. Accidently, of course, they discovered that Scotch Tape could remove a layer one carbon atom thick from a lump of graphite (graphite is the form of carbon used to make pencil lead). The procedure was not as simple as it sounds and the process needed much refining. The layer obtained was called graphene (see Figure 1).

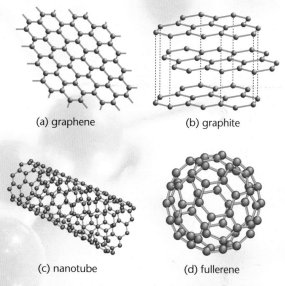

(a) graphene   (b) graphite

(c) nanotube   (d) fullerene

Figure 1 *Graphene (a) has a characteristic hexagonal structure that also occurs in graphite (b), nanotubes (c) and in fullerenes (d). The single layer of carbon atoms in graphene is 0.345 nm thick.*

A carbon atom has four electrons available for bonding. Three are used to bond covalently to other carbon atoms to make the flat hexagonal structure. In graphene, the fourth electron is not involved in making covalent bonds and is free to move around the structure and carry an electric current. Graphene is the best electrical conductor known and could be used to replace silicon in electrical components. The fourth electron can react with other substances and make new materials. Carbon's strong covalent bonds make graphene the strongest material known – more than 200 times the strength of steel. Graphene's other properties include very low mass and transparency.

Many universities have set up research departments to investigate its properties and future uses. All this is possible because of the bonding and structure of graphene.

3.1 IONIC BONDING

Figure 2 *Sea salt stockpiled at a refinery. Sodium chloride is a raw material for many industries, including the chlor-alkali industry that produces chlorine, sodium hydroxide and hydrogen – themselves the starting points for making many other chemicals.*

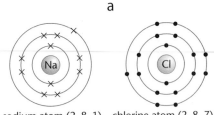

sodium atom (2, 8, 1)   chlorine atom (2, 8, 7)

The 3s electon of sodium transfers to the chlorine atom

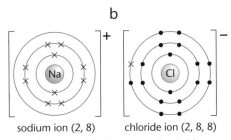

sodium ion (2, 8)   chloride ion (2, 8, 8)

As ions, sodium and chloride acquire the stable configuration of an inert gas. Opposite charges attract in ionic bonding.

Figure 3 *Ionic bonding in sodium chloride*

Sodium chloride (Figure 2) has **ionic bonds**. Figure 3a shows the electron configuration of sodium and chlorine atoms. When ionic bonds form, it is usually only the electrons in the outer shell that are involved (Figure 3b). The diagrams used to explain bonding show this outer shell only and are called **dot-and-cross** diagrams. The dots and crosses are used to represent the origin of each electron. Here, all the electrons are shown. The sodium electrons are shown with crosses and the chlorine electrons are shown with dots. Remember, all electrons are the same, regardless of where they come from.

The outer electron of a sodium atom transfers to a chlorine atom. Loss of the electron leaves a sodium ion that now has 11 protons and 10 electrons. This leaves an overall positive charge of $+1$. You write the ion as Na^+ and its electron configuration is 2,8 (Figure 3b).

The chlorine atom gains the electron to become a chloride ion. The chloride ion now has 17 protons and 18 electrons. This results in an overall charge of -1 and so the chloride ion is negatively charged. It can be written as Cl^- and its electron configuration is 2,8,8 (Figure 3b). The positive sodium ion and the neagative chloride ion are held together by **electrostatic attraction**.

Sodium ions and chloride ions have the same electron configurations as atoms of noble gases. Both have an octet of electrons (eight electrons) in the outer shell, so they both have a full outer shell. For an ion that is made from two or more atoms, you use square brackets to show that the charge on the ion acts over the whole ion, for example $[HCO_3]^-$.

Figure 4 shows dot-and-cross diagrams for some other ionically bonded compounds. In magnesium oxide, MgO, two outer electrons from magnesium's outer shell are transferred to the outer shell of the oxygen atom. The magnesium ion now has a charge of $+2$ and the oxide ion has a charge of -2.

In sodium oxide, Na_2O, one electron from each of two sodium atoms transfer to an oxygen atom to complete its outer shell and form an oxide ion, O^{2-}. In calcium fluoride, CaF_2, two electrons are transferred from a calcium atom to two fluorine atoms, forming two F^- fluoride ions.

In solid sodium chloride, vast numbers of oppositely charged sodium ions and chloride ions attract each other and form a giant structure, called a lattice. The lattice is a regular, repeating pattern of sodium ions and chloride ions (see Figure 13). Compounds consisting of ions are called ionic compounds.

Compound	Magnesium oxide	Sodium oxide	Calcium fluoride
Formula	MgO	Na$_2$O	CaF$_2$
Dot-and-cross diagrams for atoms	Mg$\overset{\times}{\times}$ O	Na $\times$ Na $\times$ O	Ca $\overset{\times}{\times}$ F F
Dot-and-cross diagrams for ions	Mg^{2+} $\left[\overset{\times}{\times} O \right]^{2-}$	Na$^+$ Na$^+$ $\left[\overset{\times}{\times} O \right]^{2-}$	Ca^{2+} $\left[\overset{\times}{\cdot} F \right]^{-}$ $\left[\overset{\times}{\cdot} F \right]^{-}$ Note: outer shells shown only

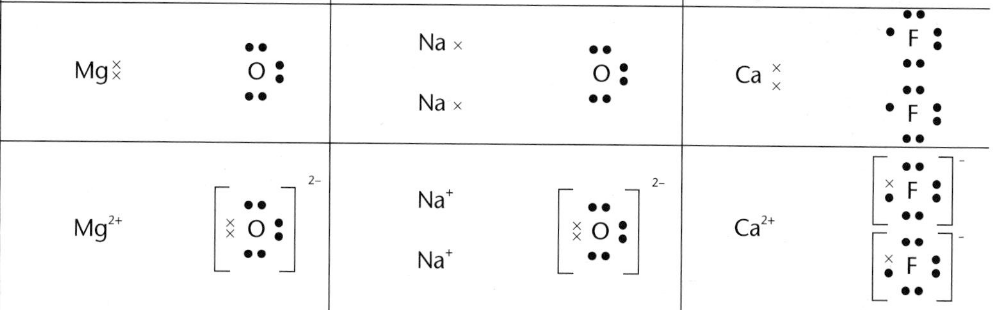

Figure 4 *Dot-and-cross diagrams for some ionically bonded compounds*

Predicting charges on simple ions

Simple ions are made from one type of element only. The position of elements in the Periodic Table can be used to predict the charge on their ions. Elements in Group 1 have one electron in their outer shell and this is transferred to form ions with a + 1 charge. Elements in Group 7(17) require one electron to complete their outer shell and form ions with a − 1 charge. Table 1 shows how an ion's position in the Periodic Table determines its charge.

Compound ions and formulae

You will also come across ions made from atoms of two or more elements. Table 2 shows some common compound ions.

QUESTIONS

1. Write the electron configurations for all the ions in Figure 4 (for example, Na$^+$ is 2,8).

2. **a.** Write the formula for:
   i. potassium chloride
   ii. calcium sulfide
   iii. magnesium chloride
   iv. lithium oxide
   v. aluminium oxide
   vi. magnesium nitride.

   b. Draw dot-and-cross diagrams to show the ionic bonding in each of i.-vi.

3. What charge would you expect on a simple ion formed from an atom of each of the following elements:
   a. nitrogen
   b. sulfur
   c. phosphorus
   d. strontium
   e. boron
   f. gallium
   g. rubidium
   h. krypton?

Group	1	2	3(13)	4(14)	5(15)	6(16)	7(17)	0
Number of electrons in outer shell	1	2	3	4	5	6	7	8
Charge on ion	+ 1	+ 2	+ 3	forms covalent bonds	−3	−2	−1	already has stable electron arrangement so does not form simple ions

Table 1 *Charges on ions formed from elements in the different groups of the Periodic Table. Note that non-metallic elements have names ending in –ide when they form ions. Chloride, nitride and sulfide are examples.*

Name of ion	Formula
hydroxide	OH^-
hydrogen carbonate	HCO_3^-
nitrate	NO_3^-
ammonium	NH_4^+
sulfate	SO_4^{2-}
sulfite	SO_3^{2-}
carbonate	CO_3^{2-}
phosphate	PO_4^{3-}

Table 2 *Formulae of complex ions*

You can use the charge on the ions, simple or complex, to construct formulae for ionic compounds.

The formula of caesium oxide
Caesium is in Group 1 and forms ions with $+1$ charge. Oxygen is in Group 6(16) and forms ions with a -2 charge.

Cs^+ O^{2-}

The positive charges and negative charges must cancel each other out. Two caesium ions are needed to give two positive charges to balance oxygen's -2 charge.

Therefore, the formula is Cs_2O.

The formula of ammonium sulfate
The ammonium ion is NH_4^+. The sulfate ion is SO_4^{2-}. Two ammonium ions are needed to balance the -2 charge on the sulfate ion.

Therefore, the formula is $(NH_4)_2SO_4$.

Note that the subscript 2 after the brackets around ammonium indicate that everything inside the brackets is doubled up.

QUESTIONS

4. Write the formulae for:

    a. strontium bromide

    b. aluminium hydroxide

    c. magnesium hydrogen carbonate

    d. ammonium hydrogen carbonate

    e. ammonium carbonate

    f. potassium sulfide

    g. barium nitrate

    h. sodium phosphide

    i. rubidium sulfide

    j. aluminium carbonate.

3.2 COVALENT BONDING

Many of the materials used every day are made from atoms held together by **covalent bonds**. The polymers and natural fibres that your clothes are made from, and most of the compounds in your body and in all other living things, contain covalent bonds.

Elements in Groups 2 to 3(13) achieve a noble gas configuration by forming positive ions. For the elements in Group 4(14), too much energy is required to remove four electrons. Instead, they achieve stability by sharing electrons in a covalent bond.

A hydrogen molecule, H_2, is made from two hydrogen atoms covalently bonded together, as shown in Figure 5. As in ionic bonding, atoms held together by a covalent bond achieve stable noble gas configurations. However, in covalent bonds, electrons are shared to complete electron shells, rather than being transferred as they are in ionic bonds.

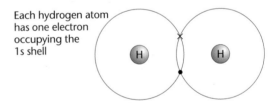

Each hydrogen atom has one electron occupying the 1s shell

The two 1s electrons are shared between the two atoms

Figure 5 *Covalent bonding in a hydrogen molecule*

In a covalent bond the positively charged nuclei of both atoms attracts the negative charge of the **bonding pair** of electrons (shared pair of electrons), as in Figure 6. You can show the covalent bond in a hydrogen molecule as H–H.

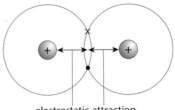

electrostatic attraction

Figure 6 *The positively charged nuclei are attracted to the negative charge of the bonding pair of electrons in the hydrogen molecule.*

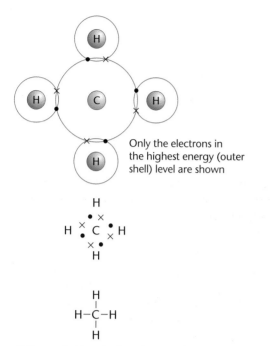

Only the electrons in the highest energy (outer shell) level are shown

$$H \overset{\bullet \times}{\underset{\times \bullet}{C}} H$$

$$H-\overset{\overset{\displaystyle H}{|}}{\underset{\underset{\displaystyle H}{|}}{C}}-H$$

Figure 7 *The covalent bonding in methane can be shown three ways.*

In a methane molecule, four covalent bonds form, as in Figure 7. The carbon atom now has the stable octet of electrons and each hydrogen atom has the electron configuration of helium.

A single covalent bond is a shared pair of electrons, one from each atom. Covalent molecules have no overall charge. They either exist as individual molecules or, if many covalent bonds are formed, they can form a giant covalent structure such as diamond or graphene. The types and numbers of atoms from which a molecule is made is given by its **molecular formula**. A molecue of methane, for example, is made from one carbon atom and four hydrogen atoms. Its molecular formula is CH_4.

QUESTIONS

5. Draw dot-and-cross diagrams (outer shells only) to show the covalent bonding in:

   a. chlorine

   b. water

   c. tetrachloromethane, CCl_4.

Double and triple covalent bonds

Some molecules contain multiple covalent bonds. Carbon dioxide has the formula CO_2, but there are not

enough electrons to make the noble gas configuration if there are single covalent bonds between the carbon and each of the oxygen atoms. Instead, a carbon dioxide molecule has **double covalent bonds** (usually just called double bonds). In a double bond four electrons are shared, so there are two bonding pairs in each double bond (Figure 8). Each carbon and oxygen atom now has the stable octet of electrons. You can write this as:

$$O = C = O$$

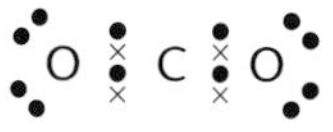

Figure 8 *Dot-and-cross diagram for carbon dioxide*

Ethyne has the formula C_2H_2. It is used as the fuel in oxyacetylene torches used to cut iron and steel (acetylene is the old name for ethyne). When ethyne burns in oxygen, the reaction is highly exothermic.

The bonding in ethyne is covalent, but again there are insufficient electrons to have single or even double bonds between the carbon atoms and attain a stable octet configuration. Ethyne therefore contains a triple covalent bond (Figure 9). The triple bond contains six electrons, to give three bonding pairs. Ethyne can be written as:

$$H-C{\equiv}C-H$$

Figure 9 *Dot-and-cross diagram for ethyne*

Dative and co-ordinate bonds

In the covalent bonds described so far, each atom donates one electron to the bonding pair. However, this is not always the case.

The ammonium ion is formed when a hydrogen ion combines with an ammonia molecule (Figure 10). In the covalent bond between the ammonia molecule and the hydrogen ion, both shared electrons come from the nitrogen atom. This is called a **co-ordinate** or **dative bond**. It can be written as shown in Figure 10, with the $\rightarrow$ representing the dative bond.

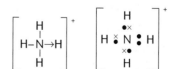

Figure 10 *Dot-and-cross diagram for the ammonium ion*

The direction of the arrow shows the origin of the pair of electrons. Since the hydrogen ion had a positive charge, the ammonium ion formed has a positive charge.

A dative bond behaves in the same way as any other covalent bond. Electrons are electrons, regardless of where they came from.

Carbon monoxide is a colourless, odourless, poisonous gas. It also contains a dative covalent bond (Figure 11). Note that there is a triple bond made up of two normal covalent bonds and one dative covalent bond. You can write carbon monoxide as:

Figure 11 Dot-and-cross diagram for carbon monoxide

The attraction between the positive metal ions and the delocalised electrons holds the metal together in the solid state.

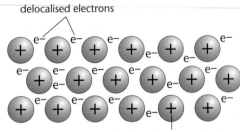

Figure 12 A simple representation of metallic bonding

3.3 METALLIC BONDS

Most elements are metals, but in their solid state they have neither ionic nor covalent bonding. The bonding that exists in metals is called **metallic bonding**.

In a lump of solid sodium metal, the sodium is present as sodium ions. Each sodium atom transfers its outer 3s electron to form a Na^+ ion to make a 'sea of electrons' that surrounds the regularly spaced sodium ions. Electrons in the 'sea' are free to move throughout the metal and are described as **delocalised electrons**.

> Dative or co-ordinate bonds are covalent bonds in which both bonding electrons originate from the same atom.

> A metallic bond is the electrostatic attraction between positively charged metal ions and the surrounding sea of delocalised electrons.

3.4 BONDING AND PHYSICAL PROPERTIES

The properties of a particular substance depend on how the atoms from which it is made bond to one another. For example:

> ionic bonding explains why sodium chloride is a crystalline solid with a high melting point

> covalent bonding explains why carbon dioxide exists as simple molecules, and the very weak attraction between molecules explains why it is a gas at room temperature.

Attraction between one molecule and another is called intermolecular bonding. You will read about it later in this chapter.

Crystal structures

The physical properties of a solid depend on the arrangement of particles in the solid and the forces of attraction between them. The structures may be **crystalline** or **amorphous**. In a crystalline structure, the different types of particle are arranged in a fixed repeating pattern. This repeating pattern gives the crystals of the substance their characteristic shape. In an amorphous solid, such as plastic or glass, there is no regular pattern to the way the particles are arranged.

You can classify crystal structures according to the type of particles and the bonding between them. The four types are:

Ionic: compounds containing ions form ionic crystals.

Metallic: metals have a metallic crystal structure when solid.

Macromolecular (giant covalent): non-metals, such as carbon and silicon, in Group 4(14) form macromolecular crystals.

Molecular: some molecules with intermolecular forces between them form molecular crystals.

Ionic crystals and their properties

Ionic crystals consist of a lattice of positive and negative ions. Sodium chloride is an example. It has a lattice of Na^+ and Cl^- ions. Each Na^+ is surrounded by six Cl^- ions, and each Cl^- ion is surrounded by six Na^+ ions.

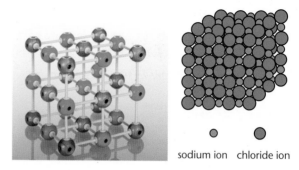

sodium ion   chloride ion

Figure 13 *Two ways of representing sodium chloride's crystal lattice – a ball-and-stick model (left) and a space-filling model (right). The shape of a sodium chloride crystal reflects the way that the ions pack together to form the sodium chloride lattice. Note: chloride ions are larger than sodium ions, but this ball-and-stick model does not show this.*

Figure 14 *Sodium chloride*

An ionic crystal melts when enough energy is transferred to it to overcome the strong electrostatic attractions between the ions. The energy is transferred by heating. Sodium and chloride ions break free from their fixed positions in the lattice and move freely. Because the electrostatic attractions are strong, the melting points are high. Sodium chloride's melting point is 801 °C.

If a compound conducts electricity, it must have ions that are free to move to carry the electrical charge. Sodium chloride does not conduct electricity in the solid state because the ions are held in a fixed position. However, when sodium chloride melts, the ions become free to move and molten sodium chloride conducts electricity.

Sodium chloride is a typical ionic solid. Ionic crystals:

> have high melting points because the ions are held firmly by strong electrostatic forces

> do not conduct electricity when solid, but when melted or dissolved in water the ions are free to move and carry the charge.

Metallic crystals and their properties

In a metal such as magnesium, metal ions are held in place by a sea of delocalised electrons. The ions pack together as closely as possible. In magnesium, each Mg^{2+} ion has 12 adjacent ions, six in the same plane, three above and three below (Figure 15). This arrangement is called hexagonal close packed.

Metal ions in a single layer:

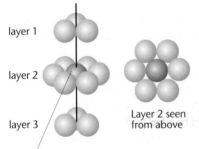

layer 1

layer 2

layer 3

Layer 2 seen from above

This metal ion has twelve atoms surrounding it: coordination number = 12

Figure 15 *The crystal structure in magnesium is due to the regular arrangement of metal ions. The crystal structure is reflected in the shape of a magnesium crystal.*

Figure 16 *A magnesium crystal*

This basic pattern of ions is repeated in the metal's structure, making up small crystals or grains. These grains are often visible on the surface of the metal.

When a metal melts, enough energy must be transferred to it by heating to overcome the strong electrostatic attractions between the ions and the sea of delocalised electrons. Metallic bonds are strong bonds. The positive metal ions are attracted to the delocalised electrons throughout the structure. Much energy is needed to break these bonds and, therefore, metals usually have high melting points.

Magnesium ions have a double positive charge. This means two electrons from each magnesium atom are delocalised in solid magnesium. In contrast, sodium ions have a single positive charge and just one electron from each sodium atom is delocalised. The sea of delocalised electrons in magnesium has twice the electron density of that for sodium. For these two reasons, the metallic bonds between the delocalised electrons and the positive metal ions are stronger in magnesium than in sodium. Also, the atomic radius of magnesium is less than that for sodium. This means that the delocalised electrons are more strongly attracted to the metal ions because they are closer together. Magnesium melts at 922 K (650 °C). Sodium melts at 371 K (98 °C).

Like all metals, magnesium and sodium are good conductors of electricity. The delocalised electrons are free to carry the charge.

Metals have crystalline structures. They:

> usually have high melting points and high boiling points

> have delocalised electrons that can move throughout the structure and, therefore, metals conduct electricity even when solid.

QUESTIONS

12. Explain why metals are malleable and ionic crystals are not.

13. Table 3 shows the relative electrical conductivity for the Period 3 elements. The conductivity of aluminium is taken as 1.00 and the other elements compared with it.

Element	Relative conductivity
sodium	0.26
magnesium	0.42
aluminium	1.00
silicon	0.10
phosphorus	0
sulfur	0
chlorine	0
argon	0

Table 3 *Relative conductivities of Period 3 elements*

a. Explain the use of the term 'relative'.

b. Draw a graph to show the electrical conductivity trend across Period 3.

c. Explain the trend from sodium to aluminium.

d. Explain the trend from silicon to argon.

Macromolecular crystals and their properties

Macromolecular crystals are also called giant covalent crystals.

The Group 4(14) elements carbon and silicon can form macromolecular crystals. Carbon is found in nature as diamond and graphite. They are allotropes. Allotropes of an element have different structures, but are in the same state.

Figures 17 and 18 *Diamond (top) and graphite (bottom) are forms of carbon, but with different structures.*

In diamond, each carbon atom forms four covalent bonds with four other carbon atoms (Figure 19). This gives a tetrahedral structure. The regular symmetrical pattern makes diamond the hardest naturally occurring substance. Diamonds can be manufactured and industrial diamonds are used on the tips of drills.

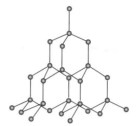

Every carbon atom bonds to four other carbon atoms

Figure 19 *The structure of diamond*

When diamond melts, covalent bonds between the carbon atoms are broken and the carbon atoms become free to move. A huge amount of energy must be transferred by heating to break the strong covalent bonds in diamond. Its melting point is very high (3500 °C).

There are no charged particles in diamond's covalent structure and diamond does not conduct electricity.

Graphite also has a giant covalent structure, but one that is quite different from that of diamond. It consists of flat sheets of carbon atoms covalently bonded together into hexagonal rings (Figure 20). Each carbon atom is bonded to three other carbon atoms. The fourth carbon electron makes up part of a cloud of delocalised electrons between the layers. The delocalised electrons hold the structure together by intermolecular forces called van der Waals forces between the layers of carbon atoms. You will read about van der Waals forces later in this chapter.

Flat sheets of carbon atoms are bonded into a hexagonal structure

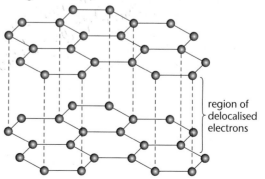

region of delocalised electrons

Figure 20 *The structure of graphite*

Like diamond, covalent bonds are broken when graphite melts. Large amounts of energy are needed to break these and graphite has a high melting point (3230 °C).

Unlike diamond, graphite conducts electricity because its delocalised electrons are free to move. The different structures give graphite and diamond different properties.

Molecular crystals and their properties

Molecular crystals contain molecules held together by intermolecular forces. These are weak bonds that form between the molecules. Iodine and ice are examples of molecular crystals.

The iodine molecule, I_2, is covalent, yet it forms crystals that contain a regular arrangement of molecules. In an iodine crystal, molecules are held together by van der Waals forces (Figure 21).

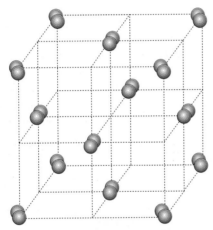

Figure 21 *Iodine crystals have a face-centred cubic arrangement of iodine molecules.*

Figure 22 *Iodine crystals*

When iodine crystals melt, the bonds between one iodine molecule and another are free to move. Because van der Waals forces are weak, the intermolecular

bonds are easily broken and iodine crystals have a low melting point (113.5 °C).

There are no charged particles in iodine molecules and they do not conduct electricity.

When water freezes, ice crystals form. Weak intermolecular forces called hydrogen bonds form between the water molecules. The water molecules are held in a regular pattern and a crystalline structure is produced.

Hydrogen bonds (a type of intermolecular force)

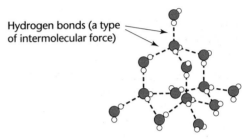

Figure 23 *The regular arrangement of water molecules in ice produces its crystalline structure. Hydrogen atoms are shown in white and oxygen atoms in red. These make up the water molecules. The dashed blue lines represent the intermolecular forces between the water molecules.*

Figure 24 *Ice crystals*

When ice melts, the intermolecular forces are broken and the water molecules can move over each other. The strong covalent bonds do not break. Intermolecular forces are weak bonds and easily broken. Ice has a low melting point (0 °C).

Ice has no charged particles to carry an electric current and is a non-conductor.

Van der Waals forces are weaker than hydrogen bonds, as you will learn later.

14. Which type of structure is likely to be present in the following?

    a. Crystalline solid that conducts electricity and has a melting point of 800 °C.

    b. Crystalline solid that does not conduct electricity and has a melting point of 84 °C.

    c. Crystalline solid that does not conduct electricity and has a melting point of more than 3000 °C.

    d. Crystalline solid that does not conduct electricity and has a melting point of 750 °C. When molten, it becomes a conductor.

15. Different types of bonds are broken when different crystalline solids melt. Which type of bonds are broken when crystals with these structures melt:

    a. ionic

    b. metallic

    c. giant covalent

    d. molecular?

Stretch and challenge

16. A_r, M_r and M_f are used to describe the relative masses of particles in a substance. Which should be used for each of the four different types of crystal?

KEY IDEAS

> Ionic crystals have high melting points, are non-conductors when solid, but conductors when molten.

> Metallic crystals have high melting points and conduct electricity both as solids and liquids.

> Macromolecular crystals (giant covalent) have high melting points and are non-conductors of electricity, whether solid or molten.

> Molecular crystals have low melting points and are non-conductors of electricity, whether solid or molten.

3.5 SHAPES OF MOLECULES AND IONS

The shapes of molecules and ions are important when finding explanations for the properties of substances. The shape of a simple molecule is determined by the number of electron pairs that surround the central atom. The pairs of electrons can be pictured as clouds of negative charge. Since like charges repel one another, electron pairs repel each other and arrange themselves to be as far apart as possible. This determines the shape of the covalently bonded molecule. In a molecule, atoms may have two types of electron pairs:

Bonding pair: pair of electrons involved in a chemical bond.

Lone pair (or non-bonding pair): pair of electrons not involved in bonding.

Table 4 shows some of the geometric shapes you can expect for different numbers of electron pairs in the outer shell of the central atom. Note that these bonds are all single covalent bonds. The shapes will be modified by the number of lone pairs and bonding pairs, as in Table 5. By using dotted lines and wedges, you can show the three-dimensional shapes of molecules.

Chemists use all this information to predict the shapes of molecules.

17. Work out the number and types of electron pairs for molecules of:

    a. ammonia

    b. water.

Lone pairs of electrons have a greater repulsion than bonding pairs because they are closer to the nucleus than bonding electron pairs. You can see this in Table 5 if you compare the bond angles of methane, ammonia and water.

Methane has four bonding pairs of electrons around its central carbon atom. These repel each other equally and all the bond angles in the molecule are 109.5°.

Ammonia has three bonding pairs of electrons and one lone pair of electrons. Because the lone pair of electrons repels more stongly than the bonding pairs, the bonding pairs are pushed together slightly and the bond angle is 107°.

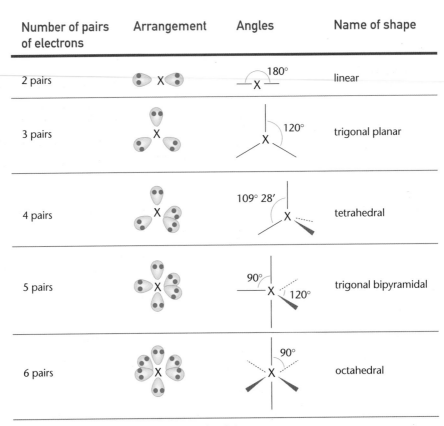

Number of pairs of electrons	Arrangement	Angles	Name of shape
2 pairs		180°	linear
3 pairs		120°	trigonal planar
4 pairs		109° 28′	tetrahedral
5 pairs		90° 120°	trigonal bipyramidal
6 pairs		90°	octahedral

Table 4 *The bond angles made by different numbers of bonding pairs of electrons*

Water has two bonding pairs and two lone pairs of electrons. The two lone pairs in water exert a greater repulsion on the bonding pairs than the single lone pair does on the bonding pairs in ammonia and the bond angle is 104.5°.

The repulsion between electron pairs around an atom increases in this order:

bonding pair–bonding pair < lone pair–bonding pair < lone pair–lone pair

This means that lone pair–lone pair repulsion is the greatest, and will produce the widest bond angles. To predict the shape of molecules you need to follow these five steps:

1. Determine the number of electron pairs (the bond structure will give you this).

2. Identify the basic shape (see Table 5).

3. Look for lone pairs (these are the electron pairs not involved in bonding).

4. Apply the rule for the order of increasing repulsion between electron pairs.

5. Draw out the shape of the molecule, adding bond angles.

Worked example 1

(MS 4.1, 4.2 and 4.2)

Work out the bond angles in the NH_4^+ ion.

Step 1 There are four bonding electron pairs around the central nitrogen atom, three from the single covalent bonds between the nitrogen atom and hydrogen atoms and one from the single co-ordinate bond between the nitrogen atom and hydrogen ion.

Step 2 The basic shape is tetrahedral.

Step 3 There are no lone pairs.

Step 4 The bond angles will be 109°.

Name of molecule	Dot-and-cross diagram	Shape	Type	Notes
Molecules with bonding electrons and no lone pairs				
Beryllium chloride		Cl—Be—Cl (180°)	linear	The Be atom has two pairs of electrons around it. The charged clouds of electrons around the chlorine atoms repel each other, so the Cl atoms are as far apart as possible
Boron trichloride		Cl, B, Cl, Cl (120°, 120°, 120°)	trigonal planar	Planar means a flat shape, and trigonal implies that the Cl atoms are at the point of an equilateral triangle
Methane	H×C×H	H—C—H (109.5°, 109.5°)	tetrahedral	A 3D structure, with all bond angles at 109.5°
Ethane		H—C—C—H (109.5°, 109.5°)	tetrahedral	The H atoms are arranged tetrahedrally around the C atoms
Ethene		C=C (120°, 120°, 120°)	trigonal planar	The double bonds hold the molecule in a flat shape. The H atoms form a trigonal planar shape around each C atom
Phosphorus(V) fluoride		P, F (90°, 120°)	trigonal bipyramidal	This shape pushes the bonding pairs further apart
Sulfur hexafluoride		S, F (90°, 90°)	octahedral	All bonding angles are 90°
Molecules with lone pairs and bonding pairs of electrons				
Ammonia	H×N×H	N, lone pair (107°, 107°)		The shape is based on a tetrahedron, but the bond angles are 107°, not 109.5°. The lone pair of electrons has a greater repulsion than a bonding pair and pushes the bonding pairs together
Water	O×H	O—H (104.5°)		The water molecule has two lone pairs. These exert a greater repulsion than one lone pair, and the hydrogen atoms are pushed together to give an angle of 104.5°

Table 5 *The shapes of molecules that have different numbers of bonding pairs and lone pairs of electrons*

Step 5 The shape of the molecule is shown in Figure 25. The overall charge on the ion does not affect the shape.

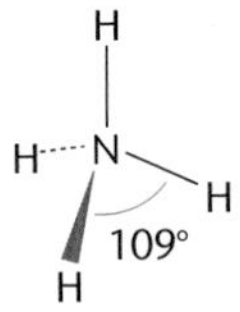

Figure 25 Shape of the ammonium ion

Worked example 2

(MS 4.1, 4.2 and 4.2)

Work out the bond angles in the hydrogen sulfide molecule, H_2S.

Step 1 There are four electron pairs around the central sulfur atom, two bonding pairs of electrons between the sulfur and hydrogen atoms and two lone pairs of electrons.

Step 2 The basic shape is tetrahedral.

Step 3 There are two lone pairs of electrons.

Step 4 The bond angle will be 104.5°.

Step 5 The molecule is described as bent (Figure 26).

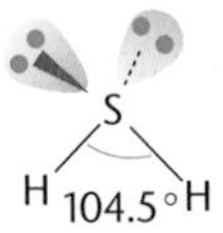

Figure 26 Shape of the hydrogen sulfide molecule

KEY IDEAS

> The shape of a molecule is determined by the bonding and lone pairs of electrons, which repel each other and arrange themselves as far apart as possible.

> Lone pair–lone pair repulsion is greater than lone pair–bond pair repulsion, which is greater than bond pair–bond pair repulsion.

> The degree of repulsion of electron pairs determines the bond angles in a molecule.

3.6 BOND POLARITY

You have read about three types of bonding: ionic, covalent and metallic. Ionic and covalent bonding are two extremes of how the electrons involved in bonding can be distributed between two atoms, either:

> in the middle of the two atoms, or

> transferred completely from one of the atoms.

However, there are compromises.

Polar covalent bonds and electronegativity

In a covalent bond between two hydrogen atoms, the electrons are distributed evenly between the two hydrogen atoms. The **electron density** is symmetrical because each hydrogen nucleus has the same power of attraction (Figure 27).

$$\delta+ \qquad \delta-$$

H   H   H   F

Figure 27 Electron distribution in hydrogen and hydrogen fluoride molecules

In a covalent bond between a hydrogen and a fluorine atom, the electrons are drawn closer to the fluorine atom. The electron density is unsymmetrical because the fluorine nucleus attracts the bonding electrons more strongly than does the hydrogen nucleus. This means that the negative charge from the bonding pair of electrons is closer to the fluorine atom. The fluorine end of the molecule is now more negative than the hydrogen end (Figure 27). You write this as:

$H^{\delta+} - F^{\delta-}$

where $\delta+$ (the Greek letter delta) means a small amount of positive charge and $\delta-$ means a small amount of negative charge.

The ability of a nucleus to attract the bonding pair of electrons is called its **electronegativity**. Fluorine is more electronegative than hydrogen, so the bonding pair of electrons lies closer to the fluorine nucleus than to the hydrogen nucleus. The bond in hydrogen fluoride is a polar covalent bond, or **polar bond** for short. In hydrogen fluoride, the polar bond makes the molecule itself polar with a **permanent dipole**. However, not all molecules with polar bonds are polar with permanent dipoles, as you will read later.

Measuring electronegativities

Electronegativities cannot be measured directly and they have units. The Nobel Prize-winning chemist Linus Pauling devised a scale that gives a numerical value to the power of an atom to attract electrons (Table 6). Fluorine is the most electronegative element, with an electronegativity of four. With values of 0.8, the metals caesium and francium (at the foot of Group 1) are the least electronegative elements. Electronegativity increases across the Periodic Table as atomic number increases and atomic radius decreases (see Chapter 4) – the larger the nuclear charge and the smaller the atomic radius, the greater the attraction for the **electron pair** in the covalent bond.

H							He
2.1							—
Li	Be	B	C	N	O	F	Ne
1.0	1.5	2.0	2.5	3.0	3.5	4.0	—
Na	Mg	Al	Si	P	S	Cl	Ar
0.9	1.2	1.5	1.8	2.1	2.5	3.0	—

Table 6 *Electronegativities for the elements in Periods 1, 2 and 3*

QUESTIONS

18. **a.** Why are there no values for the noble gases in Table 6?
    b. What are the trends in electronegativity across a period?

Stretch and challenge

19. Explain why electronegativities decrease as you go down Group 7(17).

Why should one atom be more electronegative than another? Two factors decide the electronegativity of an atom:

> the size of the atom or, more correctly, the atomic radius, and

> the **nuclear charge**.

If atoms are small, the nucleus can get close to the bonding pair of electrons and attract them more strongly. Fluorine and chlorine are both in Group 7(17), but a chlorine atom has one extra shell of electrons. A fluorine atom, being smaller than a chlorine atom, is more electronegative than the chlorine atom.

The nuclear charge is the total positive charge on the nucleus. Since the number of protons increases across a period, the nuclear charge also increases across a period. For atoms in a period, those with a higher nuclear charge will attract the bonding pair of electrons more strongly and will be more electronegative.

Lithium and fluorine are both in Period 2.

> The nuclear charge of a lithium atom is +3.

> The nuclear charge of a fluorine atom is +9.

The higher nuclear charge of the fluorine atom attracts the bonding pair of electrons more strongly, so fluorine is more electronegative than lithium.

Polar bonds but non-polar molecules

Not all molecules that have atoms with large differences in their electronegativities and, therefore, polar bonds, have permanent dipoles and are polar themselves.

Tetrachloromethane is an example. Its formula is CCl_4. The carbon atom forms four covalent bonds with the four chlorine atoms. Figure 28 shows its electronic configuration and the shape of the tetrachloromethane molecule.

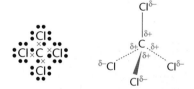

Figure 28 *The electronic configuration in tetrachloromethane and its molecular shape*

Carbon and chlorine's electronegativity values are 2.4 and 3.0 respectively. Therefore, each C–Cl bond is polar. The carbon atom has a δ^+ charge and the four chlorine atoms, have a δ^- charge.

Type of bond	Bond diagram	Explanation
pure covalent bond	$X \overset{\bullet}{\underset{\bullet}{\rule{1.5em}{0.4pt}}} Y$	X and Y have equal electronegativities.
		The bonding pair of electrons is half way between X and Y.
polar bond	$^{\delta+}X \overset{\bullet}{\underset{\bullet}{\rule{1.5em}{0.4pt}}} Y^{\delta-}$	Y is more electronegative than X.
		Y has the larger share of the electron density.
ionic bond	$\left[\,^{+}X\,\right]\ \left[\,\overset{\bullet\bullet}{\underset{\bullet\bullet}{Y}}\,\right]^{-}$	Y is a lot more electronegative than X.
		The electron pair is strongly attracted to Y and ions form.

Table 7 *The gradual change from covalent to ionic bonding*

If a molecule has a permanent dipole, one side will be slightly positive and the other side slightly negative. A molecule of tetrachloromethane does not have a positive or negative side; the outside of the molecule is uniformly $\delta-$ and tetrachloromethane is not polar. The polarities cancel each other out.

QUESTIONS

Stretch and challenge

20. With the aid of diagrams, explain which of these have a permament dipole and which do not:
    a. CH_4
    b. CH_3F
    c. CH_2F_2
    d. CF_4

ASSIGNMENT 1: INVESTIGATING THE RELATIVE STRENGTHS OF INTERMOLECULAR FORCES

(PS 1.2, 2.1, 2.4)

Figure A1 *Investigating the polarity of molecules. Water is attracted to a negatively charged rod.*

The theory
Although intermolecular forces are weak, some are stronger than others. An experiment can be set up in the laboratory to compare the relative strengths of these forces between different molecules. If a negatively charged rod is moved near to a stream of liquid consisting of molecules with a permanent dipole, the stream of liquid will be deflected.

The experiment
Six burettes were filled to the zero mark with either water (H_2O), ethanol (C_2H_5OH), propanone (CH_3COCH_3), chloromethane (CH_3Cl), tetrachloromethane (CCl_4) or hexane (C_6H_{14}). Each burette was positioned so that its tip was 5 cm above a beaker. A ruler was placed across the top of each beaker.

A plastic rod was rubbed with a piece of fur cloth, giving the surface of the rod a negative charge. In turn, the tap on each burette was opened to allow a stream of liquid to run into the beaker. The charged rod was held about 1 cm away from the stream of liquid and the deflection measured on the ruler.

Liquid	Amount of deflection/mm
Water	11
Ethanol	4
Propanone	7
Chloromethane	5
Tetrachloromethane	0
Hexane	0

Questions

A1. Which variables must be controlled in this experiment?

A2. Explain why hexane and tetrachloromethane were not deflected.

A3. Explain why water had the highest deflection.

A4. Why was chloromethane deflected but not tetrachloromethane?

A5. Why do ethanol and propanone show small amounts of deflection?

A6. Suggest why the streams of liquid were deflected.

A7. What safety precautions need to be taken when doing this experiment?

3.7 FORCES BETWEEN MOLECULES

You have seen that ionic bonding produces giant structures (lattices) and that covalent bonding can produce either giant structures or molecules. Substances consisting of covalent molecules may be solids, liquids or gases at room temperature. Most of the gases in the atmosphere consist of small, covalently bonded molecules. The forces between the molecules are very weak. In solids and liquids, the forces between the molecules must be strong enough to hold them together and prevent them from moving apart and becoming a gas.

Forces that can exist between molecules are called **intermolecular forces**. Covalent bonds in a molecule are strong, but intermolecular forces between one molecule and another are much weaker, between one-tenth and one-hundredth the strength of a covalent bond.

Types of dipole

You have already met permanent dipoles in Section 3.4 on bond polarity. Hydrogen chloride molecules have polar bonds, $H^{\delta+}-Cl^{\delta-}$, because chlorine atoms are more electronegative than hydrogen atoms (Figure 29). The molecule itself is polar with a **permanent dipole**.

Some molecules have neither polar bonds nor a permanent dipole because the atoms that bond together have the same or very similar electronegativity values. The chlorine molecule, Cl_2,

is an example. You can think of the electrons in a chlorine molecule forming a negative electron cloud. These electrons are in constant motion and, at any particular instant, may not be evenly distributed. This means that one end of the chlorine molecule may have more negative charge than the other end. The dipole lasts for only an instant and so is called an **instantaneous dipole** (Figure 30).

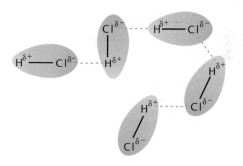

Figure 29 Hydrogen chloride molecules have a permanent dipole, resulting in weak intermolecular bonds between molecules.

Evenly distributed electron cloud; no dipole

For an instant, the electron cloud is uneven and an instantaneous dipole exists

Figure 30 Instantaneous dipole in chlorine

If a non-polar molecule is next to a polarised molecule with a permanent or instaneous dipole, an **induced dipole** is created in that molecule. The small negative charge on the polarised molecule is enough to repel the electrons in an adjacent unpolarised molecule and induce a dipole (Figure 31).

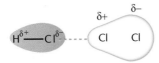

Figure 31 *Hydrogen chloride has a permanent dipole. The negative charge on the chlorine end of the molecule repels the electron cloud in an adjacent chlorine molecule and induces a dipole.*

Dipoles produce intermolecular forces

If molecules have dipoles, they will attract each other, resulting in intermolecular forces between them. There are three types of attraction:

➤ permanent dipole–permanent dipole

➤ permanent dipole–induced dipole

➤ induced dipole–induced dipole.

Induced dipole–induced dipole forces (also called van der Waals forces, dispersion forces and London forces) form when a instantaneous dipole, such as in

a chlorine molecule, induces a dipole in an adjacent non-polar molecule. The term van der Waals forces comes from the Dutch physicist Johannes van der Waals. They are very weak intermolecular forces, about one-hundredth the strength of a covalent bond.

Intermolecular forces and boiling points

Van der Waals forces explain the differences in boiling points between ethane (185 K) and pentane (309 K). Both compounds consist of non-polar molecules. Pentane molecules are longer than ethane molecules. This means that more van der Waals forces can operate between pentane molecules than between ethane molecules (Figure 32). So, the pentane molecules are more strongly attracted to each other, which makes pentane's boiling point higher than ethane's boiling point. Pentane is a liquid at room temperature and ethane is a gas.

Animal fats, such as butter, are used for cooking (Figure 34). Butter is a solid at room temperature. This is because it has long chains of carbon atoms that lie parallel to each other with van der Waals forces attracting the chains to each other. The structure of cooking oils also has long chains of carbon atoms, but the chains are in different directions and some are branched. They cannot lie close and parallel like the carbon chains in butter. Hence fewer intermolecular forces operate and less energy is needed to change the oil from a solid into a liquid. Cooking oils are, therefore, liquid at room temperature.

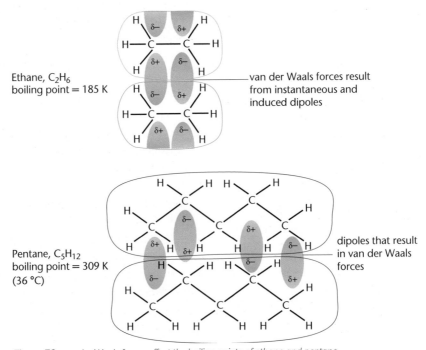

Figure 32 *van der Waals forces affect the boiling points of ethane and pentane.*

Figure 33 *Geckos can amazingly climb up walls and across ceilings. Their secret appears to be in van der Waals forces. These intermolecular forces between densely packed fine hairs on their toes and the wall seem to stop them falling off.*

Figure 34 *Fats and oils are used in cooking. Their physical properties depend on intermolecular forces.*

ASSIGNMENT 2: EXPLAINING THE PROPERTIES OF XENON AND HELIUM

(PS 1.1, 1.2)

You will use ideas about intermolecular forces to explain the different boiling points of xenon and helium.

Figure A1 *Helium is used to fill balloons.*

Xenon and helium are both noble gases. Helium is the gas used to fill party balloons and in a gas mix with oxygen for divers. Xenon is used in photographic flash bulbs. The boiling point of helium is 4 K and the boiling point of xenon is 166 K.

Questions

A1. a. How many electrons does a helium atom have?

   b. How many electrons does a xenon atom have?

A2. Draw a diagram to explain how an instantaneous dipole can arise in a xenon atom.

A3. Draw another diagram to show how this instantaneous dipole can induce a dipole in an adjacent xenon atom.

A4. What is the attraction between the instantaneous and induced dipoles called?

A5. How will the extent of this force affect the boiling point of xenon?

A6. Explain why helium has a lower boiling point than xenon.

Hydrogen bonding

Hydrogen bonds are a special type of permanent dipole–permanent dipole attraction. They are the strongest intermolecular force, typically being about one-tenth the strength of a covalent bond.

A hydrogen bond depends on three features:

> A highly polar bond between a hydrogen atom and a highly electronegative atom such as oxygen, nitrogen or fluorine.

> The small hydrogen atom can get close to other atoms.

> A lone pair of electrons on an oxgen, nitrogen or fluorine atom of one molecule can attract the positive charge of the hydrogen atom on an adjacent molecule.

The large difference in electronegativity (ability to attract electrons) between a hydrogen and a nitrogen, oxygen or fluorine atom means that strongly polar bonds are formed, which give the molecules a large permanent dipole.

When two molecules are close enough, an attraction is set up between the positive end of one and the negative end of another. This attraction is called **hydrogen bonding.** The two highly electronegative atoms in adjacent molecules and the hydrogen atoms are always in a straight line. The arrangement is: X–H–X. It is linear and hydrogen bonds are described as being directional.

A hydrogen bond is different from other dipole–dipole forces because hydrogen has no inner non-bonding electrons. An atom of nitrogen, oxygen or fluorine attracts hydrogen's single electron towards it, exposing the proton of the hydrogen nucleus. Without non-bonding electrons to repel the lone pairs of electrons in nitrogen, oxygen or fluorine, the proton is attracted strongly to their lone pair of electrons. It is embedded in the lone pair, forming a hydrogen bond. This is shown for ammonia in Figure 35.

In the water molecule, oxygen is more electronegative than hydrogen and strongly attracts the shared pair of electrons in the covalent bond (Figure 36). This makes the O–H bond highly polar and strong hydrogen bonds form between the water molecules, accounting for many of water's distinctive properties.

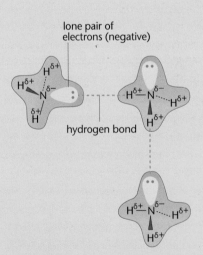

Figure 35 *Hydrogen bonding in ammonia*

The O–H bonds in water are very polar, and hydrogen bonds form between water molecules

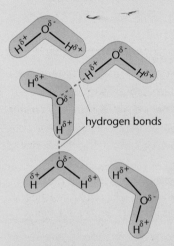

Figure 36 *Hydrogen bonding in water at room temperature. Each oxygen has two lone pairs of electrons. In each case, one pair is used to form the hydrogen bond. However, for simplicity, the lone pairs have been omitted from the diagram*

Boiling points and hydrogen bonding

Hydrides are compounds that contain hydrogen plus one other element. Water is a hydride of oxygen.

Figure 37 shows the trends for boiling points of hydrides in Groups 6(16) and 7(17). From Period 3 down each group to Period 5 there is a steady increase because the molecules get larger as you go down the group. Water and hydrogen fluoride do not follow the same pattern because the hydrogen bonds present attract the molecules to each other. This means that it takes more energy to form a vapour, making the boiling points unexpectedly high.

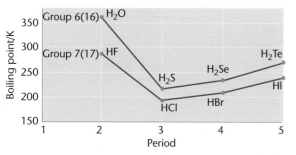

Figure 37 Boiling points of Group 6(16) and 7(17) hydrides

it becomes less dense and moves up to the surface before freezing. This still leaves liquid water below in which pond organisms can survive.

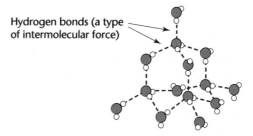

Figure 38 The 3D structure of ice

Hydrogen bonds form in ethanoic acid

Ethanoic acid has the formula CH_3COOH. Hydrogen bonds form in pure ethanoic acid as shown in Figure 39. These hydrogen bonds give higher melting and boiling points than expected.

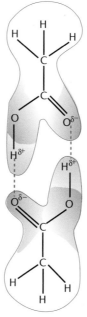

Figure 39 Hydrogen bonds in ethanoic acid

QUESTIONS

21. Imagine that hydrogen bonds did not exist. Use the graph of Figure 37 to suggest a value for the boiling points of:

   a. water
   b. hydrogen fluoride.

Water expands as it freezes

The water molecule is V-shaped, so you can think of it as having the oxygen atom at its negative apex. Either of oxygen's lone pairs attracts a positive hydrogen atom of an adjacent water molecule to form a hydrogen bond (Figure 36). The attractive force of hydrogen bonds explains its relatively high melting and boiling points. As in Figure 36, not all molecules of liquid water have hydrogen bonds at any one time. Some non-bonded molecules move more or less close than the hydrogen bonding position allows. When water is heated, the molecules gain energy, move more rapidly and become slightly further apart as their motion offsets the attraction of the hydrogen bonds. The water becomes less dense as its volume increases.

As you cool water, the reverse happens and the water becomes more dense. That is, until it reaches 277 K (4 °C) when it is at its most dense. But from 277 K (4 °C) to 273 K (0 °C) the proportion of hydrogen-bonded molecules increases, and the molecules start to form a lattice structure very similar to that of carbon in diamond (see Figure 19). The effect is that molecules are set slightly further apart and the water becomes less dense. At 273 K (0 °C), all the molecules are fixed in hydrogen-bonded positions (Figure 38) making a rigid 3D lattice – the water has now frozen and become ice.

In nature, water in a pond will cool in winter. At 277 K, it is at its most dense. But if the water cools further,

QUESTIONS

22. An experiment to find the M_r of ethanoic acid gave the value as 120. Explain why.
23. Hydrogen bonds also exist in dilute ethanoic acid between the acid molecules and the water molecules. Draw structural formulae to show the hydrogen bonds.

ASSIGNMENT 3: HYDROGEN BONDING IN PROTEINS

(PS 1.1, 1.2)

Hair consists mainly of a protein called keratin. Many thousands of keratin molecules lie side by side to make up a single hair. The top diagram shows two keratin molecules.

Like all proteins, keratin consists of a long chain of amino acids. Amino acids are organic acids that have both acidic $-COOH$ and basic $-NH_2$ groups in their molecules. The structure of hair is kept in place by three types of bonds: hydrogen bonds, ionic bonds and covalent bonds.

The ionic and covalent bonds are strong bonds and determine whether you have curly or straight hair. They hold the keratin molecules together by bonding between them. The only way to change these is by perming.

The hydrogen bonds are weak bonds and form along the length of the keratin molecules to give it a spiral shape. They are easily broken and reformed. This is why people can temporarily curl or straighten their hair.

Bonding in keratin molecules

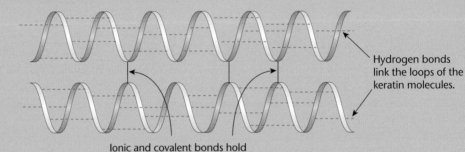

Hydrogen bonds link the loops of the keratin molecules.

Ionic and covalent bonds hold the keratin molecules together.

Breaking hydrogen bonds between keratin molecules

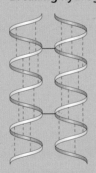

Dry hair: hydrogen bonds keep the hair in shape.

Wet hair: hydrogen bonds break and the keratin molecules can stretch.

The effect of perming

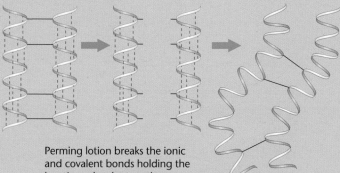

Perming lotion breaks the ionic and covalent bonds holding the keratin molecules together.

The neutralising lotion forms new covalent and ionic bonds, and so the hair curls.

When hair is wet, the hydrogen bonds are broken, but they reform when the hair is heated and dry. Using curling tongs or hair straighteners allows you to set the hydrogen bonds to give your hair a new shape. It will last until your hair gets wet again, unless you set it with hair spray. A humid day is usually enough to break the hydrogen bonds.

Perming gives hair a permanent change. The chemicals in the perming lotion break the bonds between the keratin molecules. Hair can then be wound on curlers and a second lotion used to reset the bonds to make the hair curl.

Questions

Stretch and challenge

A1. Which atoms must be in a keratin molecule for hydrogen bonds to form along its length?

A2. Explain how hair straighteners work in terms of hydrogen bonds.

A3. Why does straightened hair go frizzy in the rain?

Figure A1 *Using hair straighteners to break hydrogen bonds*

KEY IDEAS

› Covalent and ionic bonds are equally strong bonds.

› Electronegativity is an atom's ability to draw the electron density from a covalent bond towards itself.

› There are three types of intermolecular force: permanent dipole–dipole (caused by differences in electron density), hydrogen bonds (a special type of permanent dipole–dipole attraction) and induced dipole–dipole or van der Waals forces (attractions caused by the formation of instantaneous dipoles; they account for the differences in boiling points between molecules of different size).

ASSIGNMENT 4: CHOOSING THE RIGHT MIXES FOR SUMMER AND WINTER PETROL BLENDS

(PS 1.1, 1.2, 2.1, 2.4, 4.1)

Figure A1 *Petrol suppliers use different blends of alkanes in their petrol at different times of the year.*

Petrol contains a blend of alkanes, from C_5H_{12} to $C_{10}H_{22}$. In a car engine, these must be vaporised before they can be ignited with oxygen from the air in a cylinder. The ability of the petrol to vaporise is essential to start your car and keep it going.

Name	Formula	Boiling point/K
pentane	C_5H_{12}	309.2
hexane	C_6H_{14}	342.1
heptane	C_7H_{16}	371.5
octane	C_8H_{18}	398.8
nonane	C_9H_{20}	423.9
decane	$C_{10}H_{22}$	447.2

Table A1 *Boiling points of alkanes used in petrol*

Petrol suppliers need to blend their petrol so that it will vaporise, regardless of the weather (temperature in particular). This means that they use different blends of alkanes at different times of the year. In hot weather, they do not want the petrol to vaporise too easily. This would cause a pollution problem and would waste fuel. In cold weather, the petrol needs to vaporise more easily.

Questions

A1. How does the length of the alkane chain affect the boiling point?

A2. What type of bonds hold the carbon and hydrogen atoms together in pentane?

A3. Do you expect the C–H bond to be polar? Give a reason.

A4. What intermolecular forces might arise in pentane? Explain how these arise.

A5. What effect would the length of the alkane chain have on these intermolecular forces?

A6. Explain the trend in boiling points from pentane to decane.

A7. Which alkanes would you suggest are included in higher proportions for summer blends of petrol?

A8. Which alkanes would you suggest are included in higher proportions for a winter blend?

A9. An investigation is carried out to determine the evaporation rate of different alkanes at a set temperature. Outline a method and name the variables that need to be controlled and those that are measured.

PRACTICE QUESTIONS

1. Table Q1 shows the electronegativity values of the elements from lithium to fluorine.

	Li	Be	B	C	N	O	F
Electronegativity	1.0	1.5	2.0	2.5	3.0	3.5	4.0

Table Q1

  a. i. State the meaning of the term *electronegativity*.

    ii. Suggest why the electronegativity of the elements increases from lithium to fluorine.

  b. State the type of bonding in lithium fluoride. Explain why a lot of energy is needed to melt a sample of solid lithium fluoride.

  c. Deduce why the bonding in nitrogen oxide is covalent rather than ionic.

  d. Oxygen forms several different compounds with fluorine.

    i. Suggest the type of crystal shown by OF_2.

    ii. Write an equation to show how OF_2 reacts with steam to form oxygen and hydrogen fluoride.

    iii. One of these compounds of oxygen and fluorine has a relative molecular mass of 70.0 and contains 54.3 % by mass of fluorine. Calculate the empirical formula and the molecular formula of this compound. Show your working.

AQA Jan 2013 Unit 1

2. The following equation shows the reaction of a phosphine molecule (PH_3) with an H^+ ion.

$$PH_3 + H^+ \rightarrow PH_4^+$$

  a. Draw the shape of the PH_3 molecule. Include any lone pairs of electrons that influence the shape.

  b. State the type of bond that is formed between the PH_3 molecule and the H^+ ion. Explain how this bond is formed.

  c. Predict the bond angle in the PH_4^+ ion.

  d. Although phosphine molecules contain hydrogen atoms, there is no hydrogen bonding between phosphine molecules. Suggest an explanation for this.

AQA June 2012 Unit 1 Question 3

3. Fluorine forms compounds with many other elements.

  a. Fluorine reacts with bromine to form liquid bromine trifluoride (BrF_3). State the type of bond between Br and F in BrF_3 and state how this bond is formed.

  b. Two molecules of BrF_3 react to form ions as shown by the following equation.

$$2BrF_3 \rightarrow BrF_2^+ + BrF_4^-$$

    i. Draw the shape of BrF_3 and predict its bond angle. Include any lone pairs of electrons that influence the shape.

    ii. Draw the shape of BrF_4^- and predict its bond angle. Include any lone pairs of electrons that influence the shape.

  c. BrF_4^- ions are also formed when potassium fluoride dissolves in liquid BrF_3 to form $KBrF_4$. Explain, in terms of bonding, why $KBrF_4$ has a high melting point.

  d. Fluorine reacts with hydrogen to form hydrogen fluoride (HF).

    i. State the strongest type of intermolecular force between hydrogen fluoride molecules.

    ii. Draw a diagram to show how two molecules of hydrogen fluoride are attracted to each other by the type of intermolecular force that you stated in part d.i. Include all partial charges and all lone pairs of electrons in your diagram.

  e. The boiling points of fluorine and hydrogen fluoride are −188 °C and 19.5 °C, respectively. Explain, in terms of bonding, why the boiling point of fluorine is very low.

AQA Jan 2012 Unit 1 Question 1

(Continued)

4. Aluminium and thallium are elements in Group 3(13) of the Periodic Table. Both elements form compounds and ions containing chlorine and bromine.

   a. Write an equation for the formation of aluminium chloride from its elements.

   b. An aluminium chloride molecule reacts with a chloride ion to form the $AlCl_4^-$ ion. Name the type of bond formed in this reaction. Explain how this type of bond is formed in the $AlCl_4^-$ ion.

   c. Aluminium chloride has a relative molecular mass of 267 in the gas phase. Deduce the formula of the aluminium compound that has a relative molecular mass of 267.

   d. Deduce the name or formula of a compound that has the same number of atoms, the same number of electrons and the same shape as the $AlCl_4^-$ ion.

   e. Draw and name the shape of the $TlBr_5^{2-}$ ion.

   f. i. Draw the shape of the $TlCl_2^+$ ion.
      ii. Explain why the $TlCl_2^+$ ion has the shape that you have drawn in part f.i.

   g. Which one of the first, second or third ionisations of thallium produces an ion with the electron configuration $[Xe]\,5d^{10}\,6s^1$?

   AQA June 2013 Unit 1 Question 5

5. Table Q2 shows the electronegativity values of some elements.

	Fluorine	Chlorine	Bromine	Iodine	Carbon	Hydrogen
Electro-negativity	4.0	3.0	2.8	2.5	2.5	2.1

Table Q2

   a. Define the term *electronegativity*.

   b. Table Q3 shows the boiling points of fluorine, fluoromethane (CH_3F) and hydrogen fluoride.

	F–F	H–C(F)(H)H	H–F
Boiling point/K	85	194	293

Table Q3

   i. Name the strongest type of intermolecular force present in: liquid F_2; liquid CH_3F; liquid HF.

   ii. Explain how the strongest type of intermolecular force in liquid HF arises.

   c. Table Q4 shows the boiling points of some other hydrogen halides.

	HCl	HBr	HI
Boiling point/K	188	206	238

Table Q4

   i. Explain the trend in the boiling points of the hydrogen halides from HCl to HI.

   ii. Give one reason why the boiling point of HF is higher than that of all the other hydrogen halides.

   AQA January 2006 Unit 1 Question 3

6. a. i. State what is meant by the term polar when applied to a covalent bond.

   ii. Consider the covalent bonds in molecules of hydrogen and of water. State whether the covalent bonds are polar or non-polar. Explain your answers.

   b. Ammonia is very soluble in water because it is able to form hydrogen bonds with water molecules.

   i. Copy and complete Figure Q1 to show how an ammonia molecule forms a hydrogen bond with a water molecule. Include partial charges and all the lone pairs of electrons.

(Continued)

Figure Q1

ii. The bond angle in a molecule of water is about 104.5°. State the bond angle in an ammonia molecule and explain why it is different from that in water.

c. Ammonia reacts with aluminium chloride to form the molecule shown in Figure Q2.

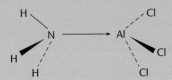

Figure Q2

Name the type of bond formed between the nitrogen and the aluminium. Explain how this bond is formed.

AQA January 2007 Unit 1 Question 3

7. a. Iodine and diamond are both crystalline solids at room temperature. Identify one similarity in the bonding, and one difference in the structures, of these two solids. Explain why these two solids have very different melting points.

   b. Graphite and diamond are both solid crystalline forms of carbon. Explain why graphite conducts electricity and diamond does not.

   c. Magnesium has a metallic structure and magnesium chloride has an ionic crystal structure.

      i. Explain why magnesium conducts electricity but solid magnesium chloride does not.

      ii. Which conditions are needed for magnesium chloride to conduct electricity?

   AQA January 2005 Unit 1 Question 5

8. a. An ammonium ion, made by the reaction between an ammonia molecule and a hydrogen ion, can be represented as shown in Figure Q3.

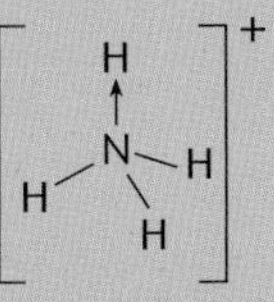

Figure Q3

   i. Name the type of bond represented in the diagram by N–H

   ii. Name the type of bond represented in the diagram by N→H

   iii. In terms of electrons, explain why an arrow is used to represent this N→H bond.

   iv. In terms of electron pairs, explain why the bond angles in the NH_4^+ ion are all 109° 28 –.

   b. Define the term *electronegativity*.

   c. A bond between nitrogen and hydrogen can be represented as $N^{\delta -} - H^{\delta +}$

      i. In this representation, what is the meaning of the symbol $\delta +$?

      ii. From this bond representation, what can be deduced about the electronegativity of hydrogen relative to that of nitrogen?

   AQA June 2002 Unit 1 Question 2

9. The hydroxide ion contains a covalent bond between the hydrogen and oxygen atoms. An extra electron on the oxygen atom gives the ion an overall + 1 charge.

   a. Explain why sodium hydroxide has the formula NaOH while aluminium hydroxide has the formula $Al(OH)_3$.

   b. Draw a dot and cross diagram to show the electron configuration of calcium hydroxide. You need only show the outer shells.

   c. Give one difference between the ionic bond in calcium hydroxide and a dative covalent bond.

(Continued)

d. An ammonia molecule, NH_3, contains three covalent N–H bonds. When an ammonium ion forms, a dative covalent bond forms between the lone pair of electrons on the nitrogen in ammonia and a hydrogen ion.

   i. Draw a dot and cross diagram to show the electron configuration of ammonia. You need only show the outer electron shells.

   ii. When ammonia dissolves in water, small amounts of ammonium hydroxide form. Draw a dot and cross diagram to show the bonding in ammonium hydroxide.

Stretch and challenge

10. Nylon is a polymer containing carbon, hydrogen, oxygen and nitrogen atoms. The repeating unit of one type of nylon is:

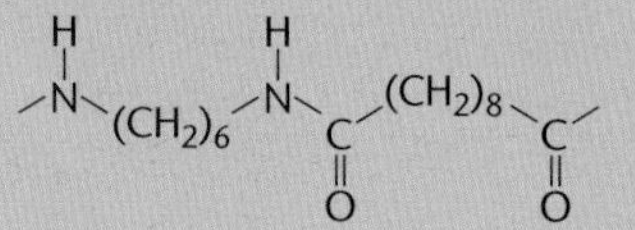

a. State two types of intermolecular forces that exist between long nylon polymer chains.

b. Copy the repeating unit above and add another repeating unit parallel to it.

   i. Annotate your diagram to show the strongest intermolecular force.

   ii. Explain the cause of this intermolecular force.

c. i. Give the bond angles between the carbon atoms in the $(CH_2)_8$ section of the polymer chain.

   ii. Give the bond angles for the carbon atom of the $C = O$ bond. Name the shape.

Multiple choice

11. What is the correct formula for ammonium carbonate?

   A. NH_4CO_3

   B. $NH_4(CO_3)_2$

   C. $(NH_4)_2CO_3$

   D. $(NH_4)_3CO_3$

12. What is the bond angle in methane, CH_4?

   A. $109° \ 28 -$

   B. $120°$

   C. $107°$

   D. $104.5°$

13. Which pairs of atoms in different compounds do not form hydrogen bonds?

   A. Hydrogen and oxygen

   B. Hydrogen and nitrogen

   C. Hydrogen and fluorine

   D. Hydrogen and chlorine

14. What is a co-ordinate bond?

   A. An ionic bond with both electrons supplied by one atom.

   B. An ionic bond with both atoms suppling one electron each.

   C. A covalent bond containing a pair of shared electrons with both electrons supplied by one atom.

   D. A covalent bond containing a pair of shared electrons with both atoms supplying one electron.

4 THE PERIODIC TABLE

PRIOR KNOWLEDGE

You have probably learnt that elements can be organised into groups and periods in the Periodic Table. You may know that the Periodic Table is a useful tool for chemists and can be used to determine an element's electron configuration. You may also know that we can use the Periodic Table to predict some properties of the elements.

LEARNING OBJECTIVES

In this chapter, you will reinforce and build on these ideas and learn more about trends in physical properties of elements and how these match the positions of the elements in the Periodic Table.

(Specification 3.2.1.1, 3.2.1.2)

All the elements in the Periodic Table up to atomic number 92 occur naturally; that is, you can find them somewhere on Earth. Elements with atomic numbers greater than 92 are artificially made. Making these heavy elements involves smashing together other atoms or nuclei to make a larger atom. This only happens in nuclear reactions or in particle accelerators.

In 2010, Russian and American scientists first reported that they had made element 117. German scientists confirmed their findings in May 2014. Element 117 has 117 protons and, as yet, no official name. It has been given the temporary name ununseptium. Its electronic configuration is:

$[Rn]5f^{14}6d^{10}7s^27p^5$

where [Rn] indicates the electronic configuration of the noble gas Radon, Rn.

The arrangement of its outermost electrons, $7s^27p^5$, means it will be in Group 7(17), the halogens.

Only a few atoms have been made and so its physical properties are unknown. Because of its high relative atomic mass, it is not expected to be a typical halogen, but it may follow the Group 7(17) trends in its melting and boiling points and in its ionisation energies.

A working party from the International Union of Pure and Applied Chemistry (IUPAC) is currently meeting to decide whether element 117 can be made official and added to the Periodic Table. Then it will be given a formal name.

4.1 CLASSIFICATION OF THE ELEMENTS IN s, p AND d BLOCKS

Developing the Periodic Table

As more and more elements were discovered in the eighteenth and nineteenth centuries, several attempts were made to arrange the elements in some sort of order. It was not until more accurate atomic mass values became available that real progress started to be made. Dmitri Mendeleev's attempt at classification was so successful that it became the basis for today's modern Periodic Table.

The modern Periodic Table

Mendeleev arranged the elements in order of relative atomic mass (then called atomic weight). In 1913, Henry Moseley showed that the order of elements in the Periodic Table could be more accurately related to another property, atomic number (Z) (see Chapter 1). In a few cases, this order is not the same as the relative atomic mass order. Iodine ($Z = 53$) has a relative atomic mass of 126.9 whereas tellurium ($Z = 52$) has a relative atomic mass of 127.6. Mendeleev's table needed a few adjustments to produce the modern Periodic Table used today (Figure 1).

The horizontal rows of elements in the Periodic Table are called **periods** and are numbered 1 to 7. These correspond to the electron shells. The vertical columns are called **groups**. Traditionally, these are numbered 1 to 0, where Group 1 contains the elements lithium, sodium, potassium and down the group to francium. Group 3 is the vertical column that contains, for example, boron and aluminium.

New IUPAC rules have renumbered the groups so that Groups 1 and 2 remain the same, but the new Group 3 is the group of transition metals from scandium down and what was called Group 3 becomes Group 13. This is because the transition elements in Periods 4 to 7 had not been assigned group numbers, but they have now. So, for example, copper, silver and gold are in the new Group 11.

In this chapter, groups will be described by their old numbers with the new numbers in brackets. The halogens, for example, are in Group 7(17).

All elements in Group 1 have one electron in their outermost shell, all elements in Group 2 have two electrons in their outermost shell and so on (Figure 2). You can now expand this idea. All the elements in a particular group have the same number of electrons in their outermost sub-shell. This characterises the group. So, Group 1 elements always have one electron in the s sub-shell and Group 2 elements have two electrons in the s sub-shell. Group 3(13) elements always have two electrons in the s sub-shell and one electron in the p sub-shell. The pattern continues to Group 7(17), with two electrons in the s sub-shell and five electrons in the p sub-shell. Group 0(18) has two electrons in the s sub-shell and six in the p sub-shell.

Using the old system, the total number of electrons in the outer shell is the same as the element's old group number. For example, nitrogen, which is in Group 5(15), has $2s^2$ and $2p^3$ electrons, totalling five electrons in its outer shell (the superscript number shows the number of electrons in the particular sub-shell).

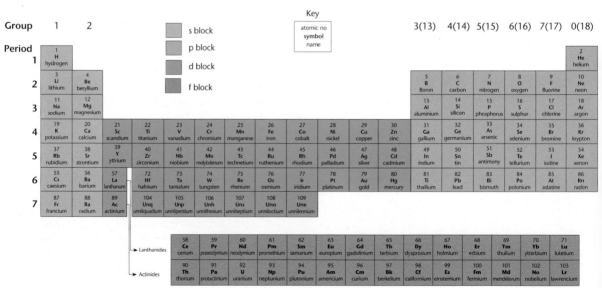

Figure 1 The modern Periodic Table.

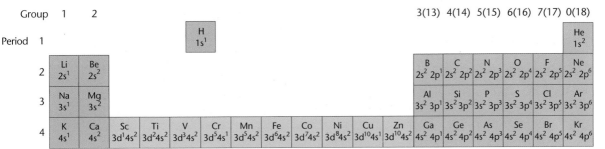

Figure 2 *The first 36 elements of the Periodic Table and their outer electronic structures.*

The Periodic Table is divided into blocks according to which sub-shell is being filled (see Figure 2). Groups 1 and 2 are called the **s block** elements because electrons are being added to s sub-shells. Groups 3(13) to 0(18) are the **p block** elements because electrons are being added to the p sub-shell.

Sandwiched between the s block and p block are the **d block** elements where it is the d sub-shells that are being filled. The d block includes the **transition** elements. For example, in Period 4, the d block elements are scandium through to zinc, but the transition elements are titanium through to copper (scandium and zinc are not transition metals).

Finally, in the **lanthanide** and **actinide** elements, electrons are filling f sub-shells and they are called the **f block** elements. Figure 1 shows these divisions of the Periodic Table.

QUESTIONS

1. For each of the elements with these atomic numbers, write the full electronic structure and state in which block of the Periodic Table it belongs:

   a. 9

   b. 20

   c. 30

2. An element has the electronic configuration $1s^2\ 2s^2\ 2p^6\ 3s^2\ 3p^6\ 3d^{10}\ 4s^2\ 4p^4$

   a. In which period is the element?

   b. In which block is the element?

   c. In which group is the element?

   d. How many electrons are there in its outer shell?

3. For each of the following elements, use the Periodic Table to name the block it is in and write its electron configuration.

   a. Strontium

   b. Fluorine

   c. Gold

   d. Aluminium

   e. Iron

   f. Germanium

4. Explain why there are ten d block elements and only six p block elements.

Stretch and challenge

5. The elements from atomic number 104 to 117 have only recently been discovered. Find out their names and suggest a reason for them.

ASSIGNMENT 1: MENDELEEV'S PERIODIC TABLE

(PS 1.1, 1.2, 2.3)

About 150 years ago, the Russian chemist Dmitri Mendeleev was a teacher at St Petersburg University. As there were no decent textbooks for his students, he decided to write one himself. He used a file of cards, one for each element, with its properties and atomic mass. By trying different arrangements of the cards, he built up a Periodic Table, which was much the same as the one we know today. The word 'periodic' means repeating.

By arranging the elements in order of atomic weight (we now use relative atomic mass) he found that, if he left a few gaps, chemically related elements occurred at regular intervals.

Mendeleev first published a version of his Periodic Table in 1869, in his textbook, *Principles of Chemistry* (Figure A1). His groupings showed the correct trends in properties and these were later confirmed by experiments. The gaps promoted a search for new elements. Mendeleev's initial table arranged the elements in columns. He later changed the layout of his table to the version that we are familiar with today.

The vertical columns correspond to the periods of elements in the modern Periodic Table and the horizontal rows are the groups of elements.

```
                            Ti=50    Zr=90     ?=180.
                            V=51     Nb=94    Ta=182.
                            Cr=52    Mo=96     W=186.
                            Mn=55    Rh=104,4 Pt=197,4
                            Fe=56    Ru=104,4  Ir=198.
                        Ni=Co=59     Pl=106,6 Os=199.
H=1                         Cu=63,4   Ag=108  Hg=200.
      Be=9,4   Mg=24        Zn=65,2   Cd=112
      B=11     Al=27,4      ?=68      Cr=116  Au=197?
      C=12     Si=28        ?=70      Sn=118
      N=14     P=31         As=75     Sb=122  Bi=210
      O=16     S=32         Se=79,4   Te=128?
      F=19     Cl=35,5      Br=80     I=127
Li=7  Na=23    K=39         Rb=85,4   Cs=133  Tl=204
               Ca=40        Sr=87,6   Ba=137  Pb=207.
               ?=45         Ce=92
               ?Er=56       La=94
               ?Yt=60       Di=95
               ?In=75,6     Th=118?
```

Figure A1 *Mendeleev's Periodic Table*

Mendeleev had left a gap below silicon (Group 4 (14)) in his Periodic Table. He gave the unknown element the name eka-silicon and predicted its properties. When the element germanium was discovered, its properties closely matched Mendeleev's predicted properties for eka-silicon (listed in Table 1). Germanium was eka-silicon. In the 20 years that followed, more of Mendeleev's predicted elements were discovered and his Periodic Table was accepted as an important tool for chemistry.

Questions

A1. Mendeleev also left a gap in his Periodic Table under aluminium in Group 3(13). What element fitted this gap?

A2. What events led to Mendeleev's table being accepted by the scientific community?

A3. The element rutherfordium (unnilquadium was renamed rutherfordium in 1997), Z = 104, is placed under hafnium in the Periodic Table. Why are scientists predicting that element 117 will be the first halogen to be made artificially?

A4. Element 118 was confirmed in 2008. Into which group is it placed?

A5. Explain the statement: *Mendeleev filled in the gaps in his Periodic Table as new elements were discovered. Today, chemists are still adding to the Periodic Table.*

Stretch and challenge

A6. Before Mendeleev's table, John Newlands had attempted to organise the elements into a table. He listed them in order of relative atomic mass and noticed that elements eight places away from each other had similar properties – he arranged them in rows of eight. His table was known as Newland's octaves. How far was his table successful, and where did it produce problems?

	Silicon, Si	Predicted properties of eka-silicon, Ek	Actual properties of germanium, Ge
Relative atomic mass	28	72	72.59
Density/ g cm^{-3}	2.3	5.5	5.3
Appearance	grey non-metal	grey metal	grey metal
Formula of oxide	SiO_2	EkO_2	GeO_2
Reaction with non-oxidising acid	none	very slow	slow with concentrated acid

Table A1 *The properties of silicon and germanium*

ASSIGNMENT 2: PREDICTING PROPERTIES OF AN UNSTABLE ELEMENT

(PS 1.1, 1.2, 2.3)

The element astatine, At, comes after iodine in Group 7(17). When Mendeleev compiled his first Periodic Table, he left a space below iodine and called the element eka-iodine. This was later identified as astatine. The *Guinness Book of Records* has dubbed astatine as *the rarest element on Earth* – at any one time there is about 25 g, less than a teaspoonful, on Earth. The Italian physicist Emilio Segrè and his team first produced it in the laboratory at the University of California in 1940. They bombarded bismuth with helium ions, He^{2+} (also called alpha particles), to produce this new element with atomic number 85. No more than 0.50 µg or 5.0×10^{-7} g has ever been made.

Its longest lived isotope has a half-life of only 8.3 hours, which means that half the astatine in a sample decays in 8.3 hours. It was named astatine, from the Greek word for 'unstable', because of this short half-life. The amount on Earth stays constant because it is formed when some heavier radioactive elements decay.

Questions

Use the data given in Table A1 to answer these questions.

A1. In which block of the Periodic Table is astatine?

A2. Give the outermost sub-shell electron structure of astatine.

A3. Predict the following values for astatine:

    a. density as a liquid

    b. melting point

    c. first ionisation energy.

A4. Give the formulae of the compounds:

    a. sodium astatide

    b. hydrogen astatide.

A5. List the scientific ideas used to make predictions about astatine's physical and chemical properties.

Stretch and challenge

A6. Explain why bombarding bismuth atoms with He^{2+} ions could, theoretically, produce astatine.

	Fluorine	Chlorine	Bromine	Iodine
Relative atomic mass	19	35.5	79.9	127
Density as a liquid/g cm^{-3}	1.11	1.56	2.93	4.93
Melting point/°C	– 220	– 101	– 7.2	113.5
Appearance at room temperature	colourless gas	green gas	orange liquid	purple solid
Formula of hydrogen halide	HF	HCl	HBr	HI
First ionisation energy/kJ mol^{-1}	1680	1260	1140	1010

Table A1 *Properties of Group 7(17) elements*

KEY IDEAS

> Elements in the Periodic Table are arranged in order of increasing atomic (or proton) number.

> The Periodic Table is divided into s, p, d and f blocks.

> Rows of elements in the Periodic Table are called periods and columns are called groups.

> Elements are classified as s, p, d or f block elements according to the electron configuration of their outer shells.

> The Periodic Table can be used to deduce the electron configuration of any element.

4.2 PROPERTIES OF THE ELEMENTS IN PERIOD 3

Moving across a period, each element has an atomic number one greater than the previous element. Successive elements have one more proton and one more electron in their atoms. This gradual change produces trends in both physical and chemical properties of the elements. These trends, or repeating patterns, are called **periodicity** – those found in one period are repeated in another period.

This section looks at the trends in atomic radius, first ionisation energy and melting point of the elements in Period 3.

Atomic radius

Atoms are extremely small. Rutherford first determined the size of an atom. He found that gold atoms have a radius of about 1×10^{-10} m.

The atomic radius is the distance from the centre of the atom to its outermost electrons, as shown in Figure 3.

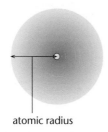

atomic radius

Figure 3 *The atomic radius of an element.*

The term atomic radius is used to measure the size of an atom, but the size of an atom depends on the space occupied by the electrons. The electron cloud of an atom has no fixed dimensions, because an atom's electrons could be found anywhere. When you use s, p and d orbitals to locate an electron, this is the most likely place to find the electron. Chemists talk about the probablity of the whereabouts of an electron. It may be elsewhere, which makes identifying the edge of an atom a problem.

Atomic radius is usually considered to be half the shortest distance from the nucleus of one atom to that of the nearest atom. Because atoms combine with different types of bonding, there are different types of atomic radii. **Covalent radius** and **metallic radius** are two types of atomic radius (Figure 4).

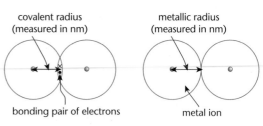

covalent radius (measured in nm)

metallic radius (measured in nm)

bonding pair of electrons

metal ion

The covalent radius is half the shortest distance from one nucleus to the next in a covalently bonded molecule.

The metallic radius is half the shortest distance from one metal ion nucleus to the next.

Figure 4 *Covalent and metallic radii.*

The nucleus in an atom is very small. The atom's size mostly depends on the space its electrons occupy. You might expect that, as the number of electrons increases, so does the size of the atom and the atomic radius. This is true for atoms of elements in the same group. Atomic radius increases down the group because an extra shell of electrons is added with each successive element. But this trend is not repeated across a period. In fact, moving across a period, atomic radius decreases, as shown in Figure 5.

Na	Mg	Al	Si	P	S	Cl	Ar
0.191	0.160	0.143	0.118	0.110	0.102	0.099	–

Figure 5 *Atomic radii of the elements in Period 3 measured in nanometres $(10^{-9}$ m).*

Note that the atomic radius decreases across the period from sodium to chlorine. Metallic radii are given for the metals Na, Mg and Al. Other values are for covalent radii. Argon is a special case.

In Period 3 all of the outer electrons are in the third energy level, $n = 3$, and have two inner shells of electrons shielding the nuclear charge. Moving across the period, a proton is added to the nucleus of each element. An electron is also added, but it goes into the same shell and the positive charge of the nucleus acts over a definite area. The increasing positive nuclear charge pulls with greater force on the negative electrons. All the electron shells are drawn closer to the nucleus. The overall effect is to pull all electrons closer to the nucleus. For example, sodium is larger than magnesium because the nuclear charge of magnesium is greater than that of sodium, but the outer electrons are in the same shell, with the same amount of shielding by the inner electrons. Electrons in magnesium are pulled closer to the nucleus than they are in sodium, resulting in a smaller radius. For the same reason, magnesium is larger than aluminium.

QUESTIONS

6. Why is there no atomic radius value for argon in Figure 5?

7. Table 1 gives the atomic radii of elements in Group 1.

Element	Number of electrons	Atomic radius/nm
lithium	3	0.152
sodium	11	0.186
potassium	19	0.227
rubidium	37	0.248
caesium	55	0.265

Table 1 *The atomic radii of elements in Group 1*

  a. What is the trend in atomic radius as you go down Group 1?

  b. Explain the trend you have described in a.

  c. Which type of radius is used to measure the atomic radii of Group 1 elements?

Stretch and challenge

  d. P^{3-} and S^{2-} have the same electron configuration, but have different atomic radii of 0.212 nm and 0.184 nm respectively. Explain why the radii are different.

First ionisation energies

The **first ionisation energy** is the energy required to remove one mole of the outermost electrons from one mole of atoms of an element in the gaseous state (see Chapter 1).

First ionisation energies depend on:

❯ the charge on the nucleus

❯ the distance of the outer electrons from the nucleus

❯ the shielding by inner electron shells

❯ whether the electron is alone in a orbital, or one of a pair.

Plotting the first ionisation energies of the Period 3 elements against their atomic numbers shows a distinct pattern (Figure 6). A similar pattern is repeated across other periods when s and p electrons are being removed. This repeating pattern of first ionisation energies across a period is an example of periodicity. The overall trend is that first ionisation energy increases across Period 3. More energy is needed to remove an electron moving across the period because as the number of protons increases, the positive charge on the nucleus increases. The force of attraction on the electrons increases.

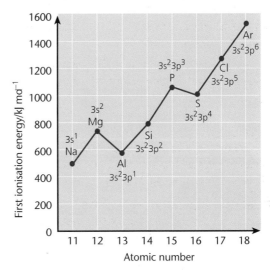

Figure 6 *First ionisation energy plotted against atomic number for Period 3 elements.*

The increase in first ionisation energy across Period 3 is not a steady rise. The characteristic 2, 3, 3 pattern of increases and dips in first ionisation energy suggests that electrons removed from the third shell are arranged in different sub-shells (see Chapter 1).

The first two values of this pattern are for electrons taken from the 3s sub-shell. The value for magnesium is greater than that for sodium because magnesium has one more proton attracting the electrons. Then there is a decrease for aluminium, where the first electron is taken from the 3p sub-shell. This electron is at a higher energy level than those in the 3s sub-shell and so is easier to remove. Aluminium, silicon and phosphorus show the expected increase in the energy required to remove each unpaired 3p electron as the increasing positive charge exerts more force on electrons in the same sub-shell. Then another dip occurs at sulfur as the first paired electron is removed. Paired electrons repel each other; there is more electrostatic repulsion and it is easier to remove the fourth electron in the 3p sub-shell. This is shown in Figure 7 for sulfur.

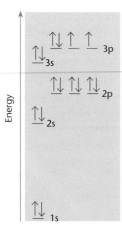

Figure 7 *The arrows represent electrons in different sub-shells of a sulfur atom.*

After sulfur, the increase in the first ionisation energy is due to the increasing positive charge attracting the electrons more strongly.

Melting points

Unlike atomic radii and first ionisation energies, the melting points across Period 3 do not show a regular pattern. Neither is this pattern repeated across any other period. Melting points are directly linked to:

> the type of bonding between atoms of the element

> the structure of the element.

The melting points across Period 3 increase from sodium to a maximum at silicon, in Group 4(14), and then decrease dramatically from silicon to argon.

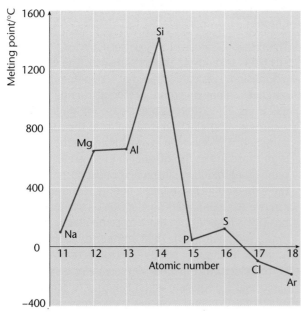

Figure 8 *Melting points of the elements in Period 3.*

When an element melts, the particles break free of the forces that hold them together. The greater the force between the particles, the higher the melting point, as energy needs to be supplied to overcome the forces between particles.

Period 3 elements with a metallic structure

The metals sodium, magnesium and aluminium are held together by metallic bonds. The electrostatic attraction between the delocalised electrons and the positive ions holds the metal crystals together. The charge on the positive ion increases from sodium to aluminium: sodium has a charge of $+1$, magnesium is $+2$ and aluminium is $+3$. The number of delocalised electrons is also increasing and the higher the charge and number of delocalised electrons, the stronger the attraction between the metal ions and delocalised electrons.

The size of the metal ion also influences the melting points of the elements sodium to aluminium. The metallic radius decreases from sodium to aluminium (see Figure 5). From sodium to aluminium, the metal ions pack closer together in the metallic structure and the forces between them increase.

The melting points increase from sodium to aluminium.

Period 3 elements with a giant covalent structure

The next element after aluminium is silicon. It is a non-metal with the highest melting point of all the Period 3 elements. Silicon has a giant covalent structure, similar to that of diamond (see Chapter 3). All the silicon atoms are held by strong covalent bonds. For silicon to melt, many of these covalent bonds must be broken. This needs a considerable input of energy, and so gives silicon a very high melting point.

Period 3 elements with a molecular structure

The non-metals phosphorus to chlorine consist of covalent molecules. The forces between molecules are van der Waals forces (weak intermolecular forces). The phosphorus molecule exists as P_4, sulfur as S_8, and chlorine as Cl_2.

When these elements melt, only enough energy has to be supplied to break the weak van der Waals forces. The melting points of the non-metals phosphorus to chlorine are low compared with the first four elements in Period 3.

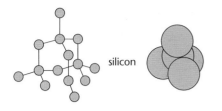

silicon

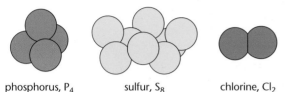

phosphorus, P_4 sulfur, S_8 chlorine, Cl_2

Figure 9 *Silicon has a giant structure (top left), with each silicon bonded to four others in a tetrahedral arrangement (top right). Phosphorus, sulfur, and chlorine have molecular structures.*

The size of the molecules accounts for the difference between the melting points of phosphorus, sulfur and chlorine. Larger molecules have stronger van der Waals forces, so more energy is needed to break

them. Their strength is in the order of $S_8 > P_4 > Cl_2$. This is the order of their decreasing melting points.

Period 3 element: Argon

Argon exists as single atoms. The atoms in solid argon are held together by very weak van der Waals forces. They are weaker than those in phosphorus, sulfur and chlorine because argon atoms are smaller. Little energy is needed to break them and argon has the lowest melting point in Period 3.

QUESTIONS

8. Look at the melting points for phosphorus and sulfur in Figure 8.

   a. Describe the type of bonding in solid phosphorus and solid sulfur.

   b. Explain the melting points in terms of the forces between molecules.

KEY IDEAS

> Periodicity is used to describe repeating patterns of properties at regular intervals in the Periodic Table.

> The atomic radius of elements decreases across Period 3 as the nuclear charge increases.

> The overall trend across Period 3 is an increase in first ionisation energy, as the nuclear charge

increases. The irregular increase across a period is due to the filling of sub-shells.

> Melting points increase from sodium to silicon, then decrease.

> Melting points depend on the bonding and structure of the element

ASSIGNMENT 3: EXPLAINING RADII

(MS 3.1; PS 1.1, 1.2, 2.3, 3.2)

Table A1 lists values for the atomic and ionic radii of some elements. The ionic radius measures the radius of an ion. For example, a sodium ion, Na^+, has a radius of 0.098 nm.

Element	Atomic radius/nm	Ionic radius/nm
lithium	0.152	0.078
sodium	0.186	0.098
magnesium	0.160	0.078
oxygen	0.073	0.124
fluorine	0.071	0.133

Table A1

You are going to use the ideas you have learnt about periodicity to explain some trends in atomic and ionic radii.

Questions

A1. Write the symbols for all the ions in Table A1.

A2. Explain why a sodium atom has a larger radius than a lithium atom.

A3. Explain why a sodium atom has a larger radius than a magnesium atom.

A4. Explain why a lithium ion is smaller than a lithium atom.

A5. Write the full electron configurations for:
  a. a sodium ion
  b. a magnesium ion.

A6. Explain why a magnesium ion is smaller than a sodium ion.

A7. Write the full electron configuration for:
  a. a fluoride ion
  b. an oxide ion.

A8. Explain why both fluoride and oxide ions are larger than their respective atoms.

A9. Positive ions are **cations**, negative ions are **anions**. Suggest how atomic radius compares to the ionic radius of:
  a. cations
  b. anions.

A10. Use data from Figure 5 to answer these questions.
  a. What is the atomic radius, in metres, of the elements from sodium to argon? Show your answer in standard form.
  b. Draw a dot-to-dot line graph of atomic number against atomic radius (nm) for the elements sodium to argon.
  c. Annotate your graph to explain the trend.

PRACTICE QUESTIONS

1. The elements in Period 2 show periodic trends.

  a. Identify the Period 2 element, from carbon to fluorine, that has the largest atomic radius. Explain your answer.

  b. State the general trend in first ionisation energies from carbon to neon. Deduce the element that deviates from this trend and explain why this element deviates from the trend.

  c. Write an equation, including state symbols, for the reaction that occurs when the first ionisation energy of carbon is measured.

  d. Explain why the second ionisation energy of carbon is higher than the first ionisation energy of carbon.

  e. Deduce the element in Period 2, from lithium to neon, that has the highest second ionisation energy.

AQA Jun 2013 Unit 1 Question 6

2. Use your knowledge of electron configuration and ionisation energies to answer this question. Figure Q1 shows the second ionisation energies of some Period 3 elements.

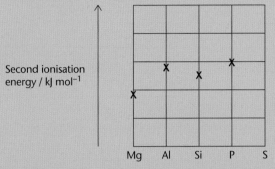

Figure Q1

a. i. Draw an 'X' on the diagram to show the second ionisation energy of sulfur.

ii. Write the full electron configuration of Al^{2+}.

iii. Write an equation to show the process that occurs when the second ionisation energy of aluminium is measured.

iv. Give one reason why the second ionisation energy of silicon is lower than the second ionisation energy of aluminium.

b. Predict the element in Period 3 that has the highest second ionisation energy. Give a reason for your answer.

c. Table Q1 gives the successive ionisation energies of an element in Period 3.

	First	Second	Third	Fourth	Fifth	Sixth
Ionisation energy/ kJ mol^{-1}	786	1580	3230	4360	16 100	19 800

Table Q1

Identify this element.

d. Explain why the ionisation energy of every element is endothermic.

AQA Jan 2013 Unit 1 Question 2

3. This question is about the first ionisation energies of some elements in the Periodic Table.

a. Write an equation, including state symbols, to show the reaction that occurs when the first ionisation energy of lithium is measured.

b. State and explain the general trend in first ionisation energies for the Period 3 elements aluminium to argon.

c. There is a similar general trend in first ionisation energies for the Period 4 elements gallium to krypton. State how selenium deviates from this general trend and explain your answer.

d. Suggest why the first ionisation energy of krypton is lower than the first ionisation energy of argon.

e. Table Q2 gives the successive ionisation energies of an element.

	First	Second	Third	Fourth	Fifth
Ionisation energy/ kJ mol^{-1}	590	1150	4940	6480	8120

Table Q2

Deduce the group in the Periodic Table that contains this element.

f. Identify the element that has a +5 ion with an electron configuration of $1s^2\ 2s^2\ 2p^6\ 3s^2\ 3p^6\ 3d^{10}$.

AQA Jun 2011 Unit 1 Question 5

4. a. State the meaning of the term *first ionisation energy* of an atom.

b. Copy and complete the electron arrangement for the Mg^{2+} ion. *$1s^2$....*

c. Identify the block in the Periodic Table to which magnesium belongs.

d. Write an equation to illustrate the process occurring when the second ionisation energy of magnesium is measured.

e. The Ne atom and the Mg^{2+} ion have the same number of electrons. Give two reasons why the first ionisation energy of neon is lower than the third ionisation energy of magnesium.

f. There is a general trend in the first ionisation energies of the Period 3 elements, Na–Ar.

i. State and explain this general trend.

ii. Explain why the first ionisation energy of sulfur is lower than would be predicted from the general trend.

AQA Jan 2006 Unit 1 Question 4

5. Figure Q2 shows the melting points of some of the elements in Period 3.

   a. Copy the diagram and use crosses to mark the approximate positions of the melting points for the elements silicon, chlorine and argon. Complete the diagram by joining the crosses.

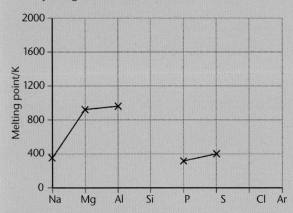

Figure Q2

   b. By referring to its structure and bonding, explain your choice of position for the melting point of silicon.

   c. Explain why the melting point of sulfur, S_8, is higher than that of phosphorus, P_4.

AQA Jun 2006 Unit 1 Question 4

6. Figure Q3 shows the values of the first ionisation energies of some of the elements in Period 3.

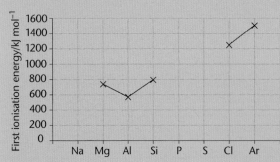

Figure Q3

a. Copy Figure Q3 and use crosses to mark the approximate positions of the values of the first ionisation energies for the elements Na, P and S. Complete the diagram by joining the crosses.

b. Explain the general increase in the values of the first ionisation energies of the elements Na–Ar.

c. In terms of the electron sub-levels involved, explain the position of aluminium and the position of sulfur in the diagram.

AQA Jan 2004 Unit 1 Question 5

7. Sodium and magnesium are both in the s block of Period 3.

   a. What is meant by s block?

   b. The atomic radius of a sodium atom is 0.191 nm. The atomic radius of a magnesium atom is 0.160 nm. Explain why a magnesium atom is smaller than a sodium atom?

   c. The first ionisation energy of sodium is 496 kJ mol^{-1}. The first ionisation energy of magnesium is 738 kJ mol^{-1}. Explain why the first ionisation of magnesium is higher than the first ionisation energy of sodium?

   d. Sodium forms Na^+ ions. Magnesium forms Mg^{2+} ions. Use this information to explain why the melting point of magnesium is higher than the melting point of sodium.

   e. i. State how the melting point of aluminium compares to the melting points of sodium and magnesium.

      ii. Explain your answer.

8. Scientists use ideas of first ionisation energy to explain models of electron arrangement in atoms.

   a. Explain what is meant by the term *first ionisation energy*.

   b. Write an equation for the first ionisation energy of sodium.

   c. State and explain the overall trend in first ionisation energies across Period 3.

   d. The increase in first ionisation energies across Period 3 is not a steady rise.

i. Which two elements show a drop in first ionisation energy across Period 3?

ii. Explain why these two elements have a drop in their first ionisation energies.

e. First ionisation energy trends across a period are an example of periodicity.

i. State what is meant by periodicity.

ii. State which element in Period 2 will have the highest first ionisation energy value. Explain your answer.

f. Write an equation to show the second ionisation energy of magnesium.

g. Explain why elements in group 1 do not usually form + 2 ions.

9. The overall trend in melting points of Period 3 elements is an increase from sodium to silicon and a decrease from silicon to argon.

a. i. Explain why melting points increase from sodium to aluminium.

ii. Explain why silicon has the highest melting point in Period 3.

b. Describe the structures of phosphorus, sulfur, chlorine and argon.

c. Explain the following:

i. Sulfur has a higher melting point than phosphorus.

ii. Melting points decrease from sulfur to chlorine to argon.

Stretch and challenge

10. Table Q3 gives the atomic radii and electronegativity values for Period 3 elements.

a. Define *electronegativity*.

b. Explain why argon does not have a value for electronegativity.

c. State and explain the trend in atomic radius across Period 3.

d. Give two reasons why electronegativity increases across Period 3.

e. Explain how electronegativity can be used to predict the type of bonding in SCl_2.

Multiple choice

11. Which block in the Periodic Table contains the transition metals?

A. s block

B. d block

C. p block

D. f block

12. Which is the correct definition of first ionisation energy?

A. The energy required to remove the outermost electron from an atom of an element in the gaseous state.

B. The energy required to produce an ion with one positive charge.

C. The energy required to remove the outermost electrons from one mole of atoms of an element in the gaseous state.

D. The energy required to remove one mole of outermost electrons from one mole of atoms of an element in the gaseous state.

13. Which element in Period 3 has a giant covalent structure?

A. Silicon

B. Phosphorus

C. Sulfur

D. Chlorine

14. Which element in Period 3 has a molecular structure?

A. Magnesium

B. Aluminium

C. Silicon

D. Phosphorus

	Na	Mg	Al	Si	P	S	Cl	Ar
Atomic radius (nm)	0.191	0.160	0.143	0.118	0.110	0.102	0.099	
Electronegativity	0.93	1.31	1.61	1.90	2.19	2.58	3.16	

Table Q3

5 INTRODUCTION TO ORGANIC CHEMISTRY

PRIOR KNOWLEDGE

You may have already studied some organic chemistry and learnt about the fractional distillation of crude oil, and about alkanes and alkenes, which are two types of hydrocarbon.

LEARNING OBJECTIVES

In this chapter you will learn about the language that chemists use when describing organic chemistry (nomenclature), different types of formulae that they use to represent molecules and how they describe how chemical changes happen (reaction mechanisms).

(Specification 3.3.1.1, 3.3.1.2, 3.3.1.3)

The market in organic products has grown rapidly in the last twenty years. We can buy organic food, organic clothes, even organic beauty products. Your shopping may even have organic packaging. Some claim that organic farming and practices produce safer and healthier products and are better for the environment.

Organic farming is controlled by the European Union (EU) organic farming standards. It relies on crop rotation, farmyard manure, compost and biological pest control. It excludes, or severely limits, the use

Figure 1 *The market in organic clothes is also expanding rapidly. This item of clothing is made from organic cotton. The cotton producer had to satisfy the EU organic standards.*

of petrochemical products such as pesticides and herbicides. It does not allow genetically modified (GM) crops or the use of plant growth regulators or antibiotics in livestock. If farmers fulfil the necessary conditions, the produce can be labelled 'organic'. About 3.5% of UK farming land is farmed organically to produce fruit and vegetables. Imported organic fruit and vegetables must also meet the same organic standards.

The public understanding of 'organic' is of a natural product, produced as nature intended. In chemistry, 'organic' has a different meaning. It is used to describe a vast group of compounds that are made from carbon atoms, with atoms of other elements, such as hydrogen, oxygen and nitrogen.

Organic chemistry is the study of carbon compounds – with a few exceptions.

5.1 ORGANIC AND INORGANIC COMPOUNDS

There are millions of different chemical compounds. Classifying them makes their study more manageable. Chemical compounds are classified according to the numbers and types of atoms that they are made from, the bonding between these atoms and the way that they are arranged.

Chemical compounds and their study are divided into two main groups: **organic compounds** and **inorganic compounds**. Organic chemistry is the study of organic compounds.

Organic compounds are compounds made from carbon atoms, together with other atoms such as hydrogen, oxygen and the halogens. They are very important in living things and in industry. It is estimated that there are about 10 million organic compounds and the number is rising all the time as chemists discover or synthesise new ones. This is more than all of the other known compounds put together. Organic compounds usually came from living things. Today, organic chemistry is the study of all carbon-based compounds, including many that are synthesised in laboratories and not found in nature. Exceptions include carbon dioxide, carbon monoxide, carbonates and hydrogen carbonates. These are usually considered to be inorganic compounds.

Inorganic chemistry is the study of compounds made from atoms other than carbon (the exceptions being those listed previously). Historically, inorganic compounds were those that melted or vaporised when heated, but returned to their original state when cooled. Early chemists thought that only inorganic compounds could be synthesised in the laboratory, because organic compounds from living things needed a vital life force. This idea was dispelled by the discovery in 1828 that urea, NH_2CONH_2, which is naturally found in urine, could be made in the laboratory.

Why carbon?

Carbon is in Group 4 (14) of the Periodic Table. It has the electronic configuration $1s^2\ 2s^2\ 2p^2$ and has four electrons available for bonding. Importantly, carbon atoms have the ability to make strong covalent bonds with other carbon atoms and with other non-metal atoms.

Chains of atoms in carbon polymers can reach thousands of atoms long and can include single and double bonds (Figures 2 and 3). Carbon can also join together to form ring and cage structures. Figure 4 shows three hydrocarbon molecules. Each is made from three carbon atoms and some hydrogen atoms. Propane has only carbon–carbon single bonds, propene has one carbon–carbon single bond and one carbon–carbon double bond and propyne has one carbon–carbon single bond and one carbon–carbon triple bond.

Carbon–carbon single bonds (C–C) are the most common. Carbon–carbon double bonds (C=C) are less common. Carbon–carbon triple bonds (C≡C) are comparatively rare.

It is carbon's ability to form strong links with itself and other non-metals that accounts for the incredibly varied and often great complexity of carbon-based structures. At the temperatures and pressures found on Earth, it is the stability of carbon-carbon bonds that has led to molecules found in living organisms being made from carbon atoms.

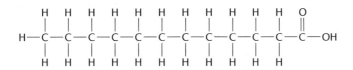

Figure 2 *Many molecules have long carbon-carbon chains, with C–C single bonds. This is a typical saturated fatty acid, lauric acid (dodecanoic acid).*

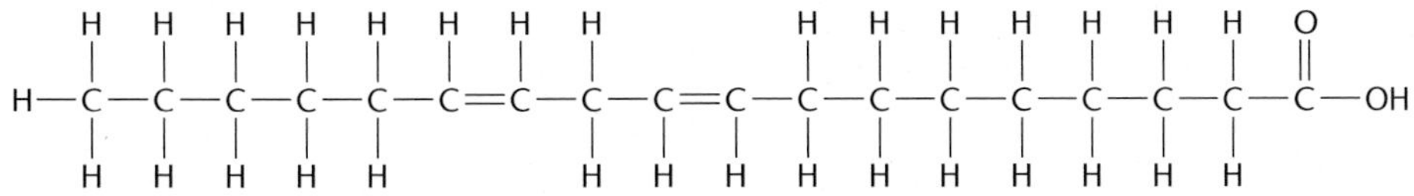

Figure 3 *This is an unsaturated fatty acid, linoleic acid. Each molecule has two C–C double bonds in the chain.*

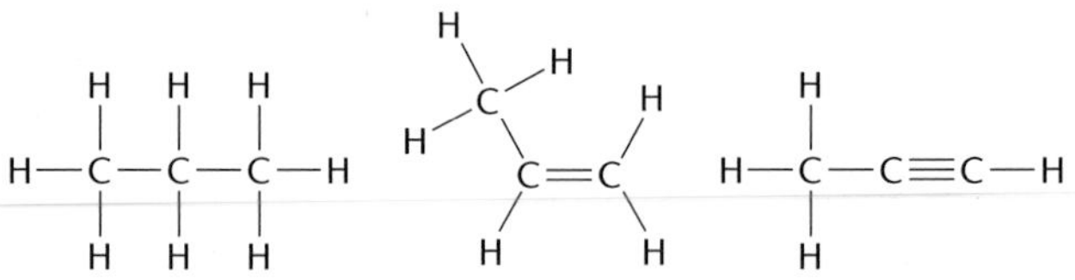

Figure 4 Compounds with single, double and triple carbon–carbon bonds: (a) propane, (b) propene, (c) propyne.

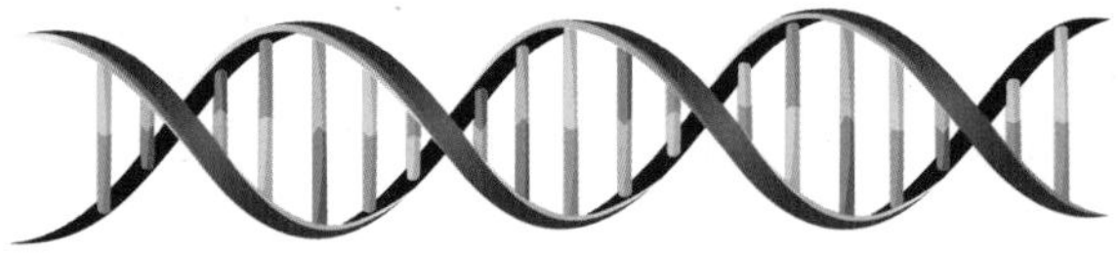

Figure 5 DNA molecules encode the genetic instructions in living material. The genetic material is bonded to long chains of carbon atoms that twist around to form two intertwined helices – the double helix structure.

The strength of carbon–carbon bonds

How strong is a carbon–carbon bond? The strength of a bond can be measured by finding out how much energy is needed to break it. The stronger the bond, the more energy is needed to break it. This is called its **bond energy** or **bond enthalpy**. Because the energy needed to break one bond is so small, bond enthalpy is measured in kilojoules per mole of bonds (kJ mol^{-1}) for gaseous molecules under standard conditions. You will learn more about bond enthalpy in Chapter 7.

Table 1 gives the mean bond enthalpies for single, double and triple carbon–carbon bonds and for other bonds between non-metal atoms. You will notice that the bond enthalpies for the carbon–carbon bonds are higher than the bond enthalpies for non-carbon–non-carbon bonds. Silicon is below carbon in Group 4 (14) of the Periodic Table, but the Si–Si bond is still weaker than the C–C bond. This helps to explain why chains made from covalently bonded carbon atoms are common, but chains made from other non-metal atoms are not.

Bond	Mean bond enthalpy/kJ mol^{-1}
C–C	347
C=C	612
C≡C	838
Si–Si	226
N–N	158
P–P	198

Table 1 Bond enthalpies

QUESTIONS

1. Draw dot-and-cross diagrams to show the bonding in a:

   a. carbon–carbon single bond

   b. carbon–carbon double bond

   c. carbon–carbon triple bond.

2. Draw dot-and-cross diagrams to show the arrangement of electrons (outer shell only) in a:

   a. carbon dioxide molecule

   b. nitrogen molecule.

Stretch and challenge

3. Explain why nitrogen and phosphorus are unlikely to form compounds that contain long chains of nitrogen and phosphorus atoms.

4. Explain why carbon atoms do not form ionic bonds.

5. Bonds between carbon atoms can be single, double or triple. The order of bond length (the distance between two nuclei) is single > double > triple. Shorter bonds are stronger and have higher bond enthalpies than longer bonds. C–C bonds are called sigma bonds. C=C and C≡C bonds consist of two types of bonds: pi (π) and sigma (σ). Explain how bond enthalpies show that:

   a. shorter bonds are stronger

   b. σ bonds are stronger than π bonds.

5.2 MOLECULAR SHAPES

The four covalent bonds around a carbon atom form a tetrahedral structure (see Chapter 3). Figure 6 shows methane, and the way that each bond is drawn indicates its three-dimensional (3D) position.

The arrangement of bonds around the central carbon

One hydrogen atom is at each of the four corners of the tetrahedron.

Figure 6 *The tetrahedral arrangement of bonds between the carbon atom and hydrogen atoms in a methane molecule*

Figure 7 shows the 3D displayed formula of propane. The chain of carbon atoms is actually a zigzag, lying in the plane of the paper, with the hydrogen atoms in front and behind the plane of the paper. The ball-and-stick model in Figure 7 gives a better idea of how the atoms are spaced in a molecule, while the space-filling model gives a more accurate picture of the arrangement of the tightly packed atoms.

Figure 7 *Different ways of showing the arrangement of bonded atoms in a propane molecule (molecular formula: $CH_3CH_2CH_3$). The space-filling model (right) of propane gives the best representation of the shape of the molecule, but the ball-and-stick model (middle) makes it easier to see the arrangement of bonds.*

5.3 TYPES OF FORMULAE

You can describe molecules in different ways. You have already come across **empirical formulae** in Chapter 2. An empirical formula gives the simplest whole-number ratio of atoms of each element in a molecule. CH_2O is the empirical formula for glucose. This tells you that there are two hydrogen atoms to each carbon atom and each oxygen atom. The **molecular formula** for glucose is $C_6H_{12}O_6$, which tells you how many of each type of atom is used to make a molecule.

There are four other types of formulae commonly used in organic chemistry: general formula, structural formula, displayed formula and skeletal formula.

General formula

A general formula allows you to find the formula of a particular molecule by substituting values for *n*. The general formula for an alcohol is $C_nH_{2n+1}OH$. The OH group is the functional group, meaning it is characteristic of an alcohol. It has no multiplication number because there is only one −OH group in any alcohol. To find the number of carbon and hydrogen atoms, you use the general formula. For example,

ethanol has two carbon atoms, so $n = 2$, which gives C_2H_5OH

propan-1-ol has three carbon atoms, so $n = 3$, which gives C_3H_7OH

You will meet general formulae as you study each group of organic compounds.

Structural formula

Like a molecular formula, a structural formula gives the numbers and types of atoms used to make a compound. But the structural formula also shows the sequence in which the atoms bond together. Propane has the molecular formula C_3H_8. Its structural formula is $CH_3CH_2CH_3$. The list of atoms is the order in which they are bonded together. If a double bond is present, structural formulae usually show this. So propene, molecular formula is C_3H_6, has the structural formula $CH_3CH = CH_2$.

Another form of structural formulae uses lines to show bonds between carbon atoms. For example, propane is sometimes shown as $CH_3{-}CH_2{-}CH_3$.

2-methylpropane has the molecular formula C_4H_{10}. A methyl group (CH_3 group) is bonded to the second carbon atom in a propane molecule and replaces a hydrogen atom. Its structural formula is $CH_3CH(CH_3)CH_3$. The brackets indicate that the CH_3 group is attached to the preceding CH group. This is sometimes shown as:

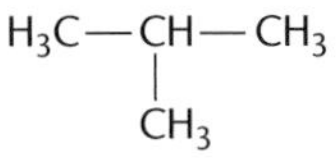

$$H_3C{-}CH{-}CH_3$$
$$|$$
$$CH_3$$

Make sure that your bond lines are between the correct atoms. You may meet both ways of showing structural formulae in different books.

Displayed formula

A displayed formula shows every bond and every atom in a compound.

The displayed formula for propane is:

$$H-\overset{\overset{\displaystyle H}{|}}{\underset{\underset{\displaystyle H}{|}}{C}}-\overset{\overset{\displaystyle H}{|}}{\underset{\underset{\displaystyle H}{|}}{C}}-\overset{\overset{\displaystyle H}{|}}{\underset{\underset{\displaystyle H}{|}}{C}}-H$$

The displayed formula for propene is:

$$H-\overset{\overset{\displaystyle H}{|}}{\underset{\underset{\displaystyle H}{|}}{C}}-\overset{\overset{\displaystyle H}{|}}{C}=\overset{}{\underset{\underset{\displaystyle H}{|}}{C}}-H$$

Skeletal formula

These are the most abbreviated types of formulae. You will probably need practice in using these. In a skeletal formula, all the hydrogen atoms have been removed, leaving a carbon skeleton with functional groups attached. The carbon atoms are not shown, but a two-dimensional (2D) version of the bond angles is.

Example: The structural formual of butan-2-ol is $CH_3CH(OH)CH_2CH_3$.

Its displayed formula is:

$$H-\overset{\overset{\displaystyle H}{|}}{\underset{\underset{\displaystyle H}{|}}{C}}-\overset{\overset{\displaystyle H}{|}}{\underset{\underset{\displaystyle O}{|}}{C}}-\overset{\overset{\displaystyle H}{|}}{\underset{\underset{\displaystyle H}{|}}{C}}-\overset{\overset{\displaystyle H}{|}}{\underset{\underset{\displaystyle H}{|}}{C}}-H$$

And its skeletal formula is:

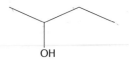

OH

In a skeletal formula, there is:

> a carbon atom at every bond junction in the chain, plus one at the end of each bond

> enough hydrogen atoms attached to each carbon to make the total number of bonds on that carbon up to four.

Ethanol

The structural formula for ethanol is CH_3CH_2OH.

The skeletal formula is:

OH

Propene

The structural formula for propene is CH_2CHCH_3.

The skeletal formula is:

Note that the double bond is shown.

Different types of formulae give us different information about an organic compound. These are summarised in Table 3.

Organic compound	Molecular formula	Empirical formula	General formula	Structural formula	Displayed formula	Skeletal formula
hexane	C_6H_{14}	C_3H_7	C_nH_{2n+2}	$CH_3CH_2CH_2CH_2CH_2CH_3$		
hex-1-ene	C_6H_{12}	CH_2	C_nH_{2n}	$CH_2=CHCH_2CH_2CH_2CH_3$		
2-methylbutane	C_5H_{12}	C_5H_{12}	C_nH_{2n+2}	$CH_3CH(CH_3)CH_2CH_3$		
2,3-dimethylpentane	C_7H_{16}	C_7H_{16}	C_nH_{2n+2}	$CH_3\,CH(CH_3)CH(CH_3)CH_3$		

(continue

Organic compound	Molecular formula	Empirical formula	General formula	Structural formula	Displayed formula	Skeletal formula
propan-2-ol	C_3H_7OH	C_3H_7OH	$C_nH_{2n+1}OH$	$CH_3CH(OH)CH_3$		
chloroethane	C_2H_5Cl	C_2H_5Cl		CH_3CH_2Cl		
benzene (a special case)	C_6H_6	CH	-	C_6H_6 (same as molecular formula)		

Table 2 *Different types of formulae for some compounds*

Type of formula	Type of atoms present	Ratio of atoms present	Number of each type of atom present	How the atoms are arranged/bonded together
empirical formula	yes	yes	no	no
molecular formula	yes	yes	yes	no
general formula	yes	yes	yes	no
structural formula	yes	yes	yes	yes
displayed formula	yes	yes	yes	yes
skeletal formula	yes	yes	yes	yes

Table 3 *What do formulae tell us?*

QUESTIONS

6. A hydrocarbon contains 240 g carbon and 44 g hydrogen.

   a. What is its empirical formula? (See examples in Chapter 2.)

   b. Its relative molecular mass is 142. What is its molecular formula?

   c. Suggest a structural formula.

7. Analysis of another hydrocarbon showed it contained 5.14 g of carbon. The total mass of the hydrocarbon was 5.98 g.

   a. What is its empirical formula?

   b. If its relative molecular mass is 42, what is its molecular formula?

   c. Draw a displayed formula for the hydrocarbon.

8. Pentane has the molecular formula C_5H_{12}.

   a. What is its empirical formula?

   b. What is its structural formula?

   c. Draw its displayed formula.

   d. Draw its skeletal formula.

Stretch and challenge

9. The types of formula in this chapter are those that we use to describe organic compounds. Which type of formula do we usually use for inorganic compounds?

10. What is the advantage of using formulae for organic compounds that show the structure?

5.4 FUNCTIONAL GROUPS AND HOMOLOGOUS SERIES

Organic compounds are classified into groups in which all of the compounds in one group have similar structures and therefore similar chemistry. A group of organic compounds is identified by its **functional group**. This is the reactive part of the molecule, and all organic compounds with the same functional group will behave in a similar way chemically.

For example, the functional group in alcohols is the hydroxyl group, $-OH$. When alcohols react, it is usually the electrons in the $-OH$ group that rearrange, and so all alcohols react similarly. The functional group for the alkenes is the carbon-carbon double bond, $C=C$. The reactions of alkenes usually involve the $C=C$ group.

Table 4 shows the functional groups of some organic compounds that you will meet in your AS or A level course.

A group of organic compounds with the same functional group and same general formula is called a **homologous series**. All of the compounds in a homologous series have similar chemical properties. Their physical properties show distinct trends. For example, all alkenes are chemically similar because

General formula	Homologous series	Name (suffix or prefix)	Functional group	Example
C_nH_{2n+2}	alkanes	suffix -ane	$C-H$	ethane, C_2H_6
C_nH_{2n}	alkenes	suffix -ene	$C=C$	ethene, C_2H_4
	haloalkanes	prefix chloro-, bromo-, iodo-	$-Cl$	chloroethane, CH_3CH_2Cl
$C_nH_{2n+1}OH$	alcohols	suffix -ol	$-OH$	ethanol, C_2H_5OH
	aldehydes	prefix hydroxy- suffix -al	$-C{\overset{H}{\underset{O}{\lessgtr}}}$	ethanal, CH_3CHO
	ketones	prefix oxo- suffix -one	$>C=O$	propanone, CH_3COCH_3 3-oxobutanoic acid, CH_3COCH_2COOH
$C_nH_{2n+1}COOH$	carboxylic acids	suffix -oic acid	$-C{\overset{O}{\underset{OH}{\lessgtr}}}$	ethanoic acid, CH_3COOH
	esters	suffix -oate	$-C{\overset{O}{\underset{O-R}{\lessgtr}}}$	methyl ethanoate, CH_3COOCH_3

Table 4 Homologous series and functional groups. We use **R** in a formula to represent a group of atoms.

they all have the same C=C functional group. The size of the alkene molecule affects its physical characteristics, such as melting point and boiling point, which increase steadily through the series. You read in Chapter 3 about the effect of intermolecular forces and the length of the carbon chain on boiling points.

Since the functional group identifies the homologous series of a compound, it is used to name the molecule. Table 4 shows the main functional groups and how they are used to name molecules. For example, if a molecule has a C=C functional group, it is an alkene and its name ends in -ene.

The naming of organic molecules is dealt with in Section 5.5.

Examples of homologous series

Alkanes are the simplest homologous series.

Alkanes have the general formula C_nH_{2n+2}. The bonds are all single bonds and alkanes are **saturated hydrocarbons**. The molecular, displayed and empirical formulae for the first six alkanes are shown in Table 5.

Name	Molecular formula	Displayed formula	Empirical formula
Methane	CH_4		CH_4
Ethane	C_2H_6		CH_3
Propane	C_3H_8		C_3H_8
Butane	C_4H_{10}		C_2H_5
Pentane	C_5H_{12}		C_5H_{12}
Hexane	C_6H_{14}		C_3H_7

Table 5 *Chemical formulae and names of the first six alkanes*

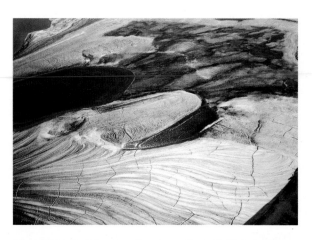

The alkenes are another homologous series of hydrocarbons. Each alkene contains a carbon–carbon double bond. A carbon atom can bond with up to four other atoms, as all carbon atoms in alkanes do. Because alkenes have carbon atoms with less than four other atoms bonded to them they are called **unsaturated hydrocarbons**. A feature of unsaturation is that further atoms or groups of atoms can be added to the molecule across the double bond so that all carbon atoms have four other atoms bonded to them.

Table 6 shows the molecular, displayed and empirical formulae of the first five alkenes. The suffix *-ene* shows that there is a double bond in the compound.

Figure 8 *Much of the ground in and near the Arctic Circle has been frozen permanently for thousands of years. The permanently frozen water in the soil is known as permafrost. With increasing global temperatures, permafrost is melting in some areas, releasing trapped methane that can destabilise trees and buildings. Methane is a powerful greenhouse gas and will contribute to more global warming.*

Name	Molecular formula	Displayed formula	Empirical formula
Ethene	C_2H_4	H H | | C=C | | H H	CH_2
Propene	C_3H_6	H H | | H—C—C=C | | | H H H	CH_2
But-1-ene	C_4H_8	H H H | | | H—C—C—C=C | | | | H H H H	CH_2
Pent-1-ene	C_5H_{10}	H H H H | | | | H—C—C—C—C=C | | | | | H H H H H	CH_2
Hex-1-ene	C_6H_{12}	H H H H H | | | | | H—C—C—C—C—C=C | | | | | | H H H H H H	CH_2

Table 6 *Chemical formulae and names of the first five alkenes*

KEY IDEAS

> Families of carbon compounds with the same functional group and general formula are called homologous series.

> Alkanes and alkenes are examples of homologous series.

> Alkanes are saturated hydrocarbons, with single bonds between the carbon atoms.

> Alkanes have the general formula C_nH_{2n+2}.

> Alkenes are unsaturated hydrocarbons, with a double bond between two of the carbon atoms.

> Alkenes have the general formula C_nH_{2n}.

ASSIGNMENT 1: GETTING THE NAME RIGHT

(PS 1.1, 1.2, 2.2)

Using the correct science terminology (words) conveys the message clearly and helps to avoid confusion. Chemistry is littered with different names for the same compound, some of them a lot more helpful than others.

Aspro™ is one of several brand names for the same medication – all of these brands contain the compound commonly known as aspirin. But it gets more complicated, because the old scientific name for aspirin is acetylsalicylic acid. Acetyl comes from *acetum*, the Latin for acetic acid, which was the old name for vinegar. Salicylic comes from *Salix*, the Latin for the willow tree, whose bark provided the extract from which a compound very similar

Figure A2 *Labels on bottles of nail polish remover for non-acrylic nails often say it contains 'aqua', perhaps because it sounds more exotic than 'water' (which is what it is). You will find aqua listed in most moisturisers and body lotions. The active ingredient in nail polish remover is shown as acetone (modern name: propanone, CH_3COCH_3).*

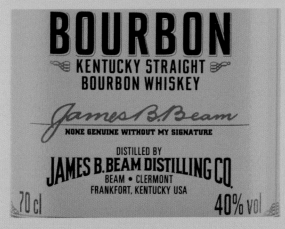

BOURBON
KENTUCKY STRAIGHT
BOURBON WHISKEY

James B. Beam
NONE GENUINE WITHOUT MY SIGNATURE

DISTILLED BY
JAMES B.BEAM DISTILLING CO.
BEAM • CLERMONT
FRANKFORT, KENTUCKY USA

70 cl   40% vol

Figure A1 *The label on this bottle of bourbon gives the alcohol content as 40%. People buying wine and spirits know that alcohol is the common name for the ingredient that adds to the enjoyment of the drink. But to a chemist, alcohols are a homologous series. The alcohol in bourbon is ethanol, C_2H_5OH.*

to aspirin was originally produced. But unless you know all that, the name acetylsalicylic acid would be meaningless. The new up-to-date name for aspirin is 2-ethanoyloxybenzenecarboxylic acid. But you wouldn't ask for that in the pharmacy.

The situation is confusing. This is why chemists now use an internationally agreed naming system for chemicals based on their structure and molecular formula. The rules of this system have been established by the International Union of Pure and Applied Chemistry, commonly known as IUPAC.

However, the correct chemical name can be equally confusing at times. Ibuprofen is commonly sold as Nurofen™ in the UK. Around the world, it has more than 100 different trade names. The medicine is an anti-inflammatory and is used to reduce inflammation and pain. All of the different brands contain the same chemical, (RS)-2-(4-(2-methylpropyl)phenyl)propanoic acid. Rather complicated for a molecular formula of $C_{13}H_{18}O_2$.

Questions

This is the skeletal formula for $C_{13}H_{18}O_2$:

A1. What do the single lines represent?

A2. What do the double lines represent?

A3. Why are some atoms shown and others not?

A4. Who decides on the chemical name for ibuprofen?

A5. Who decides on the trade names for ibuprofen sold around the world?

A6. Why is the chemical name for ibuprofen important?

Stretch and challenge

A7. Draw the full displayed formula for ibuprofen.

A8. Which functional groups can you identify?

5.5 NAMING ORGANIC COMPOUNDS

You can give compounds systematic names according to rules established by IUPAC. These rules are used throughout the world. The naming method fills several books; here you will look at just a few.

Naming organic compounds with unbranched or branched chain carbon atoms

A chain of carbon atoms in a compound is identified by the number of carbon atoms it contains. The label given to this number then forms the basis of the compound's name, as shown in Table 7.

Number of carbon atoms	Label used in name
1	meth
2	eth
3	prop
4	but
5	pent
6	hex

Table 7 Labels used to name chains of carbon atoms

These are some simple rules for naming an organic compound:

> Identify the principal functional group present, and so the homologous series. This gives the suffix (the ending of the name).

> Select the longest continuous carbon chain that contains the principal functional group and the maximum number of unsaturated bonds. This is an unbranched chain, which is sometimes called a straight chain (even if the carbon atoms are not arranged in a straight line!). This gives the main part or **stem** of the name.

> Number the carbon atoms from the end that gives the lowest possible number to the principal functional group.

> Identify additional functional groups. These provide the prefix or prefixes (the beginning of the name) and are numbered according to their position. Use commas to separate numbers and hyphens to separate numbers and letters. The rules for ordering different functional groups are covered later in this section.

Naming alkanes

You can use these rules to name alkanes.

Worked example 1

The principal functional group present is C–H. It is an alkane and will have the ending (or suffix) -*ane*.

The carbon chain is five carbons long. This gives it the prefix *pent-*.

So the name is *pentane*. All unbranched alkanes can be named in this way, but some alkanes have branched chains.

Worked example 2

$$CH_3—CH—CH_3$$
$$|$$
$$CH_3$$

The longer carbon chain has three carbon atoms in it. This gives the name *propane*. But there is a CH_3 group on the second carbon atom. This is an **alkyl** group, which is named in Table 8.

Name	Structural formula
methyl	$CH_3—$
ethyl	$CH_3CH_2—$
propyl	$CH_3CH_2CH_2—$
butyl	$CH_3CH_2CH_2CH_2—$

Table 8 *Labels used to name alkyl groups*

The branched group is a *methyl* group and the name of the branched alkane is *methylpropane*. Sometimes the molecule is called 2-methylpropane because the methyl group is attached to the second carbon atom, but in this case it is not necessary because no other carbon atom is available.

Worked example 3

| **5** | **4** | **3** | **2** | **1** | **number of carbon atoms** |

$$CH_3—CH_2—CH_2—CH_2—CH_3$$
$$|$$
$$CH_3$$

In this example, there are two possible places for the methyl group. The carbon atoms in the longest chain are numbered to give the lowest number to the carbon with the methyl group. The methyl group could be positioned on the second or fourth carbon atom. If the methyl group is on carbon number 4, this is the same as the methyl group being on carbon 2. The name needs to show the position of the methyl group.

So, the suffix is -*ane*, because this is an alkane. The prefix *pent* gives the number of carbon atoms in the longest chain. The *methyl* group is on the second carbon atom in the chain. The name is *2-methylpentane*.

Worked example 4

$$CH_3 \quad CH_3$$
$$| \quad |$$
$$CH_3—CH_2—CH_2—CH_2—CH_2—CH_3$$
$$\text{1} \quad \text{2} \quad \text{3} \quad \text{4} \quad \text{5} \quad \text{6}$$

The functional group is C–H so the suffix is -*ane*. The longest carbon chain contains six carbon atoms and gives the prefix *hex*. There are two methyl groups, one on the second carbon atom and one on the third. The name must show the positions of these two alkyl groups. The name is *2,3-dimethylhexane*. Note that numbers are separated with a comma and letters and numbers with a hyphen. The further prefix *di*- is used to show that there are two alkyl groups. Again, the carbon atoms have been numbered to give the lowest number of carbons to the first additional functional group.

QUESTIONS

13. Draw the displayed formulae of these alkanes:

    a. nonane

    b. 2-methylheptane

    c. 2,2-dimethylpentane

    d. 3,3-dimethylhexane

    e. 3-ethylheptane

Naming alkenes

Propene

$$\begin{array}{ccc} H & H & H \\ | & | & | \\ H—C— & C= & C—H \\ | \\ H \end{array}$$

The functional group is the C=C group, which gives the suffix -*ene*. The longest carbon chain is three carbon atoms, which gives the prefix *prop*. The molecule is *propene*. It may appear that the double bond could be between the other two carbon atoms, but this would be the same molecule. Try making a molecular model and turn it round.

But-1-ene

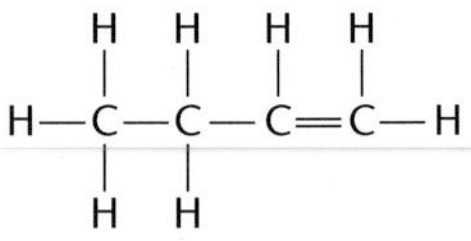

H—C—C—C=C—H (with H atoms shown)

The functional group C=C is identified using the suffix *-ene*. The carbon chain is four carbon atoms long, so the prefix *but* is used to describe it. The carbon atoms are numbered, as they are in naming alkanes, to give the lowest number for the position of the functional group, in this case, the C=C bond. In the name, the position of the C=C bond is identified by using the lowest number. In this example, the double bond is between carbon atoms 1 and 2. You use the number 1 to show the position of the bond. The name is but-1-ene.

QUESTIONS

14. Draw the displayed formulae of the following:
    a. ethene
    b. but-2-ene
    c. pent-1-ene
    d. hex-3-ene

15. Name these molecules:
    a. 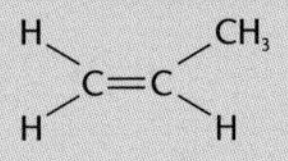
    b. 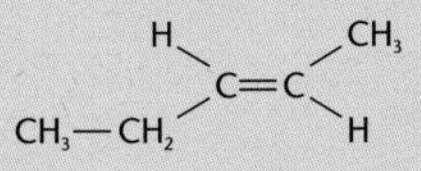
    c. 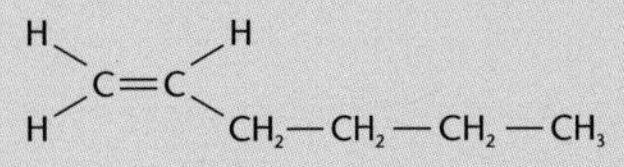
    d. 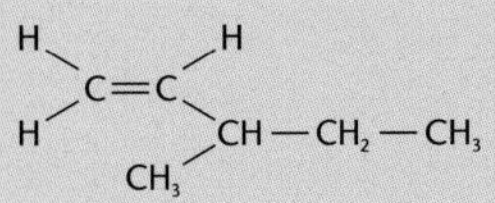
    e.

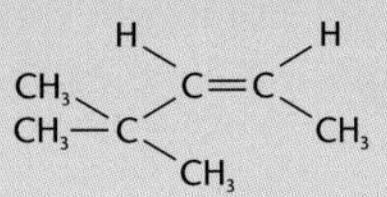

Naming halogenoalkanes

Halogenoalkanes are alkanes in which one or more hydrogen atoms have been replaced by halogen atoms: fluorine, chlorine, bromine or iodine. Figure 9 shows the displayed formulae of some halogenoalkanes.

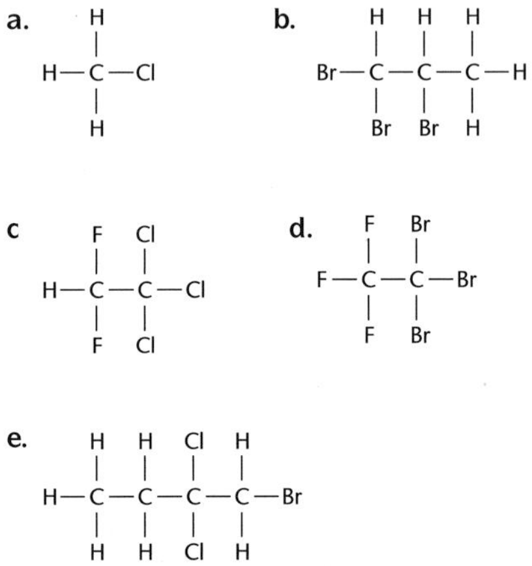

Figure 9 Displayed formulae of some halogenoalkanes

Many halogenoalkanes have traditional names. For example, chloroform, CH_3Cl, and carbon tetrachloride, CCl_4. However, you need to be able to name them systematically. Halogenoalkanes contain two functional groups, the C—H group of the alkanes and the C—X group, where X is a halogen. A hierarchy in functional groups makes some more important than others in the naming process. In the case of halogenoalkanes, the name of the alkane group gives the suffix of the name for the halogenoalkane.

Worked example 5

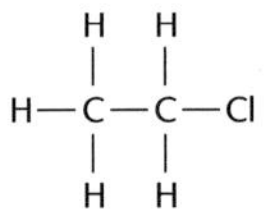

The alkane group C—H gives the suffix *-ane* to the name. The longest carbon chain is two carbon atoms long, giving the halogenoalkane the suffix *ethane*. The type of halogen atom, chlorine, becomes the prefix *chloro-*. There is one chlorine atom present, so its name is chloroethane.

The prefixes used for other halogen atoms are shown in Table 9.

Halogen	Prefix
fluorine	*fluoro-*
chlorine	*chloro-*
bromine	*bromo-*
iodine	*iodo-*

Table 9 *Prefixes used for halogen atoms*

Worked example 6

$$Cl-\overset{\overset{\displaystyle H}{|}}{\underset{\underset{\displaystyle H}{|}}{C}}-\overset{\overset{\displaystyle H}{|}}{\underset{\underset{\displaystyle H}{|}}{C}}-Cl$$

As in Worked example 5, two carbon atoms give the halogenoalkane the suffix *ethane*. But there are two chlorine atoms, one on the first carbon and one on the second carbon atom. The number and position of these must be shown in the name. You use *dichloro* to show that there are two chlorine atoms and numbers to show their position. So the name is 1,2-dichloroethane. *Di-*, *tri-*, *tetra-*, *penta-*, and so on, are used when there is more than one atom of a halogen.

Worked example 7

$$H-\overset{\overset{\displaystyle Br}{|}}{\underset{\underset{\displaystyle H}{|}}{C}}-\overset{\overset{\displaystyle H}{|}}{\underset{\underset{\displaystyle H}{|}}{C}}-\overset{\overset{\displaystyle Cl}{|}}{\underset{\underset{\displaystyle Cl}{|}}{C}}-\overset{\overset{\displaystyle H}{|}}{\underset{\underset{\displaystyle H}{|}}{C}}-H$$

The four carbon atoms give the suffix *butane*. Because this is a halogenoalkane, you now need to identify the number and position of the halogen atoms. If there is more than one type of halogen atom, they are listed in alphabetical order. The lowest numbers possible are then used to locate their position. So this halogenoalkane is 1-bromo-3,3-dichlorobutane.

QUESTIONS

16. Name the halogenoalkanes in Figure 9.

17. Draw structural formulae for:
    a. 1,1-dibromo-2,2-dichloroethane
    b. 1,2-dibromo-1,2-dichloroethane
    c. 1-bromo-1-chloro-2,2-difluoropropane
    d. 1,1,1-triiodoethane
    e. CF_3CCl_3
    f. $CH_3CHClCH_3$
    g. $CHCl_3$
    h. $CH_3CHBrCHClCCl_3$

18. Name the haloalkanes e, f, g and h in question 17.

Naming alcohols, aldehydes, ketones and carboxylic acids

Table 10 gives examples of how the names of alcohols, aldehydes, ketones and carboxylic acids are formed.

QUESTIONS

19. Name the following:
    a. $CH_3CH_2CH_2OH$ e. $CH_3CH(OH)CH_2CH_3$
    b. CH_3CH_2CHO f. CH_3CHO
    c. CH_3CH_2COOH g. $CH_3CH_2CH_2COOH$
    d. $CH_3CH_2COCH_3$ h. $CH_3CH_2CH_2COCH_3$

Stretch and challenge

20. Explain why:
    a. aldehyde and carboxylic acid groups have to be at the end of the carbon chain
    b. a ketone group cannot be on the end of a carbon chain
    c. an alcohol group can be on the end or in the middle of a carbon chain.

Homologous series	Examples of structural formula	Name
alcohols	$CH_3CH_2CH_2CH_2OH$	The functional group is −OH, which gives the suffix *-ol*. The carbon chain is four carbons long, which gives the prefix *butan-*. The −OH group is on the first carbon atom (number to give the lowest numbers). **The name is butan-1-ol**.
aldehydes	$CH_3CH_2CH_2CHO$	The functional group is −C = O, the carbonyl group. In aldehydes, a hydrogen atom is attached to the carbon with the oxygen, which gives the group −CHO. The −CHO group gives the suffix *-al*. The carbon chain is four carbon atoms long, which gives the prefix *butan-*. The aldehyde group has to be on the first carbon atom. **The name is butanal**.
ketones	CH_3COCH_3	The functional group is also the carbonyl group, C=O. In ketones the carbon atom is bonded to two other carbon atoms. The carbon chain is three carbon atoms long, which gives the prefix *propan-*. The ketone group gives the suffix *-one*. There is no other place for the carbonyl group to go so **the name is propanone**. If there are more than four carbon atoms in the chain, the carbons are numbered as low as possible to show the position of the carbonyl group.
carboxylic acids	CH_3COOH	The functional group is the −COOH group, which gives the suffix *-oic acid*. The carbon chain is two carbon atoms long, which gives the prefix *ethan-*. **The name is ethanoic acid**.

Table 10 Examples of the naming of alcohols, aldehydes, ketones and carboxylic acids

KEY IDEAS

> The naming of chemicals follows international rules agreed by the International Union of Pure and Applied Chemistry (IUPAC).

> The systematic name of an organic molecule with unbranched ('straight') or branched carbon chains tells you the homologous series to which it belongs, the number of carbon atoms it contains, the position of the functional group and the number and positions of additional functional groups.

5.6 REACTION MECHANISMS

Reaction mechanisms describe how a reaction happens. Chemical reactions involve the rearrangement of electrons. Tracking their progress through a reaction is the key to describing the mechanism of a reaction. We show reaction mechanisms by adding notation to formulae, often in an equation, to show how electrons move during a reaction. Understanding how reactions happen enables chemists to produce designer chemicals.

You will find out more about specific examples of these as you work through other sections about organic chemistry in this book.

Different mechanisms explain how different reactions happen.

Free-radical mechanisms

A radical is an atom or molecule with an unpaired electron. Unpaired electrons make a substance very reactive. We use a dot (•) alongside the formula to show the unpaired electron.

For example, chlorine gas, Cl_2, reacts explosively with hydrogen in sunlight (see Figure 10). The UV radiation in sunlight has enough energy to break the covalent bond in the chlorine molecule. This is the first stage of the reaction. Two chlorine radicals are produced, each with an unpaired electron.

$$Cl_2 \rightarrow Cl\bullet + Cl\bullet$$

In this case, the radical is simply a chlorine atom.

Chlorine radicals are very reactive. The second stage in the reaction happens when a chlorine radical collides with a hydrogen molecule:

$$Cl\bullet + H_2 \rightarrow HCl + H\bullet$$

A hydrogen radical is produced. Again, the radical is simply an atom – a hydrogen atom this time.

The reaction continues when the hydrogen radical collides with a chlorine molecule:

$$H\bullet + Cl_2 \rightarrow HCl + Cl\bullet$$

The chlorine radical produced collides with another hydogen molecule and so on. This is called a chain reaction. It will continue until two radicals collide and react. For example:

$$H\bullet + Cl\bullet = HCl$$

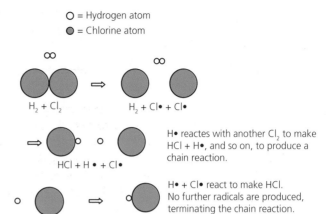

O = Hydrogen atom
● = Chlorine atom

$H_2 + Cl_2$ $H_2 + Cl\bullet + Cl\bullet$

$HCl + H \bullet + Cl\bullet$

H• reactes with another Cl_2 to make HCl + H•, and so on, to produce a chain reaction.

$H\bullet + Cl\bullet$ HCl

H• + Cl• react to make HCl. No further radicals are produced, terminating the chain reaction.

Figure 10 *A flash of light can trigger a chain reaction between chlorine gas and hydrogen gas.*

Using curly arrows to show other mechanisms

Chemical reactions involve the breaking and making of bonds. Keeping track of the electron movement during a reaction helps us to understand how the reaction happens. We use curly arrows to show the movement of the electrons.

A curly arrow with a complete arrow head with two barbs is used to show the movement of a pair of electrons:

The curly arrow with half an arrow head (one barb) is used to show the movement of a single electron:

In both cases, the arrow shows the direction of the electron movement. The tail of the arrow shows where the electron(s) come from. The head of the arrow shows where they end up.

If curly arrows are used to track a bonding pair of electrons, the arrow starts at the bond line. If the curly arrow is used to track a lone pair of electrons, we use two dots (:) to identify the electrons. You must show the electrons being tracked as either a bond line or as two dots. This is shown in the following examples.

Curly arrows: bond breaking

Imagine that you have two atoms or groups of atoms joined by a covalent bond. There are two ways the bond may break: **homolytic bond fission** and **hetereolytic bond fission**.

Homolytic bond fission

This happens when the shared pair of electrons split, with one going to one of the bonded atoms and the

117

other going to the other bonded atom. For example, when chlorine molecules form chlorine atoms in UV light:

$$Cl_2 \rightarrow Cl\bullet + Cl\bullet$$

Using curly arrows to show the movement of electrons:

Cl⌢⌢Cl $\xrightarrow{\text{UV light}}$ 2Cl•

Generalised equation for homolytic fission:

X⌢⌢Y $\longrightarrow$ X• + Y•

Heterolytic bond fission

This happens when the shared pair both move to the same bonded atom. An example:

$$H-\underset{\underset{H}{|}}{\overset{\overset{H}{|}}{C}}-Br \longrightarrow H-\underset{\underset{H}{|}}{\overset{\overset{H}{|}}{C^+}} + Br^-$$

The arrow in this diagram shows that the pair of electrons that make the carbon-bromine covalent bond are leaving carbon and moving to bromine. The methyl group becomes positively charged (CH_3^+) and bromine becomes negatively charged (Br^-). The arrow starts in the centre of the bond and ends up on the bromine atom because that is where the electron pair will end up.

Generalised equations for heterolytic fission (depending on which of X and Y has the greatest attraction for the shared pair of electrons):

X⌢Y $\longrightarrow$ X$^+$ + :Y$^-$

X⌢Y $\longrightarrow$ X:$^-$ + Y$^+$

Curly arrows: bond making

In the following example, the curly arrow shows that the electron pair on Y is now being shared between X and Y and forming a covalent bond. The arrow starts at the electron pair and points to where the new bond forms.

X$^+$ ⌢ :Y$^-$ $\longrightarrow$ X—Y

Ammonia reacts with a hydrogen ion to form an ammonium ion. The equation is: $NH_3 + H^+ \rightarrow NH_4^+$. During the reaction, the lone pair of electrons on the ammonia molecule forms a dative covalent bond with the hydrogen ion.

$H_3N:$ ⌢ H^+ $\longrightarrow$ H_3N^+—H

Because the electrons are taken from a lone pair of electrons on the nitrogen atom and not a covalent bond, they are shown with a pair of dots.

Ammonia reacts with water to form an ammonium ion and a hydroxide ion. The reaction involves the movement of two pairs of electrons, so two curly arrows are used.

The mechanism is:

The lone pair of electrons on the nitrogen atom is attracted to the positive charge of a hydrogen atom in the water molecule. The pair of electrons in the covalent O–H bond moves towards the oxygen atom and an ammonium ion and a hydroxide ion forms.

KEY IDEAS

> Mechanisms explain how organic reactions happen.

> Unpaired electrons in free radical mechanisms are shown by a dot.

> Curly arrows are used in other mechanisms to show the movement of electrons.

> A curly arrow with two barbs shows the movement of a pair of electrons.

> A curly arrow with one barb shows the movement of a single electron.

QUESTIONS

21. Use curly arrows to show how ammonia reacts with hydrogen chloride gas (HCl) to make ammonium chloride (NH_4Cl).

22. Use curly arrows to explain the reaction mechanism when a negatively charged hydroxide ion reacts with a positively charged hydrogen ion to make a water molecule.

5.7 ISOMERS

Two molecules may have the same molecular formula, but differ in the way that their atoms are arranged. These are called **isomers**. Isomers are distinct compounds with different physical properties and, sometimes, different chemical properties. Isomers are common in organic compounds because of the great variety of ways carbon can form chains and ring structures. There are two main types of isomer.

Structural isomers have the same molecular formula, but their atoms are bonded together in a different order.

Stereoisomers have the same molecular formula and their atoms are bonded together in the same order, but they are arranged differently in space. Figure 11 shows the different types of structural isomers and stereoisomers. E-Z stereoisomers used to be called, and often still are, geometric isomers.

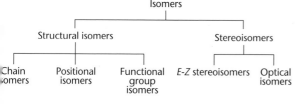

Figure 11 Different types of structural isomers and stereoisomers. You will find out about optical isomers later in your A-level studies.

Chain isomers

C₄H₁₀ and chain isomers

The molecular formula C_4H_{10} has two possible structural formulae:

butane   2-methylpropane

Butane and 2-methylpropane are structural isomers of C_4H_{10}. Both belong to the same homologous series, the alkanes. This means that they share similar chemical properties. However, they differ in their physical properties, such as boiling points (b.p.):

b.p. butane = $-0.5\ °C$

b.p. 2-methylpropane = $-11.7\ °C$

The type of isomer present can also be shown using structural formulae. Butane can be written as $CH_3CH_2CH_2CH_3$ and 2-methylpropane as $CH_3CH(CH_3)CH_3$, where the sequence of groups in the formula defines their position in the molecule.

Butane and 2-methylpropane are examples of chain isomers. Chain isomers occur in all carbon compounds more than four carbon atoms long. The longer the chain, the higher is the number of isomers possible. C_4H_{10} has two chain isomers, but $C_{40}H_{82}$ has an estimated 6.25×10^{13} chain isomers.

C₅H₁₂ and chain isomers

C_5H_{12} has three chain isomers:

Pentane
$CH_3CH_2CH_2CH_2CH_3$

2-methylbutane
$CH_3CH(CH_3)CH_2CH_3$

2,2-dimethylpropane
$CH_3C(CH_3)_2CH_3$

Note that there are only three chain isomers for C_5H_{12}.

The C–C bond rotates so that, although you may draw a carbon chain as a straight line, it actually twists and changes shape continually, so that

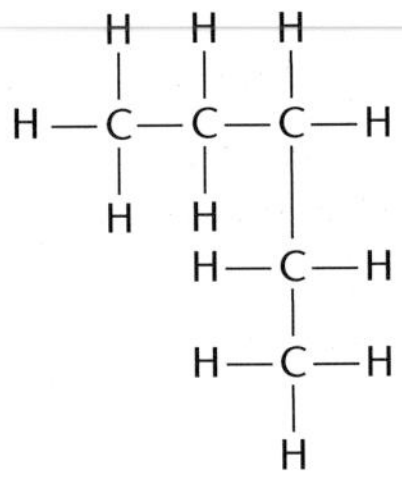

is the same as structure 1, and

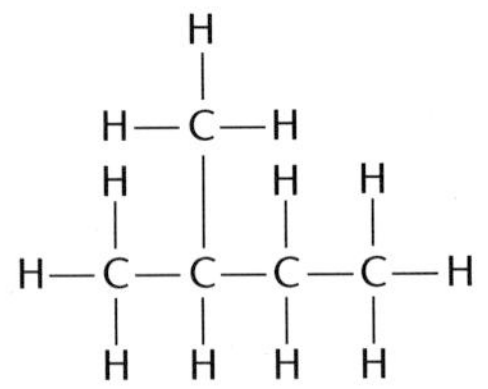

and

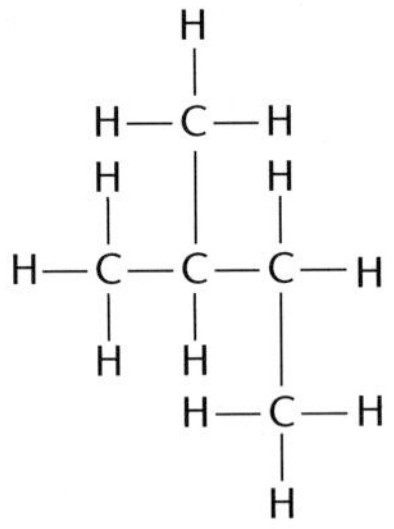

are the same as structure 2.

QUESTIONS

24. Write out the structural formulae for all the chain isomers of C_6H_{14}. Give the systematic name for each isomer.

Stretch and challenge

25. Explain why butane's boiling point is higher than methylpropane's boiling point.

Positional isomers

Positional isomers have their functional groups in different places. You have already met some of these in the naming section. C_4H_8 is an alkene with a double bond. Because there are four carbon atoms in the chain, there are two possible places for the double bond:

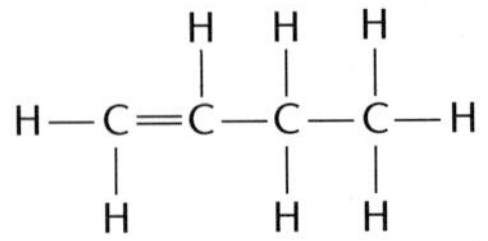

but-1-ene
$CH_2CHCH_2CH_3$

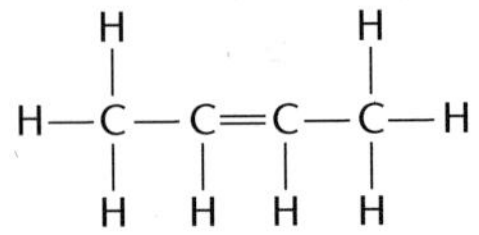

but-2-ene
$CH_3CHCHCH_3$

But-1-ene and but-2-ene are positional isomers of C_4H_8. Their chemical reactions will be the same

because they possess the same functional group. You can expect their boiling points to vary slightly.

You will meet alcohols later in this book. The functional group is –OH and the general formula is $C_nH_{2n+1}OH$. If the carbon chain is longer than two carbon atoms, alcohols can have positional isomers. The alcohol C_4H_9OH has two isomers:

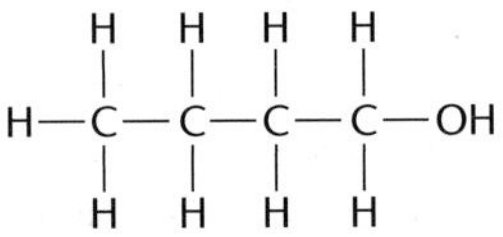

butan-1-ol
$CH_3CH_2CH_2CH_2OH$

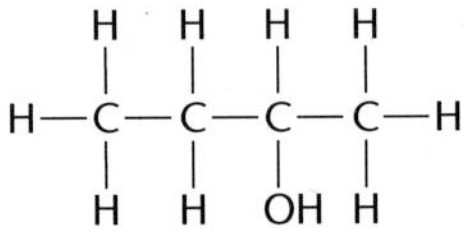

butan-2-ol
$CH_3CH_2CH(OH)CH_3$

QUESTIONS

26. Draw the positional isomers for C_5H_{10}. Name each isomer.

27. Draw the positional isomers for C_3H_7OH. Name each isomer.

Functional group isomers

Some organic compounds have the same molecular formula, but different functional groups. The atoms are arranged in a different sequence.

Ethanol has the displayed formula:

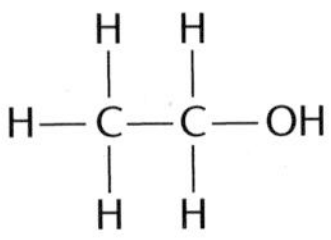

Methoxymethane is an ether and has the displayed formula:

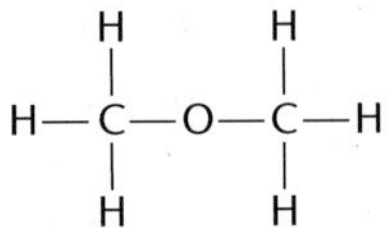

Both have the molecular formula C_2H_6O, but have different chemical and physical properties.

QUESTIONS

28. Propanal is CH_3CH_2CHO. Propanone is CH_3COCH_3. Explain why these are functional group isomers. Table 10 may help you.

Stereoisomers

Stereoisomers contain the same molecular formula and their atoms are connected in the same sequence, but the atoms are arranged differently in space. This is easier to see in 3D models of the isomers.

The type of stereoisomers you are studying here are *E-Z* stereoisomers. You may find these still called geometric isomers in some websites and books.

The C–C bond in organic compounds is free to rotate, but, because of the structure of the double bond, the C=C bond is not free to rotate. it has a flat shape and is described as being planar. This gives rise to isomers.

The fixed C=C bond means that the positions of the atoms or groups attached to it are also fixed. But, different arrangements are possible.

But-2-ene has two possible arrangements of atoms, as shown in Figure 12.

Z-but-2-ene

E-but-2-ene

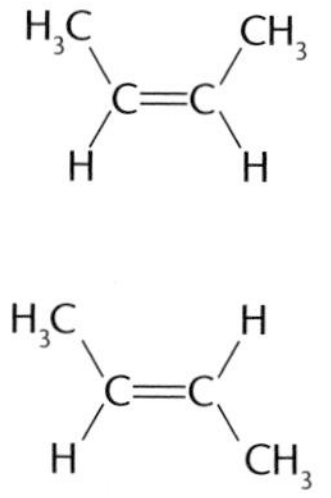

In the *Z*-isomer, the two methyl groups are on the same side of the carbon=carbon double bond. In the *E*-isomer, the two methyl groups are on opposite sides.

Figure 12 *E-Z stereoisomers of but-2-ene*

These are called ***E-Z* stereoisomers**. Molecules with the same atoms or groups on the same side of a C=C bond are called *Z*-isomers and have *Z*- added to their names as a prefix. Molecules with atoms or groups on opposite sides are *E*-isomers and have *E*- added to their names as a prefix. *Z* is used after the German word *zusammen* meaning 'together'.

There is an alternative method of naming *E-Z* stereoisomers. These are frequently referred to as *cis*- and *trans*-isomers, where *cis* means 'on the same side' and *trans*- means 'on opposite sides'. *Z*-but-2-ene also has the name *cis*-but-2-ene and *E*-but-2-ene has the name *trans*-but-2-ene. Food ingredient labels often refer to *trans* fats.

While the *cis*/*trans* system works well for most situations, its use is limited when all four groups attached to a C=C bond are different. For example, look at these two molecules:

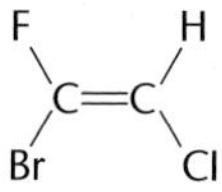

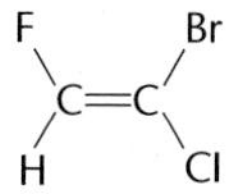

How can they be named using the *cis* and *trans* nomenclature? The answer is that they cannot. However, the *E-Z* system does enable them to be named, using **Cahn–Ingold–Prelog** (CIP) rules. These rules decide the priorites when naming the compounds.

Priority is given to the atom with highest atomic number that is attached to the carbon atoms that formed the double bond.

Looking at the previous example, the order or priority is:

Br > Cl > F > H

If the two atoms of highest priority are attached to the opposite ends of the double bond, this is the *Z*-isomer (structure (i) in the diagram). If the two atoms of highest priority are attached to the same end of the double bond, this is the *E*-isomer (structure (ii) in the diagram).

If a group is attached rather than a simple atom, then it is the atomic number of the atom of that group that is attached that is used to prioritise; for example:

H_3C- > H_2N- > HO –

QUESTIONS

29. Draw the full structural formulae for:

    a. *Z*-pent-2-ene

    b. *E*-pent-2-ene

    c. *Z*-1,2-dibromoethene

    d. *E*-1,2-dibromoethene

    e. *Z*-hex-2-ene

    f. *E*-hex-2-ene

30. 1,2-dichloroethene is used in the chemical industry.

    a. What is its molecular formula?

    b. Draw the displayed formula of its *E-Z* stereoisomers and name each isomer.

31. For two molecules to be isomers of each other, which of these properties must they have: the same functional group, the same molecular formula, the same structure, the same chemical properties, the same physical properties?

32. Which of these are isomers and which type(s) of isomerism do they possess?
   a. hexan-1-ol
   b. 2-methylbutan-1-ol
   c. 2-methylpentan-3-ol
   d. 2-methylpentan-2-ol
   e. 2,3-dimethylpropanol

ASSIGNMENT 2: AN ANTI-CANCER DRUG

(MS 4.2; PS 1.1, 1.2, 2.2)

In 1965 Dr Barnett Rosenburg, Professor of Biophysics and Chemistry at Michigan State University, was investigating the effect of an electric current on cells. He set up experiments, similar to electrolysis experiments, in which bacteria were added to the electrolyte. He used platinum electrodes because platinum is a very unreactive metal and he wanted to investigate the effect of the electric current only. When the electricity was switched on, Rosenburg realised that cell division in the bacteria had stopped. Further investigations showed that the platinum electrodes had actually reacted with the electrolyte used and produced the compound commonly known as *cis*-platin, a *Z*-stereoisomer:

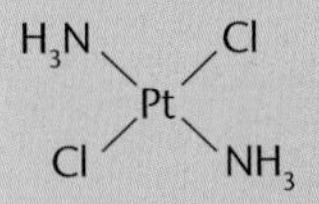

H$_3$N, Cl, Pt, Cl, NH$_3$

Figure A1 *The displayed formula of cis-platin*

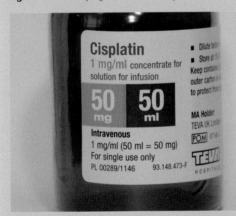

Figure A2 *Cisplatin can be delivered intravenously (as indicated on the label).*

The four groups around the platinum ion are called ligands. Each ligand forms a dative bond with the platinum ion. The whole structure is flat and the shape is described as **square planar**. Rosenburg reasoned that *cis*-platin may be effective as an anti-cancer drug.

After thorough testing, *Z*-platin became a commonly used chemotherapy drug for the treatment of different cancers, such as lung, stomach, oesophagus and others.

Cell division is a regulated process in the body, but, in cancer cells, cell division becomes unregulated and a tumour forms. *cis*-platin binds to DNA in the nucleus of cells and prevents the DNA replicating and the cells from dividing.

Living organisms are sensitive to the differences between *Z*- and *E*- geometric isomers. Only the *Z*-isomer is effective as a chemotherapy treatment. It is thought that the *E*-isomer is broken down in the body before it reaches cell nuclei.

Questions

A1. Why does platin have *E-Z* stereoisomers?

A2. Draw the structure of *E*-platin.

A3. What is a dative bond?

A4. Draw dot-and-cross diagrams to show the lone pairs of electrons in NH$_3$ and Cl$^-$.

ASSIGNMENT 3: NHS GUIDELINES

(MS 4.2, 4.3; PS 1.1, 1.2, 2.2)

Figure A1 *We need fat in our diet to provide energy, keep our nervous system working properly and help us to absorb the fat-soluble vitamins.*

We need fat in our diet, but too much of the wrong type can lead to health problems such as obesity, diabetes and heart disease. There are two main types of fat in our diet – saturated fats and unsaturated fats.

Saturated fats are thought to be linked to large amounts of cholesterol in the blood stream. These affect blood circulation and cause health problems. They are found in high amounts in dairy and meat products.

NHS guidelines of daily intake of fats

The average adult male should consume no more than 30 g of saturated fat.

The average adult female should consume no more than 20 g of saturated fat.

Adults should eat no more than 5 g of *trans* unsaturated fat.

Unsaturated fats are considered beneficial, helping to reduce damage done by saturated fats. They are found in plant oils such as olive oil. Like saturated fats, unsaturated fats consist of long carbon chains with other atoms attached, but the molecules in unsaturated fats have one or more double bond between some of the carbon atoms. As with but-2-ene, unsaturated fats have *E-Z* isomers. In the food industry, they are commonly called *cis* and *trans* isomers and give rise to the *cis* and *trans* fats (*cis* and *trans* are the old names for *E-Z* isomers).

The damaging effect of saturated fats is thought to be due to the linear shape of their long carbon chain. Trans fats (more accurately called *E-* fats) also have a linear shape and, although they are unsaturated, are thought to produce the same health risks as saturated fats. It is the *Z-* stereoisomers of unsaturated fats that are beneficial.

E- stereoisomers of unsaturated fats are rarely found in nature. But the food industry manufactures large amounts of them when converting unsaturated fats in vegetable oils to hard fats like margarine to use in baking products.

Questions

A1. What type of carbon–carbon bonds are found in molecules of saturated fats?

A2. Describe three differences between a C–C single bond and a C=C double bond.

A3. Why do unsaturated fats have stereoisomers while saturated fats do not?

A4. Food nutritionlists and health experts continue to debate whether saturated fats or unsaturated fats are beneficial or harmful. Many studies have been carried out in many different countries to provide evidence. So far, many of the results have been contradictory. Name three problems that scientists face when investigating the effects of saturated fats and unsaturated fats in diets.

Stretch and challenge

A5. Oleic acid is an unsaturated fat. Each molecule has a chain of 18 carbon atoms. There is a C=C double bond between atoms 9 and 10 in the chain and a carboxylic acid group at one end. Draw skeletal formulae to show differences between the *E-Z* isomers of oleic acid.

A6. Linoleic acid has the structural formula $C_4H_9CH_2 = CHCH_2 = CH(CH_2)_7COOH$.

Draw skeletal formulae for stereoisomers of linoleic acid.

KEY IDEAS

> Structural isomers have the same molecular formulae, but different structural formulae.

> Chain isomers have the same molecular formulae, but different arrangements of the carbon and hydrogen atoms in their chains, so you get long chains or branched chains.

> Positional isomers have the same molecular formulae, but the functional groups are in different positions on the C chain.

> Functional group isomers have the same molecular formula but different functional groups.

> Stereoisomers have the same molecular formulae and functional groups, but a different spatial arrangement of functional groups.

PRACTICE QUESTIONS

1. a. The hydrocarbon but-1-ene (C_4H_8) is a member of the homologous series of alkenes. But-1-ene has structural isomers.

   b. i. State the meaning of the term structural isomers.

   ii. Give the IUPAC name of the position isomer of but-1-ene.

   iii. Give the IUPAC name of the chain isomer of but-1-ene.

   iv. Draw the displayed formula of a functional group isomer of but-1-ene.

   AQA Jun 2013 Unit 1 Question 2a

2. a. The structure of the bromoalkane Z is:

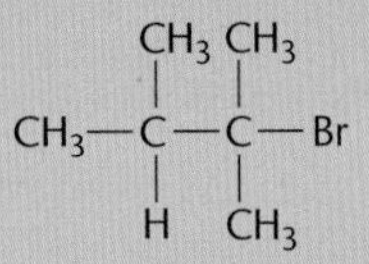

   CH₃—C—C—Br with CH₃ CH₃ groups and H CH₃

   i. Give the IUPAC name for Z.

   ii. Give the general formula of the homologous series of straight-chain bromoalkanes that contains one bromine atom per molecule.

   iii. Suggest one reason why 1-bromohexane has a higher boiling point than Z.

   b. Draw the displayed formula of 1,2-dichloro-2-methylpropane. State its empirical formula.

   AQA Jul 2013 Unit 1 Question 8

3. Compound X is:

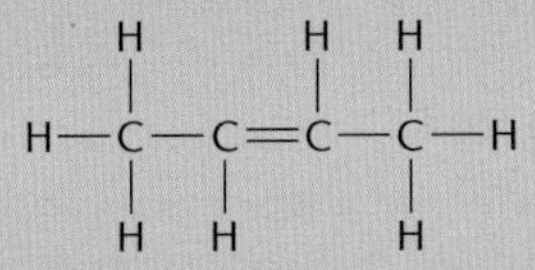

   H—C—C=C—C—H (with H atoms)

   It is a member of a homologous series of hydrocarbons.

   a. i. Deduce the general formula of the homologous series that contains X.

   ii. Name a process used to obtain a sample of X from a mixture containing other members of the same homologous series.

   b. There are several isomers of X.

   i. Give the IUPAC name of the position isomer of X.

   ii. Draw the structure of a functional group isomer of X.

   AQA Jun 2012 Unit 1 Question 2a and b

4. Octane is the eighth member of the alkane homologous series.

   a. State two characteristics of a homologous series.

   b. Name a process used to separate octane from a mixture containing several different alkanes.

(Continued

c. The structure shown below is one of several structural isomers of octane.

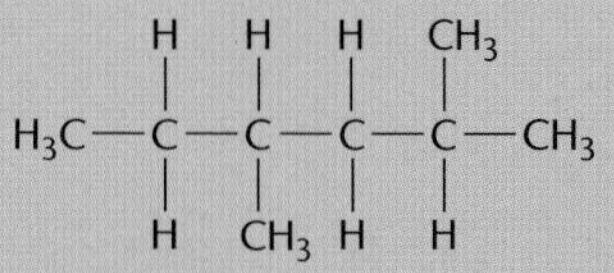

i. Give the meaning of the term structural isomerism.

ii. Name this isomer and state its empirical formula.

d. Suggest why the branched chain isomer shown above has a lower boiling point than octane.

AQA Jan 2011 Unit 1 Question 6

5. Figure Q1 gives the structures of the four isomeric alkenes with molecular formula C_4H_8.

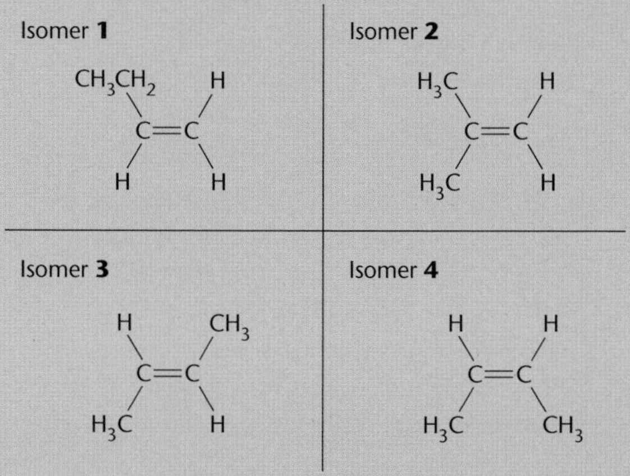

Figure Q1

a. Name Isomer 1 and Isomer 2.

b. Identify the isomer, from Figure Q1, which is a chain isomer of Isomer 1.

AQA Jan 2007 Unit 3(a) Question 3

6. Table Q1 shows some boiling points of alkanes.

Alkane	Boiling point/K
butane	272.6
pentane	309.2
hexane	342.1
heptane	371.5
2-methylpropane	261.4
2-methylbutane	301.0
2-methylpentane	333.4
2-methyhexane	363.1

Table Q1

a. 2-methyl propane and butane are structural isomers.

i. What is meant by the term structural isomer?

ii. Name the three other pairs of structural isomers in Table Q1.

iii. What type of structural isomerism is shown?

b. i. Give the molecular formula and displayed formula for 2-methylbutane.

ii. Name the type of bonding between the atoms in an alkane.

iii. What type of bonds form between the alkane molecules?

iv. Explain why butane has a higher boiling point than 2-methylpropane.

7. Alcohols are used as solvents in hair-care products. Propan-2-ol is commonly used in hair sprays because the resins needed to hold hair in place are soluble in propan-2-ol and propan-2-ol has a boiling point of 82.5 °C. It readily evaporates from the hair, leaving the resin in place.

a. What is the functional group of the alcohols?

(Continued)

b. Draw the displayed formula for propan-2-ol.

c. Propan-1-ol is a structural isomer of propan-2-ol. What type of structural isomer is this?

d. The boiling point of propan-1-ol is 97.1 °C and the boiling point of propan-2-ol is 82.5 °C. Explain this difference.

e. Suggest why propan-1-ol is less suitable as a solvent in hair sprays?

8. But-1-ene is used to make the synthetic rubber found in car types and pond liners..

   a. i. What homologous series does but-1-ene belong to?

      ii. What is the general formula of the compounds in this series?

   b. i. Give the empirical formula for but-1-ene.

      ii. Draw the displayed formula for but-1-ene.

   c. But-1-ene contains a double covalent bond. How many bonding pairs of electrons are shared in a double covalent bond?

   d. C_4H_8 has structural isomers.

      i. Name two structural isomers.

      ii. Draw their displayed formulae.

      iii. What type of structural isomerism is this?

   e. Explain why C_3H_6 does not have structural isomers.

9. Table Q2 gives some systematic names and displayed formulae of structural isomers having the molecular formula C_4H_9Cl.

Displayed formula	Systematic name
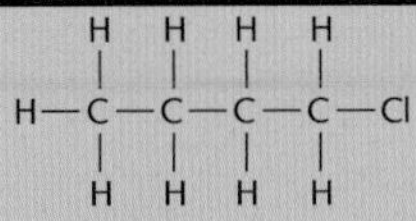 H—C—C—C—C—Cl (with H atoms shown)	
	2-chlorobutane
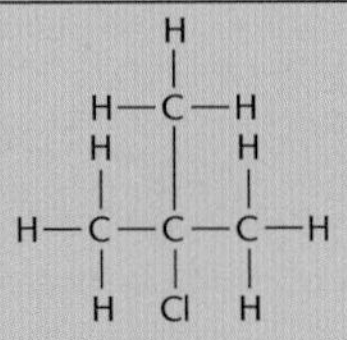 H—C—C—C—H (with CH₃ branch and Cl)	2-chloro-2-methylpropane
	1-chloro-2-methylpropane

Table Q2

a. Copy and complete the table.

b. Which homologous series do the compounds in the table belong to?

c. What is meant by the term structural isomer?

d. The C–Cl bond in $CH_3CH_2CH_2CH_2Cl$ can break heterolytically to produce a chloride ion and a positive charge on the carbon atom. Use curly arrows to describe the mechanism.

e. Draw diagrams to show the displayed formulae for:

   i. 1,2-dibromobutane

   ii. 1,1-dibromobutane

f. What is the molecular formula for 1,2-dibromobutane?

g. What is the empirical formula for 1,2-dibromobutane?

h. 1,2-dibromobutane and 1,1-dibromobutane are structural isomers. Draw the displayed formula of another structural isomer of 1,2-dibromobutane and name this isomer.

(Continued)

Stretch and challenge

10. Butenedioic acid is $HOOCCH = CHCOOH$.
It has two stereoisomers, commonly known as malic acid and fumaric acid. Malic acid is made by all living organisms and is responsible for the sour taste in fruit. Fumaric acid has a similar fruit taste, but is only made by a few living organisms. Both are used as food additives.

   a. Draw the two *E-Z* isomers of butenedioic acid and label them as *E*-butenedioic acid and *Z*-butenedioic acid.

   b. Draw the skeletal formula for each isomer.

   c. Explain why butenedioic acid has *E-Z* isomers.

   d. Suggest which part or parts of the molecule are responsible for the sour fruit taste.

Multiple choice

11. What is used to show the movement of a single electron in a reaction?

   A. Curly arrow with one barb

   B. Curly arrow with two barbs

   C. Dot after the species

   D. Two dots alongside the formula

12. Which are positional isomers?

   A. *E*-pent-2-ene and *Z*-pent-2-ene

   B. cyclohexane and cyclohexene

   C. butane and 2-methylpropane

   D. 1-chloropropane and 2-chloropropane

13. Which are functional group isomers?

   A. CH_3CH_2OH and CH_3COCH_3

   B. CH_3CH_2COH and $CH_3CH_2CH_3$

   C. CH_3CH_2CHO and CH_3COCH_3

   D. CH_3CH_2OH and CH_3CH_2COOH

14. Which of these is not shown in a skeletal formula?

   A. Carbon atoms

   B. Bonds

   C. Functional groups

   D. A 2D version of the bond angles

6 THE ALKANES

PRIOR KNOWLEDGE

You may know that crude oil is a mixture of alkanes and that we use fractional distillation to separate them. You may have learned about some of the properties of alkanes in crude oil and know how cracking is used to produce petrol and alkenes for the chemical industry.

LEARNING OBJECTIVES

In this chapter, you will build on these ideas and find out more about the chemical reactions of the alkanes. You will use some of the reaction mechanisms you learnt in Chapter 5 to describe how chlorine reacts with methane.

(Specification 3.3.2.1, 3.3.2.2, 3.3.2.3, 3.3.2.4)

Because of North Sea oil, the UK has been a major oil producer in the world. Since 2000, our oil production has fallen as reserves have been depleted and we are now a net importer of crude oil. This means that we use more than we produce and have to buy crude oil from other countries. Crude oil remains important to UK energy and contributes 37% to our total energy consumption.

Figure 1 *Photovoltaic cells are one of many alternative energy technologies. They generate electricity from the Sun's energy. 10 hectares of photovoltaic cells can produce enough electricity for 1515 homes.*

Fracking is one way to increase the amount of crude oil that the UK produces. Many areas of the UK have underground shale rocks that contain crude oil and gas, but these are not easily extracted by traditional methods.

The UK government has given the go ahead for fracking, but environmentalists are concerned that the process, which fractures underground rocks, may trigger mini earthquakes or contaminate water supplies.

Securing our future energy supply is important and tough targets are also in place for alternative energy technology. The EU wants 15% of UK energy to come from renewable supplies by 2020.

6.1 FRACTIONAL DISTILLATION OF CRUDE OIL

Crude oil is also called petroleum, a name that refers to the fact that it occurs naturally in pockets within rocks. It comes from the Latin for rock – *petra* – and oil – *oleum*.

Crude oil in its natural form is a thick, tarry substance that is difficult to ignite. In its raw state it isn't very useful. To be useful, it needs to be processed to separate out its most valuable constituents, such as petrol, lubricating oils, heating oils and fuel for power stations.

Crude oil is a very valuable resource. We use it to make detergents, plastics, paints, antifreeze, synthetic rubber and medicines. Seventy per cent of the organic chemicals we use are produced from crude oil and a massive 3000 million tonnes of crude oil products are used worldwide every year.

What does crude oil contain?

Crude oil is a mixture of about 150 different hydrocarbons. Hydrocarbons are compounds that contain only carbon and hydrogen. The majority of hydrocarbons in crude oil are **unbranched alkanes**,

Figure 2 *These are just a few of the end products from crude oil.*

but the mixture also contains cycloalkanes and arenes. Figure 3 shows the structures of some of these important hydrocarbons.

The exact composition of crude oil depends on the conditions in which it formed. Samples from different parts of the world have slightly different amounts of each type of hydrocarbon, called its 'fingerprint'. This fingerprint (Table 1) enables the source of an unknown sample of oil to be identified by analysis.

Crude oil is separated into mixtures of hydrocarbons with similar boiling points. These are called crude oil fractions and are obtained by **fractional distillation** in an oil refinery (Figure 5).

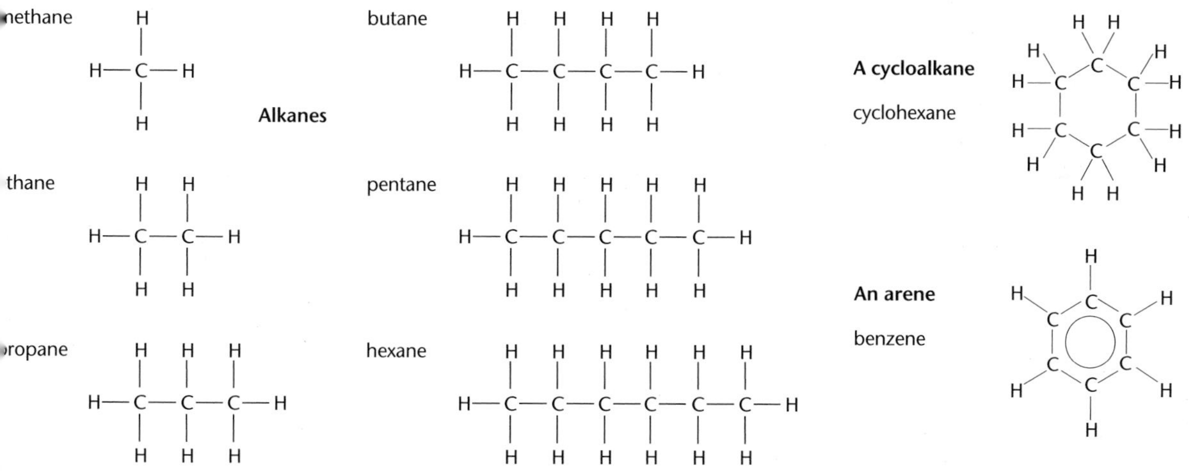

Figure 3 *Alkanes, a cycloalkane and an arene*

| Type of crude oil | Percentage composition by mass | | | |
	Petrol	Kerosene	Gas oil	Fuel oil
North Sea oil	23	15	24	38
Arabian light	18	11.5	18	52.5
Arabian heavy	21	15	21	43
Iranian heavy	21	13	20	46

Table 1 *The percentages of the fractions of petrol, kerosene, gas oil and fuel oil in different types of crude oil*

Fractional distillation

Figure 4 *The oil refinery operates 24 hours a day, seven days a week. The whole operation is monitored in control centres throughout the refinery.*

The crude oil is heated until it vaporises. The gases pass into a fractionating column (Figure 5). A temperature gradient is created between the bottom of the column and the top. The bottom of the column is kept at about 350 °C and there is a gradual cooling as vapours pass up the column. The temperature at the top is about 60 °C. Any hydrocarbons that remain liquid at temperatures as high as 340 °C fall to the bottom of the column and are removed as residue. This is not wasted, as it contains useful materials such as lubricating oil, and is used as bitumen on road surfaces.

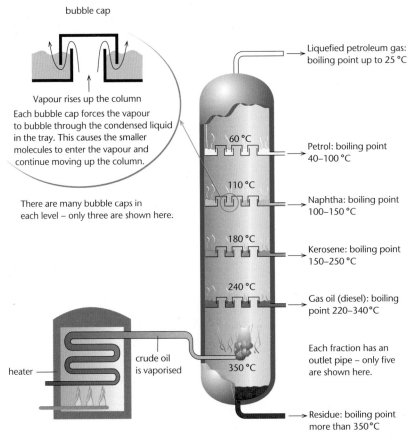

Figure 5 *A fractionating column*

The smaller hydrocarbons (lower relative molecular mass) rise up the column in their gaseous state. As a particular hydrocarbon reaches the level in the column where the temperature is equal to its boiling point, it condenses and is collected in trays. Only the most volatile hydrocarbons, those with the lowest boiling points, reach the top of the column. The major fractions obtained by fractional distillation of crude oil and their major uses are shown in Table 2.

The **primary distillation** of crude oil does not separate individual hydrocarbons. Each fraction contains a mixture of hydrocarbons that have boiling points within a particular range. Fractions can be further separated to obtain less complex mixtures, in other words, fewer different hydrocarbons. These are **secondary distillations**. The fraction that has a boiling point above 350 °C has to be distilled under reduced pressure. Lower pressure lowers its boiling point and is called **vacuum distillation**. It allows the hydrocarbons to be distilled out at lower temperatures.

QUESTIONS

1. In which fractions will the following be found:
   a. propane
   b. hexane
   c. decane?

Name of fraction	Approximate boiling range/°C	Number of carbon atoms in hydrocarbon	Uses
LPG (liquefied petroleum gas)	up to 25	1–4	car fuel, heating and cooking fuel, camping gas
petrol (gasoline)	40–100	4–12	petrol
naphtha	100–150	5–11	petrochemicals
kerosene (paraffin)	150–250	11–15	jet fuel, petrochemicals
gas oil (diesel)	220–340	15–19	central heating fuel, petrochemicals
mineral oil (lubricating oil)	over 350	20–30	lubricating oil, petrochemicals
fuel oil	over 400	30–40	fuel for ships and power stations
wax, grease	over 400	40–50	candles, grease for bearings, polish
bitumen	over 400	above 50	roofing, road surfacing

Table 2 *Fractions from crude oil*

ASSIGNMENT 1: WHY FRACTIONAL DISTILLATION WORKS

(MS 3.1; PS 1.1, 1.2, 2.2)

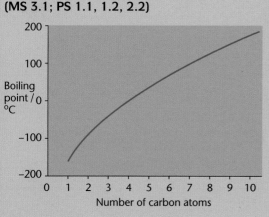

Figure A1 *Boiling point trend in alkanes*

Fractional distillation works because the alkanes that make up crude oil have different boiling points. The trends in alkane boiling points is shown in Figure A1. You may need to refer to Chapters 3 and 5 to answer the questions that follow.

Questions

A1. Alkanes are a homologous series of organic compounds.
   a. What is their general formula?
   b. What is the molecular formula of propane and decane? (Decane has 10 carbon atoms.)

c. Write the structural formulae for propane and decane.

d. Draw structural formulae for:

 i. 2,3-dimethyloctane

 ii. 3,5-dimethyloctane

e. Draw a skeletal formula for 3-ethyloctane.

f. What type of isomers of decane are the alkanes listed in questions 1d and 1e?

A2. a. The electronegativity value for carbon is 2.5 and for hydrogen is 2.1. Do alkane molecules have a permanent dipole?

b. Decane is liquid at room temperature. What forces exist between decane molecules?

c. What happens to the intermolecular bonding when decane changes from a liquid to a gas?

d. Explain why propane is a gas at room temperature and decane is a liquid.

A3. Fractional distillation can be carried out in the laboratory using a crude oil substitute (real crude oil contains a small percentage of benzene and is carcinogenic). Four fractions were collected from room temperature to 100 °C, 100 °C to 150 °C, 150 °C to 200 °C and 200 °C to 250 °C.

a. Predict the range of the number of carbon atoms in the hydrocarbons in each fraction.

b. Name the fractions from real crude oil that correspond to the fractions from the laboratory experiment.

6.2 CRACKING

Typical yields of the oil fractions obtained by primary distillation are shown in Table 3. This table also shows the relative demand for each fraction and it is immediately obvious that the two do not match. The yields of some fractions are more than enough to meet requirements, while the supply of other fractions are not enough to meet demands for them.

The most useful oil fractions tend to contain hydrocarbons of shorter chain length. Petrol for cars contains alkanes with carbon chain lengths from C_6 to C_{10}. Ideally, almost 30% of the fractions produced from fractional distillation need to be in this range to meet the demand for petrol. In practice only 16% of the fractions fall into this range.

The problem is solved by **cracking**. Cracking is a process that breaks some carbon bonds in long-chain alkanes, such as kerosene, and produces smaller molecules. Cracking is primarily carried out to produce alkanes for petrol, but the process also produces alkenes that are used by the chemical industry to make a variety of plastics and other products.

Cracking products are varied, but an overall summary of the cracking reaction is:

High M_r alkane $\rightarrow$ smaller M_r alkane + alkene

There are two types of cracking: **thermal cracking** and **catalytic cracking** (Table 4).

The overall results are similar. Cracking always produces a mixture of alkanes and alkenes, but the molecules can break up in several different ways to form a mixture of products. These can be separated by fractional distillation. Two possible fragmentations of the $C_{14}H_{30}$ molecule are:

$$C_{14}H_{30}(l) \rightarrow C_{12}H_{26}(l) + C_2H_4(g)$$

and

$$C_{14}H_{30}(l) \rightarrow C_7H_{16}(l) + C_3H_6(g) + 2C_2H_4(g)$$

Fraction	Approximate supply/%	Approximate demand/%
liquefied petroleum gas (LPG)	2	4
petrol and naphtha	16	27
kerosene	13	8
gas oil (diesel fuel)	19	23
fuel oil and bitumen	50	38

Table 3 *Supply and demand for oil fractions*

Thermal cracking

The process of thermal cracking uses heat to provide the energy required to break the C–C bonds. This is usually carried out in a **steam cracker** (Figure 6). When long-chain alkanes are heated under pressure and in the absence of air, bonds in the molecules vibrate more vigorously. This increased vibration can lead to bonds breaking and alkane molecules being split into smaller molecules. At the lower end of the temperature range, carbon chains tend to break in the centre of the molecule. At higher temperatures, carbon bonds tend to break towards the end of the chain, which leads to a higher proportion of alkenes with low molecular mass.

Ethene (C_2H_4) is the major alkene produced by thermal cracking. The double bond in an alkene makes it a reactive molecule, far more reactive than an alkane with the same chain length. Alkenes, particularly ethene, can be used as a starting point for the chemical industry (see Chapter 14). For example, polymerising ethene produces poly(ethene).

C–C bonds are strong. The mean bond enthalpy for the C–C bond is:

$$\Delta H^\theta \text{ (C–C)} = 347 \text{ kJ mol}^{-1}$$

We use the term ΔH^θ (C–C) for bond enthalpy for the bond shown in the brackets.

High temperatures (between 400 and 900 °C) and high pressures (up to 7000 kPa) are needed, but to avoid breaking too many C–C bonds and producing lots of short-chained alkanes, they are kept at these conditions for only a very short time. This exposure time, called the **residence time**, is about one second.

Steam cracking requires a lot of energy, but has the advantage that it can be used on all long-chain alkane fractions, including the residue from the bottom of the fractionating column. Mixtures of products obtained can be separated by further fractional distillation.

Catalytic cracking

Cracking can also be carried out at much lower temperatures using a catalyst. Clay catalysts were used at first, but these have been largely replaced by synthetic zeolite catalysts. These are crystalline aluminosilicates. Zeolite is a rock that contains aluminium, silicon and oxygen. Its regular atomic structure creates a network of holes (Figure 7). Zeolite is now made industrially and the pores can be made to exact sizes. This is important because the reactions involved in catalytic cracking take place in these pores (Figure 8). Long-chain alkane molecules fit into the pores and when the catalyst–alkane mixture is exposed to a slight pressure and a temperature of about 450 °C, cracking occurs.

This catalytic cracking process is efficient and produces more branched chain molecules than thermal cracking. Branched hydrocarbons burn more easily and are more useful as fuels. They make better petrol because they do not cause so many problems in car engines.

Catalytic cracking is the major cracking method used to produce alkanes for petrol. Catalytic cracking also produces aromatic hydrocarbons. These are hydrocarbons with ring structures in their overall structure. They are important in the chemical industry.

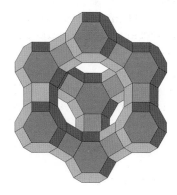

Figure 7 *The structure of zeolite*

Figure 6 *Thermal cracking is carried out at the steam cracking area of an oil refinery.*

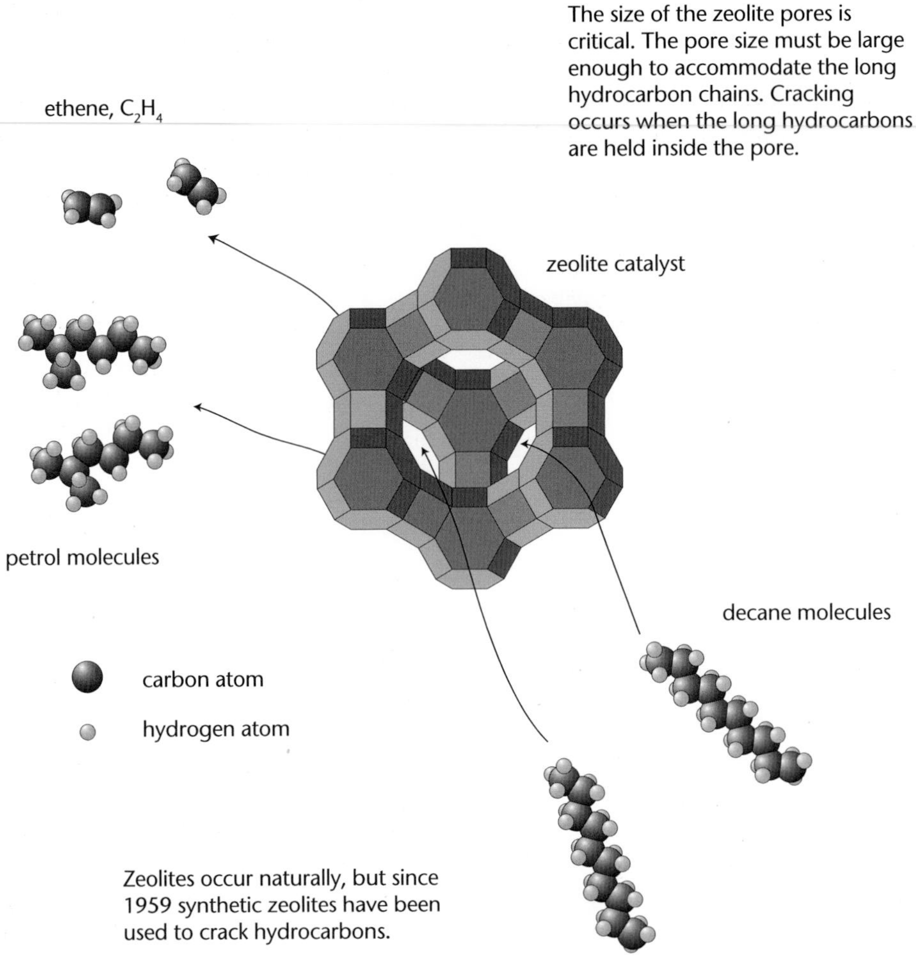

The size of the zeolite pores is critical. The pore size must be large enough to accommodate the long hydrocarbon chains. Cracking occurs when the long hydrocarbons are held inside the pore.

ethene, C_2H_4

zeolite catalyst

petrol molecules

decane molecules

● carbon atom

○ hydrogen atom

Zeolites occur naturally, but since 1959 synthetic zeolites have been used to crack hydrocarbons.

Figure 8 *Zeolite catalysts in action*

Figure 9 *Zeolites are also used in detergents to soften water. The zeolites remove the calcium ions that cause hardness in water supplies and replace them with sodium ions, which soften the water.*

	Thermal cracking	Catalytic cracking
Feedstock	All longer carbon chains, including those in the residue	Only works on a fraction from the refining of petroleum, such as diesel oil
Conditions	400–900 °C, 7000 kPa	450 °C, slightly higher than atmospheric pressure, zeolite catalyst
Products	Higher percentage yield of alkenes, mostly ethene	Lower percentage yield of ethene
	Mixture of unbranched and branched chain alkanes	Higher percentage yield of branched chain alkanes (mainly used to make petrol) and aromatic hydrocarbons

Table 4 *Thermal and catalytic cracking*

An example of catalytic cracking is shown in Figure 10. There are no set rules for the chemical reactions that happen during cracking. Some conditions will remove a hydrogen atom from an alkane chain, leaving a reactive radical that can then react with another to produce branched alkanes.

Thermal cracking is expensive, because of the high temperature and pressure required, but the high yield of ethene for the chemical industry provides a good return. Catalytic cracking is cheaper because it can operate at a lower temperature and only needs a slight pressure. Its higher yield of branched chain alkanes makes its products suitable for use in petrol.

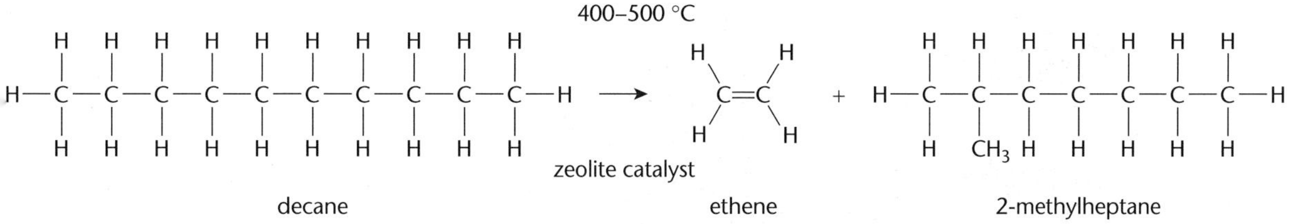

Figure 10 *Decane can be catalytically cracked using a zeolite catalyst at 400–500 °C. The products are ethene and 2-methylheptane.*

QUESTIONS

2. Write structural formulae to show two possible isomers of C_7H_{16}, both containing branched chains. Name your isomers.

3. **a.** Write an equation to show the cracking of dodecane, $C_{12}H_{26}$, into propene and 4-methyloctane.

   b. Write the structural formula for 4-methyloctane.

Stretch and challenge

4. Distilling the crude oil residue under reduced pressure in vacuum distillation lowers the boiling points and is used for fractions with boiling points > 350 °C in normal atmospheric conditions. Useful fuel oils are obtained for use in power stations and ships. Why is vacuum distillation used and not normal distillation at atmospheric pressure?

KEY IDEAS

❯ Petroleum or crude oil is a mixture consisting mainly of alkanes.

❯ The hydrocarbons in crude oil have different boiling points and can be separated out by fractional distillation.

❯ Cracking breaks down longer alkane molecules into shorter alkane molecules and alkenes.

❯ Thermal cracking provides a higher yield of alkenes, particularly ethene.

❯ Catalytic cracking provides a higher yield of branched alkanes, suitable for petrol fuel, and aromatics.

6.3 COMBUSTION REACTIONS OF ALKANES

The general formula for the alkanes is C_nH_{2n+2}. Table 5 in Chapter 5 gives the molecular and structural formulae for the first six alkanes.

Figure 11 *Combustion reactions of alkanes release much of the energy we rely on.*

Alkanes are not very reactive. The C–C and C–H bonds are strong bonds and a chemical reaction that involves breaking these requires a lot of energy.

Alkanes are not polar, so do not attract polar molecules or ions, such as OH^-. If they don't attract polar molecules, then they are unlikely to react with them. The few reactions of alkanes include combustion (Figure 12) and their reactions with the halogens. You can read about reactions with halogens in section 6.6.

Figure 12 *The match used to light the candle melts the wax. The liquid wax rises up the wick. The heat of the match vaporises the liquid wax and triggers a reaction between the candle wax (a mixture of long-chained alkanes) and oxygen. The exothermic combustion reaction releases the energy and keeps the reaction going.*

Combustion

When alkanes are heated in a plentiful supply of oxygen, combustion occurs. Initially, energy must be transferred to start the bond-breaking process, but once bond breaking has started, bond making can begin and energy is released (Figure 13). Some of this released energy is transferred to break more bonds and the reaction proceeds without further input of energy from the surroundings. Like all hydrocarbons, alkanes burn in oxygen highly exothermically. The minimum energy needed to start the reaction is called the **activation energy**, E_A. You will read more about activation energy in Chapter 8.

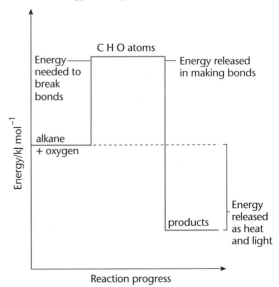

Figure 13 *Alkane combustion. The diagram shows how the enthalpies of atomisation of reactants and products relate to the energy released in the reaction. It does not show the reaction mechanism and associated activation energy.*

Complete combustion

When sufficient oxygen is present, alkanes burn to produce carbon dioxide, CO_2, and water. This is **complete combustion** (Figure 14).

$$CH_4(g) + 2O_2(g) \rightarrow CO_2(g) + 2H_2O(g)$$

$$\Delta_cH^\ominus = -890 \text{ kJ mol}^{-1}$$

$$C_4H_{10}(g) + 6\tfrac{1}{2}O_2(g) \rightarrow 4CO_2(g) + 5H_2O(g)$$

$$\Delta_cH^\ominus = -2880 \text{ kJ mol}^{-1}$$

Note that the products are always carbon dioxide and water.

The amount of energy released is shown by the $\Delta H^\ominus$ sign. It is measured in kJ mol^{-1} and a negative sign shows an exothermic reaction. A positive sign shows an endothermic reaction. The $\Delta H^\ominus$ values for alkane combustion are relatively high, which is why they are valuable fuels. They release a lot of energy when they react with oxygen.

You can find out more in Chapter 7 about energy changes when chemical reactions occur.

Incomplete combustion

If insufficient oxygen is available, carbon monoxide or carbon and water are produced. This is called **incomplete combustion** (Figure 14). Two reactions can occur:

$$2CH_4(g) + 3O_2 \rightarrow 2CO(g) + 4H_2O(g)$$

$$CH_4(g) + O_2(g) \rightarrow C(s) + 2H_2O(g)$$

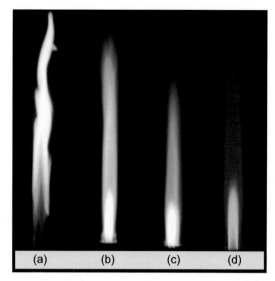

Figure 14 *When the air hole on the Bunsen burner is fully open (d), the flame is blue and the reaction between methane and oxygen is complete combustion. When the air hole is closed (a), the flame is yellow and the reaction between methane and oxygen is incomplete combustion. The yellow flame is caused by the carbon particles it contains. Flames (b) and (c) show the hole slightly open and half-open, respectively.*

QUESTIONS

5. Write equations for:

    a. the complete combustion of octane

    b. the incomplete combustion of octane.

ASSIGNMENT 2: HOW MUCH CARBON DIOXIDE?

(MS 0.0, 0.1, 0.2; PS 1.1, 1.2)

Alkane combustion produces carbon dioxide. In this assignment, you will use skills learnt from Chapter 2 to calculate the masses of carbon dioxide produced from burning some fuels in the UK and to make links between the energy we can obtain from burning fuels and the amount of carbon dioxide produced.

The UK consumption for some petroleum products is given in Table A1.

The third column gives the molecular formula of a typical constituent of the fraction. The fourth column gives the energy produced when one mole of the constituent is burnt completely in oxygen.

Questions

A1. Write balanced equations for the complete combustion of each typical constituent.

A2. For each typical constituent, calculate the energy given out when one mole of carbon dioxide is formed.

A3. Write a sentence that summarises your findings.

A4. Taking petrol to be C_8H_{18}, calculate the mass (tonnes) of carbon dioxide produced per year from petrol used in the UK.

A5. Carry out similar calculations to find the mass of carbon dioxide produced each year from the combustion of refinery gases and aviation fuel used in the UK.

A6. What is the total mass of carbon dioxide produced from refinery gases, petrol and aviation fuel in the UK?

Fraction	Thousand of tonnes	Typical constituent of fraction	$\Delta_{combustion}H/ kJ\ mol^{-1}$
refinery gases	7447	CH_4	−890.3
petrol	18 223	C_8H_{18}	−5470.2
aviation fuel	12 589	$C_{12}H_{26}$	−8086.5

Table A1 *UK consumption for some petroleum products*

6.4 PROBLEMS WITH ALKANE COMBUSTION

The major use of alkanes from the petroleum industry is as fuel, but this comes at a cost.

Use as car fuel

Petrol is the alkane mixture used to fuel cars. It usually contains a mixture of branched and unbranched alkanes between C_6 and C_{10}, plus additives, such as oxygenates. Oxygenates provide extra oxygen for the combustion reactions. When petrol burns in oxygen, the products are all gases. There is a large increase in volume, which drives the pistons and is responsible for turning the wheels of the car.

Although today's engines are the most efficient ever, the alkane combustion is never complete, however, so the exhaust fumes contain the products of incomplete combustion, as well as carbon dioxide and water.

$$2C_8H_{18}(g) + 17O_2(g) \rightarrow 16CO(g) + 18H_2O(g)$$

Carbon monoxide is poisonous. That's why running a car engine in a closed space is potentially lethal. The exhaust fumes from cars also contain a proportion of unburnt alkanes from the petrol (Figure 15).

Figure 15 *Exhaust fumes contain carbon dioxide, carbon monoxide, water, unburnt hydrocarbons, sulfur dioxide (produced from sulfur impurities), nitrogen and nitrogen oxides.*

Further, nitrogen and oxygen in the air do not normally react together – the activation energy that must be overcome to trigger the reaction is too large. But petrol burns at temperatures of about 1000 °C in an engine. At this temperature, the activation energy for the reaction between nitrogen and oxygen in the air–petrol vapour mix is overcome. There are several oxides of nitrogen, usually referred to as NO_x. Nitrogen monoxide, NO, is the one usually formed in engines, but this is rapidly oxidised to nitrogen dioxide, NO_2 (Figure 16):

$$N_2(g) + O_2(g) \rightarrow 2NO(g)$$

$$2NO(g) + O_2(g) \rightarrow 2NO_2(g)$$

Nitrogen monoxide is colourless, but nitrogen dioxide is brown. Under certain weather conditions, this can build up to give a brown haze over large cities. Nitrogen dioxide also contributes to acid rain. It reacts with water and oxygen to make nitric acid, HNO_3:

$$4NO_2(g) + 2H_2O(l) + O_2(g) \rightarrow 4HNO_3(l)$$

The acid rain can damage buildings, plants and animal life. For example, it lowers the pH of lakes and affects living things in them.

Figure 16 *The brown haze over Los Angeles is partly due to the build up of nitrogen dioxide gas.*

QUESTIONS

6. On a warm day, alkanes can evaporate from the petrol tank of a car. Explain why it is mostly the shorter chain alkanes in the petrol that evaporate.

Catalytic converters

These combustion problems have been tackled in part by fitting **catalytic converters** to cars to reduce the emissions of carbon monoxide and oxides of nitrogen. A **catalyst** increases the rate of a reaction without being used up. The catalytic converter in a car is fitted in the exhaust pipe. Most catalytic converters are three-way; three reactions occur simultaneously. Oxides of nitrogen oxidise carbon monoxide to carbon dioxide and are themselves reduced to nitrogen. For example, nitrogen monoxide is reduced by carbon monoxide:

$$2CO(g) + 2NO(g) \rightarrow 2CO_2(g) + N_2(g)$$

The catalyst used is a mixture of platinum, rhodium and palladium (all transition metals). Transition metals

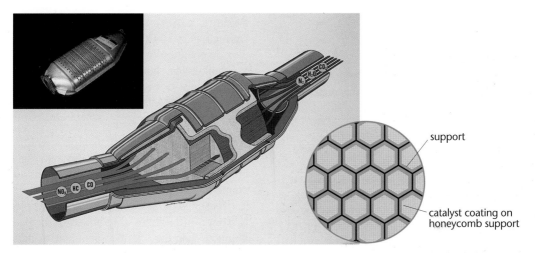

Figure 17 *Cutaway of a three-way catalytic converter showing the honeycomb filter. The gases enter the filter where platinum, rhodium and palladium catalyse reactions of exhaust gases to remove pollutants. The surface area of each filter equals that of two football pitches.*

have several oxidation states, which makes them good catalysts. You can read about oxidation states in Chapter 10. The catalyst is spread in a very thin layer over the surface of a metal oxide or ceramic honeycomb that has a large surface area (Figure 17).

Carbon monoxide and nitrogen monoxide molecules form temporary weak bonds with the surface of the catalyst. They are **adsorbed**. While they are held on the surface, old bonds break and new ones form. Cars fitted with catalytic converters should not use leaded petrol. The lead forms strong bonds with the surface of the catalyst and poisons it.

Catalytic converters remove up to 90% of the CO and NO in exhaust gases.

7. Why do you think that some scientists are suggesting that we should 'mine' road dust for platinum?

Sulfur impurities in fossil fuels

Fossil fuels contain sulfur compounds as impurities. As you have seen, some are burnt in cars and other vehicles and catalytic converters are used to tackle the problem.

However, power stations that burn coal and oil to generate electricity are a problem. Coal mined in the UK, for example, can contain up to 1.5% sulfur locked away in sulfur compounds and released as sulfur dioxide when they are heated strongly.

Before action was taken to reduce emissions (Figure 18) sulfur dioxide escaping into the atmosphere caused major pollution problems in the form of acid rain and health problems. Asthma sufferers are particularly sensitive to sulfur dioxide as it irritates the lining of the nose and throat.

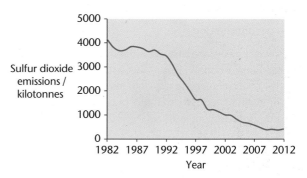

Figure 18 *In the UK, sulfur dioxide emissions have fallen by 96% since 1970. This graph shows the decrease in sulfur dioxide emissions since 1982.*

Sulfur dioxide and desulfurisation

Sulfur-containing fuels burn to produce sulfur dioxide, SO_2:

$$S(s) + O_2(g) \rightarrow SO_2(g)$$

Sulfur dioxide is oxidised in the atmosphere to make sulfur trioxide, SO_3.

$$2SO_2(g) + O_2(g) \rightarrow 2SO_3(g)$$

Sulfur trioxide then dissolves in water to produce sulfuric acid, $H_2SO_4(aq)$.

$$SO_3(g) + H_2O(l) \rightarrow H_2SO_4(aq)$$

The overall reaction can be summarised:

$$SO_2(g) + O_2(g) + H_2O(l) \rightarrow H_2SO_4(l)$$

The sulfuric acid formed is a major contributor to acid rain. Alternatively, sulfur dioxide may dissolve in water (for example, rainwater) to form sulfurous acid, $H_2SO_3(aq)$.

$$SO_2(aq) + H_2O(l) \rightarrow 2H_2SO_3(aq)$$

Sulfurous acid is a solution of SO_2 molecules in dynamic equilibrium with hydrogen ions, $H^+(aq)$, and hydrogensulfite ions, HSO_3^{2-}. You will find out more about chemical equilibria in Chapter 8. Sulfuric acid is an example of a strong acid and sulfurous acid is an example of a weak acid. You can find out more in *Student Book 2*. The sulfur dioxide from car engines only accounts for 2% of the total sulfur dioxide emissions. Most sulfur dioxide emissions potentially come from power stations and industrial processes.

The significant reductions in sulfur dioxide emissions (Figure 18) have been achieved due to action taken by power stations and industry. Sulfur dioxide is removed from waste gases by a process called **flue gas desulfurisation**.

The waste gases are treated with either calcium oxide or calcium carbonate. Both react with sulfur dioxide to produce calcium sulfate. Calcium sulfate is then used in the building industry to make plaster and plaster board. The reactions are:

$$SO_2(g) + CaCO_3(s) \rightarrow CaSO_3(aq) + CO_2(g);$$
then $2CaSO_3(s) + O_2(g) \rightarrow 2CaSO_4(s)$

and

$$SO_2(g) + CaO(s) \rightarrow CaSO_3(s);$$
then $2CaSO_3(s) + O_2(g) \rightarrow 2CaSO_4(s)$

Figure 19 shows how gas flue desulfurisation is used in a coal-fired power station.

Tracking acid rain

During the 1970s and 1980s when levels of sulfur dioxide pollution were high and acid rain was a problem, it was thought that much of the UK's sulfur dioxide emissions were being blown by the prevailing south-westerly winds and causing acid rain in Norway and Sweden.

The Met Office conducted an experiment to find out if UK emissions were responsible for acid rain problems in the Scandinavian countries. They monitored a plume of emission from a single power station on the north-east coast of the UK by taking samples over two days. Trace gases were added to the emission plume to help the scientist identify it. Wind speeds and directions were used to predict the movement of the plume. An aircraft equipped with electronic apparatus measured its sulfur dioxide, nitrogen oxide and ozone content as well as the products of any chemical reactions. Samples of the atmosphere containing no emissions were taken at the same time. They concluded that the acid rain that was impacting on Scandinavia was not just from the UK, and that pollution from across Europe was having an effect.

QUESTIONS

8. When monitoring the drift of gaseous emissions from power stations, what properties must the added trace gases have?

9. Why is it necessary to take measurements from air containing no emissions?

10. What chemical changes from the release into the atmosphere of

    a. sulfur dioxide and

    b. nitrogen oxides

    might scientists expect to detect after two days?

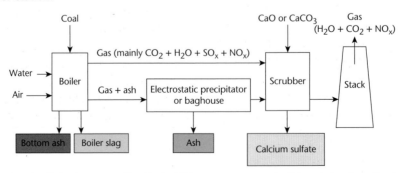

Figure 19 By using gas flue desulfurisation systems like this, the UK is on target to meet the EU regulations for sulfur dioxide reduction by 2020

6.5 REACTIONS OF ALKANES WITH CHLORINE

Methane, CH_4, reacts with chlorine, Cl_2, in the presence of ultraviolet light (usually sunlight) to produce a mixture of chloroalkanes. It is an example of a **photochemical reaction**. The chloroalkanes formed depend on the number of hydrogen atoms substituted in the methane molecule.

If one hydrogen atom is substituted by a chlorine atom, chloromethane forms:

$$CH_4(g) + Cl_2(g) \rightarrow CH_3Cl(g) + HCl(g)$$

But the reaction doesn't stop there. If a second hydrogen atom is substituted, then dichloromethane forms:

$$CH_3Cl(g) + Cl_2(g) \rightarrow CH_2Cl_2(g) + HCl(g)$$

Similarly, a third hydrogen atom can be substituted:

$$CH_2Cl_2(g) + Cl_2(g) \rightarrow CHCl_3(g) + HCl(g)$$

And a fourth:

$$CHCl_3(g) + Cl_2(g) \rightarrow CCl_4(g) + HCl(g)$$

Chlorine reacts with methane in a **free-radical substitution** reaction. In Chapter 5 you learnt how we use dots to help explain the free radical substitution reaction between chlorine and hydrogen.

Chlorine's reaction with methane follows a similar pattern. This is clearly defined in three steps – **initiation, propagation** and **termination**.

Initiation

The reaction only takes place in light. The UV wavelengths of light provide enough energy to split the chlorine molecule into two chlorine atoms:

$$Cl_2 \xrightarrow{\text{UV radiation}} Cl\bullet + Cl\bullet$$

The dot-and-cross diagram is:

$$\overset{\bullet\bullet}{\underset{\bullet\bullet}{Cl}} \overset{}{\underset{\times\times}{\times}} \overset{\times\times}{\underset{\times\times}{Cl}} \xrightarrow{\text{UV radiation}} 2\,\overset{\bullet\bullet}{\underset{\bullet\bullet}{Cl}}\bullet$$

This is an example of **homolytic fission**. Homolytic fission occurs when a covalent bond breaks and each atom receives one electron. You read about this in Chapter 5.

The chlorine atoms formed are extremely reactive because their unpaired electrons readily pair up with electrons from other molecules. They are called radicals. This fission is called the initiation step, the first or initial stage of the reaction.

Propagation

A chlorine radical can now react with a methane molecule:

$$CH_4 + Cl\bullet \rightarrow CH_3\bullet + HCl$$

$CH_3\bullet$ is a methyl radical and is also highly reactive. This is a **propagation** step because it produces another radical. When a methyl radical collides with a chlorine molecule, chloromethane is produced, along with another chlorine radical:

$$CH_3\bullet + Cl_2 \rightarrow CH_3Cl + Cl\bullet$$

The chlorine radical is free to react with more methane molecules. The above two reactions are propagation steps since they produce radicals that can now take part in another propagation step.

If plenty of chlorine is present, further substitution of hydrogen atoms can occur:

$$CH_3Cl + Cl\bullet \rightarrow CH_2Cl\bullet + HCl \qquad \text{Equation 1}$$
$$CH_2Cl\bullet + Cl_2 \rightarrow CH_2Cl_2 + Cl\bullet \qquad \text{Equation 2}$$

The process can continue until tetrachloromethane is produced. It is a chain reaction.

Termination

Propagation steps continue until two radicals collide. When a methyl radical collides with a chlorine radical, chloromethane is produced. There is no radical in the products:

$$CH_3\bullet + Cl\bullet \rightarrow CH_3Cl$$

This is a **termination** step. Other termination reactions are possible:

$$CH_3\bullet + CH_3\bullet \rightarrow C_2H_6$$

and

$$Cl\bullet + Cl\bullet \rightarrow Cl_2$$

Only very small amounts of chlorine gas are formed in termination steps.

Initiation, propagation and termination steps are easily identified from equations because initiation steps produce radicals, propagation steps have radicals as both reactants and products and termination steps only have radicals as reactants.

QUESTIONS

11. What step in free-radical substitution is shown by Equations 1 and 2?

    $CH_3Cl + Cl\bullet \rightarrow CH_2Cl\bullet + HCl$ Equation 1

    $CH_2Cl\bullet + Cl_2 \rightarrow CH_2Cl_2 + Cl\bullet$ Equation 2

12. Write equations to show the reactions that produce $CHCl_3$ from CH_2Cl_2 and Cl_2:

    a. initiation

    b. propagation

    c. termination.

13. Chloroethane is formed in a similar free-radical substitution reaction between ethane and chlorine gases. Write equations to show the reactions:

    a. initiation

    b. propagation

    c. termination.

ASSIGNMENT 3: A DANGEROUS PESTICIDE

(MS 4.1, 4.2, 4.3; PS 1.1, 1.2)

Bromomethane (CH_3Br) was extensively used as a pesticide and to fumigate soil until it was phased out in the early 2000s. Along with the notorious CFCs, bromomethane was found to break down ozone in the higher atmosphere. CFCs are chlorofluorocarbons that break down in the upper atmosphere with UV radiation to produce reactive chlorine radicals that break down ozone. In fact, bromine radicals are more destructive to ozone than chlorine radicals. You can read more about ozone destruction in Chapter 13.

Figure A1 *Although we still manufacture bromomethane for use in industry, it is also naturally produced by marine organisms (an estimated 1–2 billion kilograms a year). Although this source of bromomethane also has a destructive effect on ozone, it is naturally balanced by new ozone being made in the upper atmosphere.*

Questions

A1. Use dotted lines and wedges to show the 3D structure of bromomethane.

A2. Write an equation to show bromomethane forming a bromine radical and a methyl radical. Use dots to show the mechanism.

A3. Why are bromine radicals very reactive?

A4. Explain why the manufactured bromomethane used in agriculture has helped deplete the ozone layer, while bromomethane has been naturally produced by marine organisms for thousands of years and has has no such effect.

A5. Bromomethane reacts with bromine to form dibromomethane (CH_2Br_2) in a free radical substitution mechanism. The reaction is similar to the reactions of chlorine with methane. Write equations to show the reaction of bromomethane with bromine to form dibromomethane:

    a. initiation step

    b. first propagation step

    c. second propagation step.

A6. Write equations for two possible termination steps during the free radical substitution of bromomethane to produce dibromomethane.

A7. Reacting bromine with bromomethane will produce a mixture of products, including tribromomethane and tetrabromomethane. Write an overall equation for the reaction between bromine and bromomethane to produce tetrabromomethane.

KEY IDEAS

❯ Alkanes are non-polar and are not very reactive.

❯ Alkanes are used as fuels. One reason is because they react exothermically with oxygen.

❯ Complete combustion produces carbon dioxide and water.

❯ Incomplete combustion produces carbon monoxide or carbon and water.

❯ Combustion of fossil fuels in combustion engines and industry produces pollutants.

❯ Catalytic converters remove carbon monoxide and oxides of nitrogen from car exhaust gases.

❯ Gas flue desulfurisation removes sulfur dioxide from industrial waste gases and has contributed to tackling environmental pollution.

❯ Methane reacts with chlorine in a free radical substitution reaction.

PRACTICE QUESTIONS

1. a. But-1-ene burns in a limited supply of air to produce a solid and water only.

    i. Write an equation for this reaction.

    ii. State one hazard associated with the solid product in part a. i.

    b. One mole of compound Y is cracked to produce two moles of ethene, one mole of but-1-ene and one mole of octane (C_8H_{18}) only.

    i. Deduce the molecular formula of Y.

    ii. Other than cracking, give one common use of Y.

    c. In cars fitted with catalytic converters, unburned octane reacts with nitrogen monoxide to form carbon dioxide, water and nitrogen only.

    i. Write an equation for this reaction.

    ii. Identify a catalyst used in a catalytic converter.

    AQA Jul 2013 Question 2 b, c, d

2. Table Q1 shows the boiling points of some straight-chain alkanes.

	CH_4	C_2H_6	C_3H_8	C_4H_{10}	C_5H_{12}
Boiling point/°C	− 162	− 88	− 42	− 1	36

Table Q1

    a. State a process used to separate an alkane from a mixture of these alkanes.

    b. Both C_3H_8 and C_4H_{10} can be liquefied and used as fuels for camping stoves. Suggest, with a reason, which of these two fuels is liquefied more easily.

    c. Write an equation for the complete combustion of C_4H_{10}.

    d. Explain why the complete combustion of C_4H_{10} may contribute to environmental problems.

    e. Balance this equation that shows how butane is used to make the compound called maleic anhydride:

    _____$CH_3CH_2CH_2CH_3$ + _____ $O_2 \rightarrow$ _____$C_2H_2(CO)_2O$ + _____ H_2O

    f. Ethanethiol (C_2H_5SH), a compound with an unpleasant smell, is added to gas to enable leaks from gas pipes to be more easily detected.

    i. Write an equation for the combustion of ethanethiol to form carbon dioxide, water and sulfur dioxide.

    ii. Identify a compound that is used to react with the sulfur dioxide in the products of combustion before they enter the atmosphere. Give one reason why this compound reacts with sulfur dioxide.

    iii. Ethanethiol and ethanol molecules have similar shapes. Explain why ethanol has the higher boiling point.

    AQA Jan 2013 Unit 1 Question 4 a–f

3. a. There is a risk of gas explosions in coal mines. This risk is mainly due to the presence of methane. If the percentage of coal mine methane (CMM) in the air in the mine is greater than 15%, the explosion risk is much lower. CMM slowly escapes from the mine into the atmosphere.

   i. Write an equation to show the complete combustion of methane.

   ii. Suggest one reason why there is a much lower risk of an explosion if the percentage of CMM is greater than 15%.

   iii. State why it is beneficial to the environment to collect the CMM rather than allowing it to escape into the atmosphere.

   b. Methane can be obtained from crude oil. Some of this crude oil contains an impurity called methanethiol (CH_3SH). This impurity causes environmental problems when burned.

   i. Write an equation to show the complete combustion of methanethiol.

   ii. State why calcium oxide can be used to remove the sulfur-containing product of this combustion reaction.

   iii. State one pollution problem that is caused by the release of this sulfur-containing product into the atmosphere.

   AQA Jun 2011 Unit 1 Question 6

4. Chlorine can be used to make chlorinated alkanes such as dichloromethane.

   Write an equation for each of the following steps in the mechanism for the reaction of chloromethane (CH_3Cl) with chlorine to form dichloromethane (CH_2Cl_2):

   a. Initiation step

   b. First propagation step

   c. Second propagation step

   d. The termination step that forms a compound with empirical formula CH_2Cl

   AQA Jun 2013 Unit 2 Question 7a

5. The fractions obtained from petroleum contain saturated hydrocarbons that belong to the homologous series of alkanes.

   a. Any homologous series can be represented by a general formula.

   i. State two other characteristics of homologous series.

   ii. Name the process that is used to obtain the fractions from petroleum.

   iii. State what is meant by the term saturated, as applied to hydrocarbons.

   b. Decane has the molecular formula $C_{10}H_{22}$

   i. State what is meant by the term *molecular formula*.

   ii. Give the molecular formula of the alkane which contains 14 carbon atoms.

   iii. Write an equation for the incomplete combustion of decane, $C_{10}H_{22}$, to produce carbon and water only.

   c. When petrol is burned in an internal combustion engine, some nitrogen monoxide, NO, is formed. This pollutant is removed from the exhaust gases by means of a reaction in a catalytic converter.

   i. Write an equation for the reaction between nitrogen and oxygen to form nitrogen monoxide.

   ii. Identify a catalyst used in a catalytic converter.

   iii. Write an equation to show how nitrogen monoxide is removed from the exhaust gases as they pass through a catalytic converter

   AQA Jan 2006 Unit 3(a)

6. The fuels used most frequently in car engines are mixtures of alkanes obtained from petroleum. In car engines, fuels undergo combustion reactions, which can lead to the formation of pollutants.

   a. i. Write an equation for the complete combustion of pentane, C_5H_{12}.

   ii. Identify a solid pollutant formed when pentane undergoes incomplete combustion.

   iii. Give one reason why sulfur dioxide gas may be found in the exhaust gases of cars.

   iv. Give one reason why sulfur dioxide is considered to be a pollutant.

(Continued)

b. Sulfur dioxide is also produced in many industrial processes. Calcium carbonate is sometimes used to remove sulfur dioxide from waste gases before they enter the atmosphere.

  i. Name one other compounds that is used to remove sulfur dioxide from waste gases.

  ii. Write an equation for the reaction between sulfur dioxide and calcium carbonate.

  iii. Give a use for the solid formed by the reaction between sulfur dioxide and calcium carbonate.

AQA Jan 2007 Unit 3(a)

7. a. Thermal cracking of large hydrocarbon molecules is used to produce alkenes. Write an equation for the thermal cracking of $C_{21}H_{44}$ in which ethene and propene are produced in a 3:2 molar ratio together with one other product.

  b. Write equations, where appropriate, to illustrate your answers to the questions below.

  i. Explain why it is desirable that none of the sulfur-containing impurities naturally found in crude oil are present in petroleum fractions.

  ii. The pollutant gas NO is found in the exhaust gases from petrol engines. Explain why NO is formed in petrol engines but is not readily formed when petrol burns in the open air.

  iii The pollutant gas CO is also found in the exhaust gases from petrol engines. Explain how CO and NO are removed from the exhaust gases and why the removal of each of them is desirable.

AQA Jan 2002 Unit 3(a) Question 6

8. Ethane reacts with chlorine in a free radical substitution reaction in a similar way to methane.

  a. The reaction only takes place in light. Write an equation for the initiation step.

  b. i. Write two equations to show the propagation step when one chlorine atom replaces one hydrogen atom in ethane.

  ii. Name the organic product formed.

  iii. Explain why this is called a propagation step.

  c. i. Write an equation to show the termination step when C_4H_{10} is produced.

  ii. Give the names of two other substances that can be produced in termination steps.

Stretch and challenge

9. Cyclopropane used to be used as an anaesthetic, but its tendency to explode when mixed with oxygen led to its eventual phasing out. Cyclopropane has the displayed formula:

Cyclopropane is an example of a cycloalkane.

  a. Give the name, displayed formulae and skeletal formulae for the first six cycloalkanes, and the general formula for cycloalkanes.

  b. Give the displayed formula of an isomer of cyclopropane and name the type of isomerism.

  c. Calculate the C–C bond angles for each cycloalkane.

  d. Explain why the larger cycloalkanes are less reactive than the smaller cycloalkanes.

  e. Given that cycloalkanes have higher melting and boiling points than alkanes with the same number of carbon atoms, compare the van der Waals forces between cycloalkanes and alkanes. Give a reason for your answer.

(Continued)

Multiple choice

10. Why can hydrocarbons in crude oil be separated by fractional distillation?

    A. Because they are liquid at room temperature

    B. Because alkanes with higher M_r values can have chain isomers

    C. Because they have different melting points

    D. Because they have different boiling points

11. Which is the best reason for cracking longer alkane chains in crude oil?

    A. The demand for petrol and alkenes exceeds the supply from the initial fractional distillation of crude oil

    B. The fractional distillation of crude oil provides an excess of short-chained alkanes

    C. The demand for fuel oil outstrips the demand for motor fuel

    D. The demand for longer chained alkanes is in excess of the demand for shorter chained alkanes

12. Which compound cannot be produced by the incomplete combustion of methane?

    A. Carbon dioxide

    B. Carbon monoxide

    C. Carbon

    D. Water

13. Which of these compounds is removed by a catalytic converter?

    A. Sulfur dioxide

    B. Chloromethane

    C. Carbon dioxide

    D. Nitrogen oxide

7 ENERGETICS

PRIOR KNOWLEDGE

You may know that energy changes usually accompany chemical reactions and that reactions may be exothermic or endothermic. You may also have carried out experiments and calculations to measure the energy transferred when foods like crackers or peanuts are burnt.

LEARNING OBJECTIVES

In this chapter, you will find out how we accurately measure some energy changes and how we can use known values to calculate other energy changes. You will also learn about bond enthalpy (or energy) and how it is used.

(Specification 3.1.4.1, 3.1.4.2, 3.1.4.3, 3.1.4.4)

Figure 1 *Since 2010, London buses on the RV1 route have used hydrogen gas as a fuel.*

Petroleum is a finite resource, meaning it will run out eventually. We will need alternative ways to power vehicles. London transport has already carried out experiments to see if hydrogen fuel can be used to power buses. As a result, several hydrogen buses are now used on the tourist route between the Tower of London and Covent Garden (Figure 1).

Hydrogen reacts exothermically with oxygen, releasing energy to power a vehicle. But unlike fossil fuels, burning hydrogen does not produce carbon dioxide, so it does not contribute to global warming. Neither does hydrogen fuel produce sulfur dioxide, unburnt hydrocarbons or large amounts of nitrogen oxides. The problem is that hydrogen production uses fossil fuels.

Hydrogen is a highly flammable gas and it is difficult to store and transport. Industrial chemists can accurately measure the amount of energy given out when a fuel burns and hydrogen fuel stores less energy than the same volume of petrol. Large volumes of hydrogen need to be carried on the vehicle and there are only a handful of hydrogen fueling stations in the UK.

Hydrogen gas can be used in two ways to power a vehicle: it can be reacted with oxygen in an internal combustion engine as petrol is, or it can be used in a fuel cell where it reacts with oxygen to produce an electric current that drives an electric motor.

7.1 EXOTHERMIC AND ENDOTHERMIC REACTIONS

When methane (natural gas) burns, it reacts with oxygen to produce carbon dioxide and water. Every time we use natural gas to cook our food, this is part of the chemistry that we are using. Combustion releases energy stored in the methane and oxygen. This energy is transferred to our food, heating it. The reaction is **exothermic**.

An exothermic reaction is one in which energy is transferred from the **system** to the **surroundings**, heating them. The system is the methane, oxygen, carbon dioxide and water. The surroundings are the food, the wok, the stove, the air – in fact, everything except the methane, oxygen, carbon dioxide and water (Figure 2).

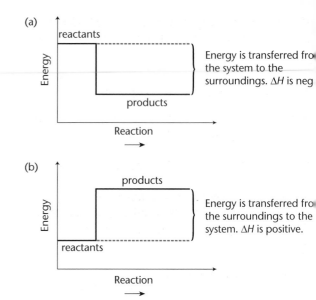

(a) Energy / Reaction — reactants, products. Energy is transferred from the system to the surroundings. ΔH is neg

(b) Energy / Reaction — products, reactants. Energy is transferred from the surroundings to the system. ΔH is positive.

Figure 3 *Energy level diagrams for exothermic reactions (top) and endothermic reactions (bottom) showing the energy content of reactan compared with the energy content of products.*

Figure 2 *We rely on the energy transferred from combustion reactions to cook our food.*

In photosynthesis, carbon dioxide and water react to form glucose and oxygen. The reaction only happens in sunlight. Energy from the Sun is transferred to the reactants (carbon dioxide and water). The reaction mixture absorbs energy and is said to be **endothermic**. Energy is transferred from the surroundings to the system and the surroundings cool down. This is shown in the energy level diagrams in Figure 3.

7.2 ENTHALPY CHANGE

All substances store energy. It is the sum of:

  ❯ the **potential energy** stored by the electrostatic forces within and between particles

  ❯ the **kinetic energy** stored by movement of particles from place to place (translational energy) and when bonds in them vibrate and rotate (vibrational energy and rotational energy).

In chemical reactions, energy transfers happen when bonds are broken and new bonds form. Some reactants transfer energy to the surroundings (exothermic reactions); other reactants absorb energy from the surroundings (endothermic reactions). This is shown in Figure 3.

The energy stored in a compound is called its **enthalpy** and given the symbol H. We cannot measure or calculate this, but we can make measurements that enable changes in enthalpy during chemical reactions to be calculated. In other words, the quantity of energy transferred when a chemical reaction happens. This is described as the **enthalpy change** and given the symbol ΔH. Delta, Δ, is a Greek letter and is used to say 'a change in'.

When methane burns, the enthalpy of methane and oxygen is higher than the enthalpy of the products (carbon dioxide and water). The difference in the enthalpy values is the energy transferred to the surroundings. We write this as:

$$CH_4(g) + 2O_2(g) \rightarrow CO_2(g) + 2H_2O(g)$$

Under standard conditions (100 kPa and 298 K) the enthalpy change is -890 kJ mol^{-1}. We write this as:

$$\Delta H^{\ominus} = -890 \text{ kJ mol}^{-1}$$

The symbol $^{\ominus}$ indicates standard conditions.

You will often see the chemical equation and enthalpy change written together; for example:

$$CH_4(g) + 2O_2(g) \rightarrow CO_2(g) + 2H_2O(g) \quad \Delta H^{\ominus} = -890 \text{ kJ mol}^{-1}$$

Since this is a combustion reaction, the enthalpy change is called the standard enthalpy of combustion, $\Delta_c H^{\ominus}$. The subscript 'c' denotes combustion.

$$\Delta_c H^{\ominus} = -890 \text{ kJ mol}^{-1}$$

The negative sign shows that the reaction is exothermic. A positive sign is used for endothermic reactions. The enthalpy change is measured in kilojoules per mole of methane, kJ mol^{-1}.

This information can also be shown on an energy level diagram (Figure 4).

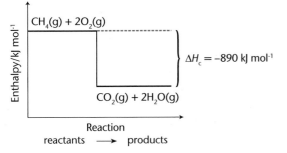

Figure 4 *Energy level diagram for the combustion of methane*

7.3 STANDARD ENTHALPY OF COMBUSTION, $\Delta_c H^{\ominus}$

Fuels react with oxygen in combustion reactions. Measuring the amount of energy given out in the reactions allows us to compare the energy content of different fuels. This is important when choosing a suitable fuel, for example, for a car or a heating boiler in a home.

We use the term **standard enthalpy of combustion** to describe the enthalpy change when a substance burns (combusts) under standard conditions. Its symbol is $\Delta_c H^{\ominus}$.

$\Delta_c H^{\ominus}$ is the enthalpy change when 1 mole of a substance burns completely in oxygen under standard conditions with all reactants and products in their standard states (solid, liquid or gas under standard conditions).

Substances usually burn at a higher temperature than 298 K, so adjustments are made to the experimental values to give data-book or web-page values. Some enthalpy of combustion values for fuels are given in Table 1.

QUESTIONS

1. Draw an energy level diagram for the standard enthalpy of combustion of hydrogen.

Fuel	Main constituent	Formula and standard state	$\Delta_c H^{\ominus}$ of main constituent/kJ mol^{-1}
hydrogen, compressed	hydrogen	$H_2(g)$	-286
natural gas, liquid (CNG or compressed natural gas)	90% methane	$CH_4(g)$	-890
liquid petroleum gas (LPG)	95% propane	$C_3H_8(g)$	-2219
methanol	methanol	$CH_3OH(l)$	-726
ethanol	ethanol	$C_2H_5OH(l)$	-1367
petrol	Octane	$C_8H_{18}(l)$	-5470

Table 1 *The standard enthalpy of combustion values of various fuels. Methanol and ethanol fuels are used in some racing cars. The majority of petrol sold in the UK contains 5% ethanol.*

7.4 STANDARD ENTHALPY OF FORMATION, $\Delta_f H^\Theta$

Like the standard enthalpy of combustion, the standard enthalpy of formation $\Delta_f H^\Theta$ is a specific type of enthalpy change. The subscript 'f' denotes formation.

$\Delta_f H^\Theta$ is the enthalpy change when 1 mole of a compound is formed from its elements under standard conditions. All reactants and products are in their standard states.

Notice that:

> the standard enthalpy of combustion is the enthalpy change when 1 mole of a substance is burnt, whereas

> the standard enthalpy of formation is the enthalpy change when 1 mole of a substance is formed.

The standard enthalpy of formation of sulfuric acid can be written as:

$$H_2(g) + S(s) + 2O_2(g) \rightarrow H_2SO_4(l)$$

$$\Delta_f H^\Theta = -814.0 \text{ kJ mol}^{-1}$$

When writing equations for enthalpies of formation, you may need to use fractions on the left-hand side of the formula. You must end up with one mole of product, so using fractions is allowed. For example,

The standard enthalpy of formation of water:

$$H_2(g) + \tfrac{1}{2}O_2(g) \rightarrow H_2O(l) \quad \Delta_f H^\Theta = -285.8 \text{ kJ mol}^{-1}$$

The standard enthalpy of formation of nitric acid:

$$\tfrac{1}{2}H_2(g) + \tfrac{1}{2}N_2(g) + 1\tfrac{1}{2}O_2(g) \rightarrow HNO_3(l)$$

$$\Delta_f H^\Theta = -207.4 \text{ kJ mol}^{-1}$$

Most standard enthalpies of formation are not directly measurable in a laboratory, unlike standard enthalpies of combustion. But they are useful. Since we cannot measure the enthalpy of a compound, standard enthalpies of formation give us a reference point and enable us to calculate enthalpy changes for any reaction.

By definition, the standard enthalpy of formation of an element is 0. After all, if you make one mole of hydrogen gas from one mole of hydrogen gas, the enthalpies before and after are the same.

QUESTIONS

2. Write equations, including state symbols, to show these enthalpy changes:

   a. The standard enthalpy of combustion of hydrogen, $H_2(g)$

   b. The standard enthalpy of formation of ethanol, $C_2H_5OH(l)$

   c. The standard enthalpy of combustion of ethanol, $C_2H_5OH(l)$

   d. The standard enthalpy of formation of pentane, $C_5H_{12}(g)$

   e. The standard enthalpy of combustion of pentane, $C_5H_{12}(g)$

   f. The standard enthalpy of formation of glucose, $C_6H_{12}O_6(s)$

   g. The standard enthalpy of combustion of glucose, $C_6H_{12}O_6(s)$

KEY IDEAS

> Enthalpy, H, is the energy content of a substance.

> Enthalpy change, ΔH, of a reaction is the energy transferred to or from the number of moles specified in the equation.

> In an exothermic reaction, energy is transferred from the reactants to the surroundings. ΔH is negative. The surroundings become hotter.

> In an endothermic reaction, energy is transferred from the surroundings to the reactants. ΔH is positive. The surroundings become cooler.

> Standard enthalpy changes are measured under standard conditions of 100 kPa and 298 K, with all substances in their standard states.

> The standard enthalpy of combustion, $\Delta_c H^\Theta$, is the enthalpy change when 1 mole of a substance is completely burnt in oxygen under standard conditions (100 kPa and 298 K), with all reactants and products in their standard states.

> The standard enthalpy of formation, $\Delta_f H^\Theta$, is the enthalpy change when 1 mole of a substance is formed from its elements. All reactants and products are in their standard states at 100 kPa and 298 K.

ASSIGNMENT 1: COMBUSTING ALKANES

(MS 3.1, 3.2; PS 1.1, 1.2, 3.1, 3.2)

Alkanes are a homologous series with the general formula C_nH_{2n+2}. Alkanes react exothermically with oxygen and make good fuels.

$$C_nH_{2n+2} + \frac{(n+1)}{2}O_2(g) \rightarrow nCO_2(g) + (n+1)H_2O(l)$$

The state symbol for the alkane has been omitted as it depends on the relative molecular mass of the alkane.

Table A1 gives the enthalpies of combustion for the first eight unbranched alkanes.

Alkane	$\Delta_cH^{\ominus}$/kJ mol^{-1}
methane, CH_4	−890
ethane, C_2H_6	−1560
propane, C_3H_8	−2219
butane, C_4H_{10}	−2877
pentane, C_5H_{12}	−3509
hexane, C_6H_{14}	−4163
heptane, C_7H_{16}	−4817
octane, C_8H_{18}	−5470

Table A1 *Enthalpies of combustion for the first eight unbranched alkanes*

Questions

A1. Plot a graph of the number of carbon atoms in the alkane (x-axis) against the enthalpy of combustion (y-axis).

A2. Use your graph to predict the $\Delta_cH^{\ominus}$ for nonane, C_9H_{20}.

A3. a. What is the difference in enthalpy values between each pair of successive alkanes?

    b. What conclusion can you make about the differences in enthalpy values between successive alkanes?

    c. Explain the shape of your graph.

7.5 MEASURING ENTHALPY CHANGES

The technique used to work out enthalpy changes of reactions is called **calorimetry**. The apparatus used is a **calorimeter** and it enables temperature changes to be measured when energy is transferred during a chemical reaction.

Measuring enthalpy of combustion

A known mass of fuel is burnt. The amount of energy transferred to the surroundings is determined by measuring the temperature rise of a known mass of water in a calorimeter.

The energy required to raise the temperature of 1 g of water by 1 K is 4.2 J g^{-1} K^{-1} (joules per gram per Kelvin). This is the **specific heat capacity** of water.

Knowing the specific heat capacity of water we can calculate the amount of energy needed to heat the known mass of water by the recorded temperature rise. This is equal to the energy transferred from the mass of fuel burnt:

energy released in the reaction = energy transferred to the surroundings

The specific heat capacity of a substance is the energy required to raise the temperature of 1 g of the substance by 1 K.

If their **specific heat capacities are known**, liquids other than water could be used, although this is very unusual. Also, the energy transferred to the reaction vessel can be calculated if its **heat capacity** is known. This will depend on the specific heat capacities and quantities of the materials from which it is made.

Measuring the enthalpy of combustion of ethanol

The apparatus used to measure the enthalpy change of combustion of ethanol is shown in Figure 5.

151

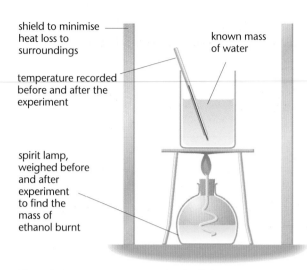

Figure 5 *Apparatus used to determine $\Delta_c H^\theta$ of ethanol*

shield to minimise heat loss to surroundings

known mass of water

temperature recorded before and after the experiment

spirit lamp, weighed before and after experiment to find the mass of ethanol burnt

Initial mass ethanol burner	55.56 g
Final mass ethanol burner	54.80 g
Mass of ethanol burnt	0.76 g
Final temperature	309.5 K
Initial temperature	294.0 K
Temperature rise	15.5 K
Volume of water in calorimeter	200 cm^3

Table 2 *Sample results of the enthalpy change of combustion of ethanol*

The sample results are shown in Table 2.

Calculations

To calculate the energy that was transferred to the water we use the equation:

$q = mc\Delta T$

where:

q = energy transferred/J

m = mass of water/g

c = specific heat capacity of water (4.2 J g^{-1} K^{-1})

ΔT = temperature change/K

Substituting the experimental data (Table 2)

energy transferred to the water (J) = 200 × 15.5 × 4.2

= 13 020 J

= 13.02 kJ

To calculate the moles of ethanol burnt:

$$\text{moles of ethanol burnt} = \frac{\text{mass of ethanol burnt/g}}{M_r}$$

M_r ethanol = 46; therefore moles of ethanol burnt:

$$= \frac{0.76}{46}$$

= 0.0165 mol

To calculate $\Delta_c H^\theta$ for ethanol:

The combustion of 0.0165 moles of ethanol releases 13.02 kJ of energy.

The enthalpy of combustion of ethanol is the energy released when 1 mole of ethanol burns completely.

$$\Delta_c H^\theta \text{ (C}_2\text{H}_5\text{OH(l))} = \frac{-13.020}{0.0165} \times 1$$

$$= -789.1 \text{ kJ mol}^{-1}$$

The value has a negative sign since it is an exothermic reaction that releases energy.

Because the value is for 1 mol of substance combusting, it is also called the **molar enthalpy of combustion**.

Accuracy

The $\Delta_c H^\theta$ for C$_2$H$_5$OH(l) given in data books is -1367 kJ mol^{-1}. The laboratory result is much lower because of procedural errors.

Most significant is that not all of the energy released in the combustion reaction is used to heat the water. Some is transferred to the copper calorimeter, the stirrer, the air around the apparatus and so on, warming all of them.

There are also errors from the measuring cylinder, thermometer and balance, but these are insignificant compared to the procedural errors.

QUESTIONS

3. Calculate the percentage errors on the measurements resulting from the apparatus used (Figure 5). The apparatus errors are:

   Balance ±0.05 g

   Thermometer ±0.1 °C

   Measuring cylinder ±0.5 cm^3

4. $\Delta_c H^\theta$ for (C$_2$H$_5$OH(l)) is -1367 kJ mol^{-1}. Why is the calculated result much lower?

Measuring the enthalpy of neutralisation

Hydrochloric acid reacts with sodium hydroxide solution in a neutralisation reaction.

$$HCl(aq) + NaOH(aq) \rightarrow NaCl(aq) + H_2O(l)$$

The reaction can be carried out in a polystyrene cup with a lid (Figure 6) and the results used to calculate the enthalpy change. For example, 50 cm^3 1.00 mol dm^{-3} hydrochloric acid is neutralised by 50 cm^3 1.00 mol dm^{-3} sodium hydroxide and the temperature is recorded every minute. A graph of the results shows a maximum measured temperature rise of 5.5 °C (as shown in Figure 7).

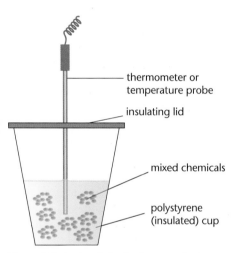

Figure 6 *A simple calorimeter for measuring temperature changes can use a thermometer or a temperature probe.*

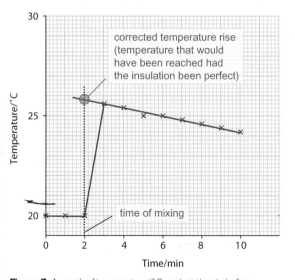

Figure 7 *A graph of temperature/°C against time/min for a neutralisation reaction. The temperature taken at three minutes was the maximum observed temperature rise (5.5 °C). However, a more accurate value can be estimated by extrapolating the cooling line back to the time of mixing.*

Calculations

To calculate the energy transferred to the water:

$$q = mc\Delta T$$

The volume of water is 100 cm^3. Since the density of water is 1 g cm^{-3}, $m = 100$ g.

Energy transferred = $100 \times 4.2 \times 5.5 =$ 2310 J or 2.31 kJ

(Note: this example uses the maximum observed temperature rise of 5.5 °C)

To calculate the amount of acid and alkali used in moles:

1000 cm^3 1.00 mol dm^{-3} hydrochloric acid contains 1.00 mol.

50 cm^3 1.00 mol dm^{-3} hydrochloric acid contains 0.050 mol.

The same number of moles of sodium hydroxide are used.

To calculate the enthalpy change.

$$HCl(aq) + NaOH(aq) \rightarrow NaCl(aq) + H_2O(l)$$

Therefore, 1 mol hydrochloric acid produces 1 mol water.

0.050 mol hydrochloric acid produces 0.050 mol water.

When 0.050 mol water is formed, 2.31 kJ is given out.

So, energy given out when 1 mol water is formed

$$= \frac{2.31}{0.050} \times 1 \text{ kJ}$$

$$= 46.2 \text{ kJ mol}^{-1}$$

QUESTIONS

5. The data-book value for this experiment is 57.2 kJ mol^{-1}. Suggest why the actual result is lower.

In the required practical you will need to show that you can measure an enthalpy change. The enthalpy change may involve a combustion reaction, a dissolution (dissolving a solute in a solvent to produce a solution), a neutralisation reaction or a displacement reaction.

Examples of the apparatus and techniques are given in the two required practical activities:

› Measurement of enthalpies of combustion.

› Measurement of the enthalpy changes in solution.

REQUIRED PRACTICAL ACTIVITY 2: APPARATUS AND TECHNIQUES (PART 1)

(MS 0.2; PS 3.3, 4.1; AT a, d and k)

Measurement of an enthalpy change (enthalpies of combustion)

Making these types of measurements in the laboratory gives you the opportunity to show that you can

> use appropriate apparatus to record mass, volume and temperature

> safely and carefully handle solids and liquids, including flammable substances.

Apparatus

Various calorimeters are available to measure an enthalpy of combustion. They range from simple to sophisticated, but the principle is the same. When a known mass of substance is burned the energy released is transferred to the calorimeter and water contained in it. They become hotter. The temperature rise is measured and this enables the amount of energy released to be calculated.

In the calculations, it is assumed that:

energy released when the substance burns = energy transferred to the calorimeter and the water it contains.

To obtain accurate results, the energy transfer from the calorimeter to the surroundings must be minimised.

A spirit burner can be used to combust liquids. Solid substances need to be supported in some way (the simplest method is to hold a piece with tongs).

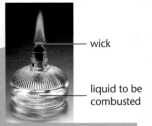

Figure P1 *A spirit burner. A ground glass cap is used to prevent the liquid evaporating from the wick when the spirit burner is not in use.*

The simplest calorimeter is a glass beaker containing a known mass of water. However, heat losses are significant. Insulating the beaker, placing a loose lid on it and shielding it from draughts can reduce these losses. A copper can is an alternative to a beaker. Copper is a good thermal conductor and the energy released from combustion heats the water quickly, reducing heat loss to the surrounding air.

Temperature changes may be measured with a thermometer or a temperature probe (thermistor). The advantage of the latter is that it can be connected to a data-logger.

Even when heat losses are minimised, it is still assumed that the energy released increases the temperature of the water in the calorimeter. However, the container (for example, a beaker or copper can) also warms up. This can be allowed for by calibrating the calorimeter by burning a known mass of substance that has a known enthalpy of combustion.

Technique

A typical experimental set up for determining an enthalpy of combustion was shown in Figure 5.

The required volume of water is measured into the calorimeter using a measuring cylinder. A dropper pipette can be used to add the final few drops until the bottom of the water's meniscus is level with the required graduation mark. The initial temperature of the water is measured.

The mass of the spirit burner, cap and liquid is measured to the appropriate resolution (usually the nearest 0.01 g). The cap is kept on the spirit burner at all times when the liquid is not being burned. This prevents loss through evaporation.

Heat shields positioned around the apparatus help to reduce heat loss. These are usually sheets of aluminium metal that reflect radiated heat. They must enclose the calorimeter as much as possible, but still allow you to work.

When the spirit burner is ignited, it is important to start heating the calorimeter immediately so that as much energy as possible is transferred to the water. To reduce heat losses, the top of the flame should be almost touching the bottom of the calorimeter. The water should be stirred gently, but constantly.

After a suitable temperature rise, say 15–20 °C, the spirit burner is extinguished by replacing the cap and the final temperature of the water taken (this is the highest temperature reached, which will be a short time after heating was stopped). The final mass of the alcohol burner, cap and liquid is measured.

QUESTIONS

P1. Which has the largest percentage error: measuring 25 cm^3 in a 25 cm^3 measuring cylinder with an error of ± 0.5 cm^3, or measuring 25 cm^3 in a 50 cm^3 measuring cylinder with a percentage error of ± 1.0 cm^3?

P2. Why is it important to prevent loss by evaporation from an uncapped spirit burner?

P3. If you are measuring the enthalpies of combustion of a series of alcohols, what practical steps must you take before measuring the enthalpy of combustion of the second alcohol?

REQUIRED PRACTICAL ACTIVITY 2: APPARATUS AND TECHNIQUES (PART 2)

(PS 4.1; AT a, d and k)

Measurement of an enthalpy change (enthalpy changes in solution)

Making these types of measurements in the laboratory gives you the opportunity to show that you can:

> use appropriate apparatus to record mass, volume and temperature

> safely and carefully handle solids and liquids, including flammable substances.

Apparatus

Reactions that happen in aqueous solution or between a solid and an aqueous solution can be carried out in glass or plastic beakers or polystyrene cups. Energy is transferred from the reactants to the water (exothermic reaction) or from the water to the reactants (endothermic reaction).

It is essential to minimise heat losses. This may be done by insulating the reaction vessel, such as a beaker, or by using a vessel made from an insulating material such as an expanded poly(styrene) cup. In both cases, a lid further reduces heat losses. The more effective the insulation, the smaller the heat losses and the more accurate the measurement of energy transferred.

A measuring cylinder can be used to measure volumes of solutions or liquids. One with the lowest capacity possible reduces the percentage error.

A balance with suitable precision (usually the nearest 0.01 g) is used for weighing solids. A weighing bottle and the 'tare' button on the balance can be used. The weighing bottle is placed on the balance and the 'tare' button pressed (the digital readout should show 0.00 g). The required amount of solid can now be measured, taking care not to spill any on the balance pan as this will affect the results.

Temperature changes can be measured with a thermometer or a thermistor. This can be inserted through a hole in the lid.

The reactants should be stirred. Options include using a thermometer (if there is no lid), a loop stirrer, a mechanical stirrer or a magnetic stirrer (examples are shown in Figure P1).

Figure P1 *A mechanical stirrer (left) and a magnetic stirrer (right)*

Technique

The liquid or one of the solutions is placed in the calorimeter and the initial temperature measured. The second solution or the solid is then added and the mixture stirred to mix the reactants.

The largest temperature change is recorded. It is important to monitor the temperature for a few minutes because although the reaction may be immediate, the temperature will take a short time to register.

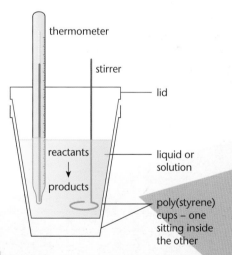

Figure P2 *A typical experimental set-up for determining the enthalpy change in solution or between a solid and a liquid or solution*

Figure P3 *In this experimental set-up the temperature is measured using an electronic temperature probe and the reaction mixture is stirred using a magnetic stirrer bar (often called a flea).*

QUESTIONS

P1. In Figure P2, why are two poly(styrene) cups used?

P2. A copper can may be used as a calorimeter in enthalpy of combustion experiments. Why is it not used for measuring enthalpy changes in solution?

Stretch and challenge

P3. As soon as a solution begins to warm up as energy is transferred from a reaction, it also cools as some energy is transferred to the surroundings. This is reduced by insulation. However, a technique called a cooling correction can also be useful. Find out how to carry out a cooling correction for an enthalpy change when two solutions are mixed and react.

ASSIGNMENT 2: THE BOMB CALORIMETER

(MS 0.0, 0.1, 0.2, 2.1,2.3, 2.4; PS 1.1, 1.2, 3.2)

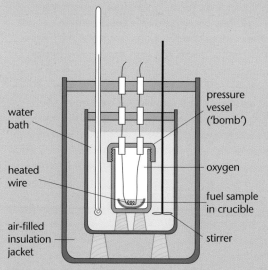

Figure A1 *A bomb calorimeter*

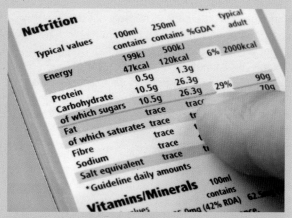

Figure A2 *The energy content of foods is measured in the laboratory using a system of standardised chemical tests and an analysis of the recipe. A 250 ml serving of this orange juice can provide you with 500 kJ of energy. (Note: 250 ml = 250 millilitres = 250 cm³.)*

The bomb calorimeter is used to calculate enthalpies of combustion. An electric current is used to ignite the fuel sample in a pressure vessel (the 'bomb'). The burning fuel transfers energy to the air around it, which in turn transfers energy to the water that surrounds the bomb. In some models, the heated air escapes through a coiled copper tube submerged in water, warming the water.

The temperature rise in the water is used to calculate the enthalpy change of the fuel. The air-filled insulation jacket reduces heat loss. In more modern bomb calorimeters, the whole calorimeter is submerged in a known volume of water, the temperature of which is monitored and the change used to produce a more accurate enthalpy value.

Bomb calorimeters can be used to measure the energy contents of foods. The food industry calls the energy content of a food its energy value. In our body cells, energy is produced from glucose in respiration. Glucose is the product of carbohydrate digestion. Respiration is a similar chemical process to combustion, but the rate is controlled by enzymes and is much slower. The problem with using the bomb calorimeter to measure food values is that it tends to overestimate the amount of energy the human digestive system can extract from foods. We cannot digest fibre in our diets, but in the bomb calorimeter dietary fibre is burnt along with other digestible carbohydrates, and the energy released is included in the measurement of food energy.

Questions

A1. 0.86 g of hexane was completely burnt in a bomb calorimeter. A temperature rise of 19.4 K was recorded. The bomb calorimeter contained 500 g of water. The specific heat capacity of water is 4.2 J g^{-1} K^{-1}.

   a. Calculate the energy released (kJ) from the burning fuel.

   b. How many moles of hexane were burnt?

   c. Calculate the enthalpy change of combustion of hexane.

A2. A breakfast cereal contains 13.4 g glucose per 100 g cereal.

$$\Delta_c H^\theta [C_6H_{12}O_6(s)] = -2802.5 \text{ kJ mol}^{-1}$$

How much energy can be obtained from the glucose in a 50 g helping?

Stretch and challenge

A3. Explain why a bomb calorimeter will provide a more accurate measurement of the energy transferred than burning hexane in the type of calorimeter shown in Figure 6.

7.6 USING HESS'S LAW TO MEASURE ENTHALPY CHANGES

Many reactions are difficult to carry out under normal laboratory conditions and so their enthalpy changes cannot be measured directly in the laboratory. For example, we cannot react carbon with hydrogen easily to produce methane and, therefore, collect data to enable the calculation of the enthalpy of formation of methane. The reaction doesn't happen under laboratory conditions.

We need an indirect method. One that can be used relies on the law of conservation of energy and involves energy cycles and calculations. The law of conservation of energy states:

Energy cannot be created or destroyed, only changed from one form to another.

This is applied to enthalpy cycles as **Hess's law**.

Hess's law states that if a reaction can take place by more than one route, the overall enthalpy change is the same, regardless of the route, as long as the initial and final conditions are the same.

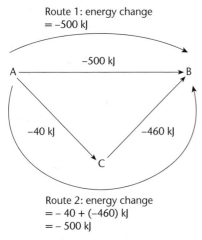

Route 1: energy change
= −500 kJ

−500 kJ

Route 2: energy change
= − 40 + (−460) kJ
= − 500 kJ

Figure 8 *Energy cycle diagram with imaginary values*

Figure 8 shows a simple energy cycle, complete with imaginary ΔH values. Note that these values are not for 1 mole amounts of A, B and C, therefore mol^{-1} is omitted from the units. The cycle obeys Hess's law because the enthalpy change from A to B is − 500 kJ, and from A to C to B is also − 500 kJ.

Hess's law enables us to calculate enthalpy changes that cannot be measured directly. Figure 9 shows how Hess's law is used to calculate the enthalpy of formation of methane. The top line of the cycle in Figure 9 is labelled ΔH_1 and represents the enthalpy change for the formation of methane. The reactions used to provide the alternative route are the combustion reactions of carbon, hydrogen and methane. ΔH_2 is the enthalpy change of the combustion of carbon; ΔH_3 is the combustion of hydrogen; and ΔH_4 is the enthalpy change of combustion of methane. We can determine the enthalpy changes of combustion of carbon, hydrogen and methane experimentally. The unknown enthalpy change of formation of methane can be calculated as it is the only unknown in the cycle. From Figure 9:

$$\Delta H_1 + \Delta H_4 = \Delta H_2 + \Delta H_3$$

$$\Delta H_1 = \Delta H_2 + \Delta H_3 - \Delta H_4$$

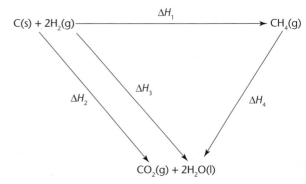

Figure 9 *Two ways to represent the enthalpy cycle for the formation of methane (see Worked example 1: Calculate the enthalpy of formation of methane). A value that cannot be measured directly in the laboratory is determined indirectly using Hess's law.*

Worked example 1

(MS 2.4)

Calculate the enthalpy of formation of methane.

Enthalpy change	kJ mol^{-1}
$\Delta_c H^{\ominus}[C(s)]$	-393
$\Delta_c H^{\ominus}[H_2(g)]$	-286
$\Delta_c H^{\ominus}[CH_4(g)]$	-890

Table 3 *Values for Worked example 1*

Step 1 Draw an enthalpy cycle diagram. This could be either of the types shown in Figure 9.
To do this you need:

> the equation for the formation of methane from its elements:

$$C(s) + 2H_2(g) \rightarrow CH_4(g)$$

This is the enthalpy change we are calculating. Make this reaction the top line, represented by ΔH_1. Make the arrow point to the products.

Use the enthalpy change values in Table 3 to complete the enthalpy diagram (Figure 9).

Step 2 Arrange the enthalpy changes to give an expression for ΔH_1.

$$\Delta H_2 + \Delta H_3 = \Delta H_1 + \Delta H_4$$

$$\Delta H_1 = \Delta H_2 + \Delta H_3 - \Delta H_4$$

Step 3 Substitute values for ΔH_2, ΔH_3 and ΔH_4.

$$\Delta H_2 = \Delta_c H^{\ominus}[C(s)]$$

$$= -393 \text{ kJ mol}^{-1}$$

$$\Delta H_3 = 2 \times \Delta H_c[H_2(g)]$$

(Note that there are two moles of hydrogen in the equation.)

$$= 2 \times -286$$

$$= -572 \text{ kJ mol}^{-1}$$

$$\Delta H_4 = \Delta_c H^{\ominus}[CH_4(g)]$$

$$= -890 \text{ kJ mol}^{-1}$$

Step 4 Substitute these values into the equation for ΔH_1.

$$\Delta H_1 = -393 + (-572) - (-890)$$

$$= -75 \text{ kJ mol}^{-1}$$

The enthalpy change of formation of methane is -75 kJ mol^{-1}.

Worked example 2

(MS 2.4)

Calculate the enthalpy of formation of nitromethane, CH_3NO_2

Figure 10 *Nitromethane makes a good fuel because it already contains some of the oxygen needed for combustion.*

Enthalpy change	kJ mol^{-1}
$\Delta_c H^{\ominus}[C(s)]$	-393
$\Delta_c H^{\ominus}[H_2(g)]$	-286
$\Delta_c H^{\ominus}[CH_3NO_2(l)]$	-709

Table 4 *Values for Worked example 2*

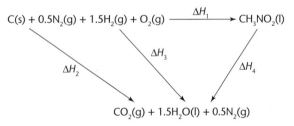

Figure 11 *The enthalpy cycle for the formation of nitromethane from its elements (see Example 2)*

Step 1 Draw the enthalpy cycle diagram (shown in Figure 11).

Step 2 Arrange the enthalpy changes to give an expression for ΔH_1.

$$\Delta H_1 + \Delta H_4 = \Delta H_2 + \Delta H_3$$

$$\Delta H_1 = \Delta H_2 + \Delta H_3 - \Delta H_4$$

Step 3 Substitute values for ΔH_2, ΔH_3 and ΔH_4.

$\Delta H_2 = \Delta_c H^\ominus[C(s)]$

$\quad = -393 \text{ kJ mol}^{-1}$

$\Delta H_3 = 1.5 \times \Delta_c H^\ominus[H_2(g)]$

$\quad = 1.5 \times -286 \text{ kJ mol}^{-1}$

$\Delta H_4 = \Delta_c H^\ominus[CH_3NO_2(l)]$

$\quad = -709 \text{ kJ mol}^{-1}$

Step 4 Substitute these values into the equation for ΔH_1.

$\Delta H_1 = -393 + (1.5 \times -286) - (-709)$

$\quad = -113 \text{ kJ mol}^{-1}$

$\Delta_f H^\ominus[CH_3NO_2(l)] = -113 \text{ kJ mol}^{-1}$

Worked examples 1 and 2 both involve using standard enthalpies of combustion to calculate standard enthalpies of formation that are difficult to measure or do not occur under normal laboratory conditions. Data books and web pages give standard enthalpy changes of formation for many compounds. We can use these to calculate the enthalpy changes for other reactions, such as the reaction between carbon dioxide and water to make glucose in photosynthesis.

Worked example 3

(MS 2.4)

Calculate the enthalpy of the reaction to form one mole of glucose from carbon dioxide and water.

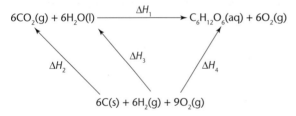

Figure 12 *The enthalpy cycle for photosynthesis*

Enthalpy change	kJ mol⁻¹
$\Delta_f H^\ominus[CO_2(g)]$	−393
$\Delta_f H^\ominus[H_2O(l)]$	−286
$\Delta_f H^\ominus[C_6H_{12}O_6(s)]$	−1273

Table 5 *Values for Worked example 3*

Step 1 Draw the enthalpy cycle diagram (Figure 12). Note:

ΔH_1 is to be calculated.

ΔH_2 is the enthalpy change of formation of carbon dioxide and must be multiplied by six since six moles of CO_2 are formed.

ΔH_3 is the enthalpy change of formation of water and must be multiplied by six since six moles of water are formed.

ΔH_4 is the enthalpy change of formation of glucose. We do not include oxygen here, since it is an element.

You will need the data in Table 5.

Step 2 Arrange the enthalpy changes to give an expression for ΔH_1.

$\Delta H_4 = \Delta H_1 + \Delta H_2 + \Delta H_3$

$\Delta H_1 = \Delta H_4 - \Delta H_2 - \Delta H_3$

Step 3 Substitute values for ΔH_2, ΔH_3. and ΔH_4

$\Delta H_2 = 6 \times (-393) = -2358$

$\Delta H_3 = 6 \times (-286) = -1716$

$\Delta H_4 = -1273 \text{ kJ mol}^{-1}$

Step 4 Substitute these values into the equation for ΔH_1:

$\Delta H_1 = -1273 - (-4074)$

$\quad = +2801 \text{ kJ mol}^{-1}$

$\Delta_f H^\ominus[C_6H_{12}O_6(l)] = +2801 \text{ kJ mol}^{-1}$

QUESTIONS

6. Use standard enthalpy of combustion values, Hess's law and the data in Table 5 to calculate $\Delta_f H^\ominus$ of:

   a. ethyne, C_2H_2

   b. ethanol, C_2H_5OH

   c. ethane, C_2H_6

   d. carbon monoxide, CO

7. The standard enthalpy change of combustion of hexane is -4163 kJ mol^{-1}.

   a. What is the M_r of hexane?

   b. How much energy is transferred when:

      i. 100 g of hexane is completely burnt

      ii. 1 kg of hexane is completely burnt?

8. Use $\Delta_f H^\ominus$ values to calculate the enthalpy changes for the following reactions (use data in Table 6):

   a. the combustion of ethanal, CH_3CHO

   b. $NH_3(g) + HCl(g) \rightarrow NH_4Cl(s)$

   c. the combustion of methanol, CH_3OH

Standard enthalpy change	kJ mol^{-1}
$\Delta_f H^\ominus[CO_2(g)]$	-393
$\Delta_f H^\ominus[H_2O\,(l)]$	-286
$\Delta_f H^\ominus[CH_3CHO(l)]$	-166
$\Delta_f H^\ominus[NH_3(g)]$	-46.2
$\Delta_f H^\ominus[HCl(g)]$	-92.3
$\Delta_f H^\ominus[NH_4Cl(s)]$	-314
$\Delta_f H^\ominus[CH_3OH(l)]$	-239

Table 6 *Standard enthalpies of formation*

9. Calculate the enthalpy change for the reaction of anhydrous copper(II) sulfate with water to produce hydrated copper(II) sulfate, $CuSO_4.5H_2O$, using the standard enthalpy changes of formation in Table 7.

Standard enthalpy change	kJ mol^{-1}
$\Delta_f H^\ominus[CuSO_4(s)]$	-771
$\Delta_f H^\ominus[CuSO_4(s)]$	2280
$\Delta_f H^\ominus[H_2O(l)]$	286

Table 7 *Standard enthalpies of formation*

A shortcut

In Worked example 3, we used $\Delta_f H^\ominus$ values to calculate the enthalpy change when 1 mol of glucose is formed in photosynthesis. Overall, we subtracted the total standard enthalpies of formation of the reactants from the total standard enthalpies of formation of the products. We can use this as a general rule. $\Delta_r H^\ominus$ is the enthalpy of a reaction. It must refer to the amounts in an equation. The Σ sign means 'the sum of'.

Figure 13 *Anhydrous copper(II) sulfate is white. When water is added, blue hydrated copper(II) sulfate forms.*

For any reaction, the standard enthalpy change of reaction, $\Delta_r H^\ominus$, is:

$$\Delta_r H^\ominus = \Sigma\Delta_f H^\ominus[\text{products}] - \Sigma\Delta_f H^\ominus[\text{reactants}]$$

For example, $CH_4(g) + 2O_2(g) \rightarrow CO_2(g) + 2H_2O(l)$

$$\Delta_r H^\ominus = [(-393) + (2 \times -286)] - (-74.9)$$

$$= 890.1 \text{ kJ mol}^{-1}$$

This is the standard enthalpy of combustion of methane.

Similarly, we can use the values for the standard enthalpy changes of combustion of products and reactants to work out standard enthalpy changes of reaction, but this time we subtract the value for the products from the value for the reactants.

$$\Delta_r H^\ominus = \Sigma\Delta_c H^\ominus[\text{reactants}] - \Sigma\Delta_c H^\ominus[\text{products}]$$

Worked example 4

(MS 2.4)

Figure 14 *Hydrazine is one of the fuels used in space shuttles.*

Hydrazine, N_2H_4, can be used as rocket fuel. It reacts with oxygen to produce nitrogen and water:

$$N_2H_4(l) + O_2(g) \rightarrow 2H_2O(g) + N_2(g)$$

Use standard enthalpy changes of formation to calculate the enthalpy change for the combustion of hydrazine (use the data in Table 8).

Standard enthalpy changes of formation	kJ mol^{-1}
$\Delta_fH^\ominus[N_2H_4(l)]$	+50.6
$\Delta_fH^\ominus[H_2O(l)]$	−286
$\Delta_fH^\ominus[N_2(g)]$	0
$\Delta_fH^\ominus[O_2(g)]$	0

Table 8

Step 1 $\Delta_rH^\ominus = \Sigma\Delta_fH^\ominus[\text{products}] - \Sigma\Delta_fH^\ominus[\text{reactants}]$

Remember that the enthalpy change of formation of any element is 0.

Step 2 $\Delta_rH^\ominus = 2(-286) - (+50.6)$

$= -623$ kJ mol^{-1} (to three significant figures)

KEY IDEAS

- Calorimetry can be used to determine enthalpy changes in the laboratory. The energy released is usually transferred to a known mass of water and the temperature change used to calculate the enthalpy change

- The energy transferred/J = specific heat capacity of water/J g^{-1} K^{-1} × mass of water/g × temperature change/K.

- The specific heat capacity is the energy needed to raise the temperature of 1g of a substance by 1K. The specific heat capacity of water is 4.2 J g^{-1} K^{-1}.

- Enthalpy changes can be calculated using Hess's law.

- Hess's law states that if a reaction takes place by more than one route, the enthalpy change of the reaction is the same, regardless of the route, providing that the conditions are the same.

7.7 BOND ENTHALPIES

When we measure or calculate enthalpy changes, we are measuring the energy taken in or given out in a reaction. Since chemical reactions involve the breaking of chemical bonds and the making of new ones, we are measuring the difference between the energy needed to break bonds and the energy given out when new bonds are made.

Breaking bonds requires energy and different bonds require different amounts of energy to break them. A strong bond is one that requires more energy to break than a weak bond. The same amount of energy will be given out when the same bond is formed.

The energy needed to break a particular bond in a gaseous molecule is called its **bond enthalpy**. It is measured per mole of bonds and its units are kJ mol^{-1}. For a $H_2(g)$ molecule

$$H_2(g) \rightarrow H(g) + H(g)$$

$$\Delta H^\ominus = +436 \text{ kJ mol}^{-1}$$

Bond enthalpy has the symbol *E*. It is written as: $E(H-H) = +436$ kJ mol^{-1}.

436 kJ of energy are needed to break one mole of H−H bonds in gaseous molecules. This is also the energy given out when one mole of H−H bonds form.

Methane has four C−H bonds. But as soon as the methane molecule begins to break down, the remaining C−H bonds exist in a different environment. When the first C−H bond is broken, it is from a CH_4 molecule, but the second C−H bond is broken from a CH_3 fragment. The third bond to break is in a CH_2 fragment. The energy needed to break the bonds is affected by the environment of the bond, so the energy needed to break each C−H bond in a methane molecule is slightly different. The eventual figure for the enthalpy change for the breaking or **dissociation** of C−H bonds in methane will be the mean of four values. How much does this affect the value?

When methane is broken into its atoms, 1664 kJ mol^{-1} of energy is absorbed:

$$CH_4(g) \rightarrow C(g) + 4H(g)$$
$$\Delta H^\ominus = +1664 \text{ kJ mol}^{-1}$$

This process involves breaking four C−H bonds, so the mean C−H bond enthalpy in methane is:

$\dfrac{1664}{4}$ kJ mol^{-1}

or 416 kJ mol^{-1}

We write this as:

$\Delta H^{\ominus}$[C–H in CH$_4$] = 416 kJ mol^{-1}

However, considering the first part of this reaction, to break one bond only:

CH$_4$(g) $\rightarrow$ CH$_3$(g) + H(g)

$\Delta H^{\ominus}$ = +423 kJ mol^{-1}

This is different from the mean value of 416 kJ mol^{-1}. Since bond enthalpies vary with their molecular environment, it is often more useful to use **mean bond enthalpy** when applying data to other compounds. Note that the values for C–H bonds in data books are not just the mean of the values for methane, but the mean of C–H bonds in a wide variety of molecules. This average value is 413 kJ mol^{-1}.

The mean bond enthalpy is the average enthalpy change when one mole of bonds of the same type are broken in gaseous molecules under standard conditions.

QUESTIONS

10. Water is H–O–H.
    Breaking the first O–H bond:
    H–O–H(g) $\rightarrow$ H(g) + O–H(g)
    $\Delta H^{\ominus}$ = +502 kJ mol^{-1}

    Breaking the second O–H bond:
    O–H(g) $\rightarrow$ O(g) + H(g) $\Delta H^{\ominus}$ = +427 kJ mol^{-1}

    a. Calculate the mean bond enthalpy for the O–H bonds in water.

    b. Explain any difference between the value you have calculated in part a. and the value given in data books.

Using bond enthalpies to calculate the enthalpy change of a reaction

So far in this chapter we have calculated enthalpy changes from experiments and by using Hess's law. Another way to calculate the enthalpy change of a reaction is to consider which bonds are broken and which are made. When bonds are broken, energy is required, so breaking bonds is endothermic. When bonds are made, energy is released, so making bonds is exothermic.

Table 9 gives some mean bond enthalpy values.

Bond	Mean bond enthalpy/kJ mol^{-1}
C–H	413
C–C	348
C–O	360
C=C	612
C=O	743
O–H	463
O=O	498
N=N	409
N≡N	944
N–N	391

Table 9 *Some mean bond enthalpies*

Worked example 5

(MS 2.4)

Calculating bond enthalpies in the combustion of ethanol.

Remember that bond enthalpies apply to molecules in the gaseous state – the reactants and products must be gases.

Step 1 Write out the equation using structural formulae:

Step 2 For each reactant, list the type and number of bonds broken. Look up the mean bond enthalpy for each bond type and calculate the enthalpy change. Do this for the bonds broken and then for the bonds made.

Breaking bonds is an endothermic process. Therefore, the enthalpy changes (in kJ mol^{-1}) when bonds are broken have positive values:

5 $\times$ (C–H) = 5 $\times$ 413 = +2065

1 $\times$ (C–C) = 1 $\times$ 348 = +348

1 $\times$ (C–O) = 1 $\times$ 360 = +360

1 $\times$ (O–H) = 1 $\times$ 463 = +463

3 $\times$ (O = O) = 3 $\times$ 498 = +1494

Total enthalpy change when all of the bonds in 1 mol ethanol and 3 mol oxygen are broken = +4730 kJ.

Making bonds is an exothermic process. The enthalpy changes when bonds are made (kJ mol^{-1}) have negative values:

$4 \times (C=O) = 4 \times -743 = -2972$

$6 \times (O–H) = 6 \times -463 = -2778$

Total enthalpy change when all of the bonds in 3 mol carbon dioxide and 3 mol water $= -5750$ kJ.

Step 3 Calculate the overall enthalpy change.
Overall enthalpy change $= +4730 - 5750 = -1020$ kJ.

Remember that the enthalpy of combustion refers to 1 mol of the compound that is combusting. Therefore, the enthalpy of combustion of ethanol is -1020 kJ mol^{-1}.

This is the theoretical standard enthalpy change of combustion for gaseous ethanol to form gaseous products using mean bond enthalpies. It will not be the same or even similar to the $\Delta_c H^\ominus$ value in the data book because enthalpies of combustion apply to reactants in their standard state in standard conditions.

When using bond enthalpies to calculate enthalpies of reaction, you can use the formula:

$\Delta H^\ominus = \Sigma \Delta H^\ominus[\text{bonds broken}] - \Sigma \Delta H^\ominus[\text{bonds made}]$

Figure 15 is a diagram of energy levels for the molecules and atoms at different steps in the reaction. It shows that the products are at a lower energy level than the reactants, so overall the enthalpy change is negative.

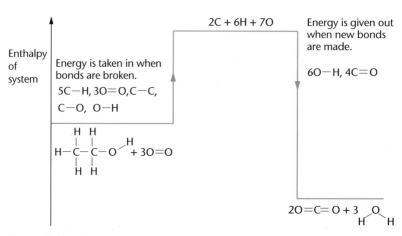

Figure 15 *Mean bond enthalpies and the combustion of ethanol*

QUESTIONS

11. Use the mean bond enthalpies in Table 9 to calculate:

    a. the enthalpy change of combustion for ethene

    b. the enthalpy change of combustion for propan-1-ol gas

    c. the enthalpy change of combustion for hydrazine burning in oxygen to produce nitrogen gas and water

    d. the enthalpy change of combustion for butan-1-ol gas.

    Hints: you must write a balanced equation before calculating the enthalpy changes.

    Hydrazine is:

Stretch and challenge

12. Use Table 9 to calculate the amount of energy needed to change 1 mol gaseous ethanol into its individual gaseous atoms.

13. Use ideas about evaporation to explain why $\Delta_c H^\ominus[C_2H_5OH(l)] = 1367$ kJ mol^{-1}, but calculating the enthalpy of combustion from bond enthalpies gives a value of -1020 kJ mol^{-1}.

Butan-1-ol is:

KEY IDEAS

> Bond enthalpy is the enthalpy change when one mole of bonds in gaseous molecules are broken, under standard conditions.

> Bond enthalpies depend on the molecular environment of the bond, and the values quoted in data books are mean bond enthalpies.

> Bond enthalpies can be used to calculate enthalpy changes for reactions by considering the total enthalpy change when the bonds are broken and the total enthalpy change when the bonds are made.

ASSIGNMENT 3: WHICH FUEL?

(MS 0.0, 0.1, 0.2, 2.1, 2.3, 2.4; PS 1.1, 1.2, 3.2)

Figure A1 *Indy racing cars may get less than three miles to a gallon, but they use fuel containing 85% ethanol. Ethanol fuel produces less pollution than conventional petrol fuel.*

Figure A2 *In the US, ethanol fuel is an important alternative to petrol. Blends of ethanol and petrol (gasoline) are common.*

Fuel	Constituent or typical constituent	$\Delta_c H^{\ominus}$/kJ mol^{-1}	Density/g cm^{-3}
liquid petroleum gas (LPG)	C_4H_{10}	-2876	0.57
petrol	C_8H_{18}	-5470	0.70
bioethanol	C_2H_5OH	-1367	0.79
hydrogen gas	H_2	-286	0.000089

Table A1 *Data for four fuels*

The fuels we use depend on many factors. In this assignment, you are comparing the amounts of energy obtainable from 1 dm^3 (one litre) of different fuels used in a combustion engine. Remember that the amounts you calculate will not be the amounts of energy that we are able to use to drive the engine and, therefore, move the car, because the combustion engine is very inefficient.

The technology to run cars on ethanol fuel is not new – Brazil, for example, has used alcohol fuel (ethanol fuel) since 1975. Many governments consider that ethanol fuel is the way ahead to compensate for dwindling oil reserves. Ethanol can be produced by fermentation of the sugars in plant material, such as corn. The enzyme zymase in yeast converts the sugar into carbon dioxide and ethanol. As a fuel, it burns exothermically to release energy.

The UK Central Science Laboratory claims that ethanol fuel produces 65% fewer greenhouse gases than petrol does. This is mainly because the amount of carbon dioxide emitted during the production and consumption of ethanol is almost equal to the amount removed from the atmosphere when crops for conversion are being grown. Ethanol is renewable and the UK government is aiming for 10% of all car fuel to be from renewable sources by 2020.

Fuel is sold in litres (1 litre = 1 dm³). Table A1 displays some data for four fuels.

Questions

A1. **a.** What is the mass, in grams, of 1 dm³ of LPG (C_4H_{10})?

    b. What is the M_r of C_4H_{10}?

    c. How many moles are there in 1 dm³ of C_4H_{10}?

    d. Calculate the energy that can be obtained from the complete combustion of 1 dm³ of C_4H_{10} fuel.

A2. Calculate the energy that can be obtained from the complete combustion of:

    a. 1 dm³ of C_8H_{18} fuel

    b. 1 dm³ of C_2H_5OH fuel

    c. 1 dm³ hydrogen gas.

A3. Write a sentence to summarise your answers to A1 and A2.

Stretch and challenge

A4. Why is the combustion engine so inefficient?

PRACTICE QUESTIONS

1. A student carried out an experiment to determine the enthalpy change when a sample of methanol was burned. The student found that the temperature of 140 g of water increased by 7.5 °C when 0.011 mol of methanol was burned in air and the heat produced was used to warm the water.

Use the student's results to calculate a value, in kJ mol⁻¹, for the enthalpy change when one mole of methanol was burned.

(The specific heat capacity of water is 4.18 J K⁻¹ g⁻¹.)

AQA Jun 2012 Unit 2 Question 2e

2. Hydrogen reacts with oxygen in an exothermic reaction, as shown by the following equation:

$H_2(g) + \frac{1}{2}O_2(g) \rightarrow H_2O(g)$ $\Delta H = -242$ kJ mol⁻¹

Use the information in the equation and the following data to calculate a value for the bond enthalpy of the H–H bond.

	O–H	O=O
Mean bond enthalpy / kJ mol⁻¹	+463	+496

AQA Jan 2010 Unit 2 Question 1d

3. Hess's law is used to calculate the enthalpy change in reactions for which it is difficult to determine a value experimentally.

    a. State the meaning of the term enthalpy change.

    b. State Hess's law.

    c. Consider the data in Table Q1 and the scheme of reactions.

Reaction	Enthalpy change/ kJ mol⁻¹
$HCl(g) \rightarrow H^+(aq) + Cl^-(aq)$	−75
$H(g) + Cl(g) \rightarrow HCl(g)$	−432
$H(g) + Cl(g) \rightarrow H^+(g) + Cl^-(g)$	+963

Table Q1

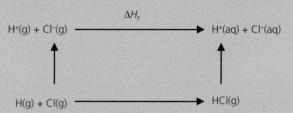

Calculate a value for $\Delta_r H$, using the data in the table, the scheme of reactions and Hess's law.

AQA Jan 2010 Unit 2 Question 2

(Continued)

4. Methanol (CH_3OH) is an important fuel that can be synthesised from carbon dioxide.

   a. Table Q2 shows some standard enthalpies of formation.

	$CO_2(g)$	$H_2(g)$	$CH_3OH(g)$	$H_2O(g)$
ΔH/kJ mol^{-1}	-394	0	-201	-242

   Table Q2

   i. Use these standard enthalpies of formation to calculate a value for the standard enthalpy change of this synthesis.

   $$CO_2(g) + 3H_2(g) \rightarrow CH_3OH(g) + H_2O(g)$$

   ii. State why the standard enthalpy of formation for hydrogen gas is zero.

   AQA Jun 2012 Unit 2 Question 2a

5. a. Write an equation, including state symbols, for the reaction with enthalpy change equal to the standard enthalpy of formation for $CF_4(g)$.

   b. State why the standard enthalpy of formation of fluorine is zero.

   c. CF_4 has a bond angle of $109.5°$. Justify this statement.

   d. Table Q3 gives some values of standard enthalpies of formation ($\Delta_f H^{\ominus}$).

Substance	$F_2(g)$	$CF_4(g)$	$HF(g)$
$\Delta_f H^{\ominus}$/kJ mol^{-1}	0	-680	-269

   Table Q3

   The enthalpy change for the following reaction is -2889 kJ mol^{-1}.

   $$C_2H_6(g) + 7F_2(g) \rightarrow 2CF_4(g) + 6HF(g)$$

   Use this value and the standard enthalpies of formation in the table to calculate the standard enthalpy of formation of $C_2H_6(g)$.

   e. Hydrogen fluoride can be made by the reaction shown.

   $$H_2(g) + F_2(g) \rightarrow 2HF(g)$$

Use the bond enthalpies in Table Q4 to calculate the enthalpy of formation of HF.

Bond		H–H	F–F	H–F
Bond enthalpy/kJ mol^{-1}		436	158	562

Table Q4

AQA Specimen paper 1 2014 Question 3

6. a. Explain the meaning of the term *enthalpy change* of a reaction.

   b. Write the equation for the reaction for which the enthalpy change is the standard enthalpy of formation of the gas nitrous oxide, N_2O.

   c. The equation for the formation of nitrogen trifluoride is given below.

   $$\tfrac{1}{2}N_2(g) + 1\tfrac{1}{2}F_2(g) \rightarrow NF_3(g)$$

   i. Using the mean bond enthalpy values given in Table Q5, calculate a value for the enthalpy of formation of nitrogen trifluoride.

Bond		N–F	N≡N	F–F
Mean bond enthalpy/kJ mol^{-1}		278	945	159

   Table Q5

   ii. A data-book value for the enthalpy of formation of nitrogen trifluoride is -114 kJ mol^{-1}. Give one reason why the answer you have calculated in part c.i. is different from this data-book value.

   d. Some standard enthalpies of formation are given in Table Q6.

Substance	$NH_3(g)$	$F_2(g)$	$NF_3(g)$	$NH_4F(s)$
$\Delta_f H^{\ominus}$/kJ mol^{-1}	-46	0	-114	-467

   Table Q6

   i. State why the enthalpy of formation of fluorine is zero.

   ii. Use these data to calculate the enthalpy change for the following reaction.

   $$4NH_3(g) + 3F_2(g) \rightarrow NF_3(g) + 3NH_4F(s)$$

   AQA Jan 2007 Unit 2 Question 1

(Continued)

7. a. Explain the meaning of the terms *mean bond enthalpy* and *standard enthalpy of formation*.

b. Some mean bond enthalpies are given in Table Q7.

Bond	N–H	N–N	N≡N	H–O	O–O
Mean bond enthalpy/ kJ mol^{-1}	388	163	944	463	146

Table Q7

Use these data to calculate the enthalpy change for the following gas-phase reaction between hydrazine, N_2H_4, and hydrogen peroxide, H_2O_2.

$$N_2H_4 + 2H-O-O-H \rightarrow N \equiv N + 4H-O-H$$

c. Some standard enthalpies of formation are given in Table Q8.

	N_2H_4(g)	H_2O_2(g)	H_2O(g)
$\Delta_f H^\ominus$/kJ mol^{-1}	+75	–133	–242

Table Q8

These data can be used to calculate the enthalpy change for the reaction in part b.

$$N_2H_4(g) + 2H_2O_2(g) \rightarrow N_2(g) + 4H_2O(g)$$

i. State the value of $\Delta_f H^\ominus$ for N_2(g).

ii. Use the $\Delta_f H^\ominus$ values from the table to calculate the enthalpy change for this reaction.

d. Explain why the value obtained in part b. is different from that obtained in part c.ii.

AQA Jun 2005 Unit 2 Question 1

8. a. Define the term *standard enthalpy of formation*, $\Delta_f H^\ominus$.

b. Use the data in Table Q9 to calculate the standard enthalpy of formation of liquid methylbenzene, C_7H_8.

$$7C(s) + 4H_2(g) \rightarrow C_7H_8(l)$$

Substance	C(s)	H_2(g)	C_7H_8(l)
Standard enthalpy of combustion/kJ mol^{-1}	–394	–286	–3909

Table Q9

c. An experiment was carried out to determine a value for the enthalpy of combustion of liquid methylbenzene using the apparatus shown in Figure Q1.

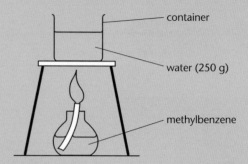

Figure Q1

Burning 2.5 g of methylbenzene caused the temperature of 250 g of water to rise by 60 °C. Use this information to calculate a value for the enthalpy of combustion of methylbenzene, C_7H_8.

The specific heat capacity of water is 4.18 J K^{-1} g^{-1}. (Ignore the heat capacity of the container.)

d. A 25.0 cm^3 sample of 2.00 mol dm^{-3} hydrochloric acid was mixed with 50.0 cm^3 of a 1.00 mol dm^{-3} solution of sodium hydroxide. Both solutions were initially at 18.0 °C.

After mixing, the temperature of the final solution was 26.5 °C.

Use this information to calculate a value for the standard enthalpy change for the following reaction.

$$HCl(aq) + NaOH(aq) \rightarrow NaCl(aq) + H_2O(l) \text{ b}$$

In your calculation, assume that the density of the final solution is 1.00 g cm^{-3} and that its specific heat capacity is the same as that of water. (Ignore the heat capacity of the container.)

e. Give one reason why your answer to part d. has a much smaller experimental error than your answer to part c.

AQA Jun 2006 Unit 2 Question 1

(Continued)

Practical skills question

9. Students are measuring the enthalpy change when propan-1-ol combusts. They are using the apparatus in Figure Q2.

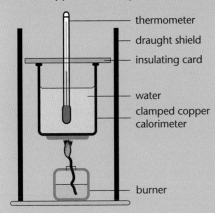

thermometer
draught shield
insulating card
water
clamped copper calorimeter
burner

Figure Q2

a. What measurements do they need to take in order to calculate the enthalpy of combustion of propan-1-ol?

b. What is the advantage of using a copper calorimeter instead of a glass beaker?

c. Explain why the students used insulating card over the calorimeter and draught shields.

d. Give the formula for calculating the energy transferred from the burning propan-1-ol to the water.

e. i. Give the major procedural error (errors due to the method used).

   ii. Give the precision errors (errors due to the apparatus used).

   iii. The result calculated by the students was less than half the value quoted in a data book. Which type of error was the main cause of this?

Stretch and challenge

10. Table Q10 gives standard enthalpy of combustion values for five alcohols.

a. i. Write equations to show the enthalpy of combustion for each alcohol.

ii. Deduce a general equation to represent the combustion of alcohols where n = number of carbon atoms.

iii. As the number of carbon atoms in a molecule of each alcohol in the series increases by one, what happens to the number of carbon dioxide molecules produced and the number of water molecules produced?

iv. Explain the relationship between the number of carbon atoms in a molecule of the alcohol and its enthalpy of combustion.

Multiple choice

11. Which is the correct definition for the standard enthalpy of combustion?

A. The standard enthalpy of combustion is the enthalpy change when one mole of a substance burns completely in oxygen under standard conditions with all reactants and products in their standard state.

B. The standard enthalpy of combustion is the enthalpy change when a substance burns completely in one mole of oxygen under standard conditions with all reactants and products in their standard state.

C. The standard enthalpy of combustion is the enthalpy change when a substance burns completely in oxygen to form one mole of carbon dioxide under standard conditions with all reactants and products in their standard state.

D. The standard enthalpy of combustion is the enthalpy change when one mole of a substance burns completely in oxygen.

Alcohol	methanol	ethanol	propan-1-ol	butan-1-ol	pentan-1-ol
$\Delta_c H^\ominus$/kJ mol^{-1}	− 726.0	− 1367.0	− 2021.0	− 2675.6	− 3328.7

Table Q10

(Continued)

12. Which is the correct definition for the standard enthalpy of formation?

A. The standard enthalpy of formation is the enthalpy change when a substance forms from one mole of reactants under standard conditions. All reactants and products are in their standard state.

B. The standard enthalpy of formation is the enthalpy change when one mole of a substance is formed from its elements under standard conditions. All reactants and products are in their standard state.

C. The standard enthalpy of formation is the enthalpy change when one mole of a substance reacts under standard conditions with all reactants and products in their standard state.

D. The standard enthalpy of formation is the enthalpy change when one mole of a substance is formed in a combustion reaction under standard conditions. All reactants and products are in their standard state.

13. What is bond enthalpy?

A. The energy change in a reaction

B. The energy given out when a particular bond breaks

C. The energy needed to break a particular bond in a molecule in its standard state at 20 °C

D. The energy needed to break a particular bond in a gaseous molecule

14. What are standard conditions?

A. 100 kPa, 0 K, concentration of any solutions: 1 mol dm^{-3}

B. 100 kPa, 298 K, concentration of any solutions: 1 mol dm^{-3}

C. 100 kPa, 298 K, concentration of any solutions: 0.1 mol dm^{-3}

D. 1 kPa, 298 K, concentration of any solutions: 1 mol dm^{-3}

8 KINETICS

PRIOR KNOWLEDGE

You may have learnt about rates of reaction and activation energy. You may have carried out experiments to investigate the effects of changing temperature, concentration or particle size on the rate of reaction. You may also know how catalysts can affect the rate of a reaction.

LEARNING OBJECTIVES

In this chapter, you will reinforce ideas about rates of reaction and activation energy and find out how we can manipulate variables to control chemical reactions. You will find out how the energy of molecules varies at a set temperature and how to show this graphically. You will also find out how catalysts work.

(Specification 3.15.1, 3.1.5.2, 3.15.3, 3.1.5.4, 3.1.5.5)

The public want to buy very fresh produce from supermarkets and other shops. However, the food industry has a problem. When fresh fruits and vegetables are harvested, they start to decay. Microbes on their surfaces start to feed on the fruit and vegetables, which deteriorate. Fresh fruit and vegetables spend three to seven days in retail distribution. A number of strategies have been developed to deal with the problem.

Refrigeration and freezing are two solutions. Lowering the temperature in refrigeration means that the

Figure 1 *Cooking food needs temperatures greater than room temperature to increase the kinetic energy of the food molecules.*

particles in the food have less kinetic energy and move more slowly and the chemical reactions causing decay will slow down. The rate of deterioration decreases and the produce keeps fresh for longer.

Freezing offers a longer-lasting solution. It does not destroy microbes, but it does deactivate them, meaning that food can be frozen for several months. The rate of reaction of the chemical reactions causing decay are almost zero.

The reverse happens when food is cooked. Cooking food requires heating it. If they are not heated, your baked potato will not become soft or your hard-boiled egg hard. Increasing the temperature of the food increases the kinetic energy of molecules in the food and the speed at which they move. The greater the speed of molecules, the more they will collide with one another. These collisions can produce new molecules that give the cooked food different colours, flavours and textures to those of the uncooked food.

8.1 WHAT AFFECTS THE RATE OF A REACTION?

Different reactions proceed at different rates. The decomposition of trinitrotoluene (TNT) and the combustion of fuel in the cylinder of a car engine happen very quickly. The rusting of iron is an example of a reaction with a slow rate of reaction.

The rate of a reaction is the number of moles of reactant used up or the number of moles of product produced in a set time, usually one second.

$$\text{rate} = \frac{\text{change in concentration}}{\text{time}}$$

The units depend on the units in the equation. If a sample of nitroglycerine produces 500 moles of gaseous products in two seconds, it is producing gaseous products at an average rate of 250 mol s^{-1} during the two seconds.

Figure 2 *Fine-powdered magnesium is used in military decoy flares. They are intended to confuse infrared homing missiles. The reaction causing the flare must happen very quickly.*

The reactants in the flare (Figure 2) must have a fast reaction rate.

The rate of a reaction is affected by four factors:

1. **Concentration and pressure.** Higher concentrations of reactants lead to faster reactions. Chlorine atoms in the stratosphere can react with ozone and produce oxygen gas. If the concentration of chlorine atoms in the stratosphere increases, then the rate at which ozone breaks down will increase. Increasing gas pressure effectively increases the concentration of a gas. This leads to an increase in the rate of reaction.

2. **Temperature.** Most reactions go faster at higher temperatures. At higher temperatures, particles have more kinetic energy, so collisions between particles are more frequent. There is more chance of successful collisions.

3. **Particle size of a solid.** Smaller particles have a larger surface area than the same mass of larger particles. The smaller the size, the greater the surface to volume ratio and the faster the reaction. Powdered magnesium reacts faster than larger lumps of magnesium. The magnesium used in flares contains powdered magnesium to produce an intense light from a fast reaction (Figure 2).

4. **Presence of a catalyst.** Catalysts increase the rate of reaction. They take part in the reaction but can be recovered, chemically unchanged, after the reaction has happened.

Before we can understand why each of these affects the reaction rate, we need to look at how individual particles react.

8.2 COLLISION THEORY

For a reaction to occur between particles such as atoms, molecules and ions, the particles must collide. This is the key idea behind **collision theory**.

In a gas, molecules are in constant motion, colliding with one another and with the walls of their container. For example, in one cubic centimetre of a gas at 100 kPa pressure and room temperature, there are about 10^{27} collisions every second. If all of the collisions between the molecules of two reacting gases resulted in a reaction, then the reaction would be over in microseconds. However, not all collisions do result in a reaction. For example, at one extreme, oxygen and nitrogen molecules coexist in the air around us without reacting.

The key to whether two colliding particles react is how much kinetic energy they store.

Figure 3 shows collisions between two types of gas molecules that can react with one another. The speed at which the molecules are moving affects how much kinetic energy they are storing.

The collision energy depends on the speed of two particles and the angle at which they collide. The energy transfer when a head-on collision happens is greater than the energy of a collision at an angle. So head-on collisions are more likely than glancing collisions to result in a reaction.

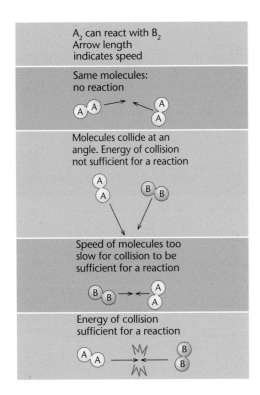

Figure 3 *Colliding molecules must have sufficient kinetic energy to produce an effective collison, in other words, one that leads to reaction.*

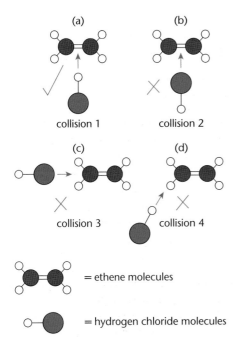

Figure 4 *Ethene reacts with hydrogen chloride to form chloroethane, but only if the two molecules collide with sufficient energy and in the right direction (orientation). (a) shows this orientation.*

Molecules must also have the correct orientation when they collide. This is not the same as whether a collision is head-on rather than glancing. The correct

orientation means that the molecules are aligned correctly – they approach one another at the right angle and from the right direction. For example, when ethene reacts with hydrogen chloride, the hydrogen end of hydrogen chloride must collide with the double bond. If it doesn't, the reaction will not happen (as shown in Figure 4).

If changing the conditions of the reaction changes the number of effective collisions (those resulting in reaction) that occur in a set time, then the rate of reaction changes. So, an increase in concentration increases the number of collisions per second that have the necessary energy to react, and the rate of reaction increases.

An increase in temperature increases the particles' kinetic energy and speed, so they collide more frequently. Increasing the proportion of effective collisions increases the rate of reaction.

8.3 ACTIVATION ENERGY

As you read in Chapter 7, molecules store energy in different ways. The total energy stored is the sum of the molecule's potential energy and its kinetic energy.

When a reaction happens, existing bonds in the reacting molecules are broken and then new bonds form to make the products of reaction. For this to happen, two molecules must collide with sufficient combined kinetic energy. The combined kinetic energy of two molecules is called the **collision energy**. Colliding particles with insufficient energy do not react – they simply bounce off each other.

So, why do some collisions result in reaction while others do not? The answer is **activation energy**. This energy barrier to reaction has the symbol E_A.

The activation energy is the minimum energy needed by colliding molecules for them to react. The energy transferred enables bonds in the reactants to be broken before new ones can form in the products.

For a collision to lead to a reaction, the collision energy must be equal to or greater than the activation energy.

We can show this on an **energy profile**. Figure 5 shows the activation energy in an enthalpy profile diagram for the reactants AB + C. When reactants collide with energy E_A, the bond A–B is broken and the reactants are in the **transition state**. The bond B–C can now form. The enthalpy of the products is lower than the enthalpy of the reactants, and the reaction is exothermic.

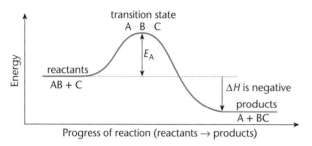

Figure 5 *An enthalpy profile showing activation energy*

When methane reacts with oxygen, carbon dioxide and water are formed. The reaction needs additional energy, such as from a burning match or lighter, to overcome the activation energy barrier (Figure 6). This energy breaks the existing C–H and O = O bonds in the reactants. New bonds can now form to produce carbon dioxide and water. The reaction is exothermic, so the energy given out provides the activation energy for the reaction to continue (Figure 7).

Figure 6 *Methane does not react with oxygen until sufficient energy from a burning match is applied to overcome the activation energy barrier.*

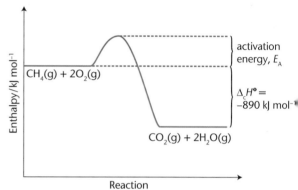

Figure 7 *An enthalpy profile diagram for the reaction between methane and oxygen*

ASSIGNMENT 1: EXPLOSIVES

(PS 1.1, 1.2, 2.2)

Figure A1 *Nitroglycerine or trinitrotoluene?*

Nitroglycerine was discovered in 1847. Its explosive properties were used extensively in the mining industry. However, nitroglycerine is very unstable. Its explosive reaction with oxygen has a very low-activation energy and the smallest shock can cause it to explode. In 1866, a package sent via financial company Wells Fargo exploded in its San Francisco office, killing many and demolishing a whole street. In the same year, Alfred Nobel's brother died in an explosion in their family nitroglycerine factory. A year later, Alfred Nobel developed dynamite. Dynamite is powdered silica rock impregnated with nitroglycerine. This gives it more stability and it can be handled more safely.

Trinitrotoluene (TNT) was first developed in 1863. It is a yellow powder at room temperature and was first used as a dye. Its explosive properties were not realised at first because it is difficult to detonate. It wasn't until 1902 that TNT was first used as an explosive. TNT has a high-activation energy (Figure A2). This means that initially energy has to be supplied to start the reaction, so TNT can be handled relatively safely.

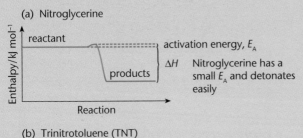

(a) Nitroglycerine

Nitroglycerine has a small E_A and detonates easily

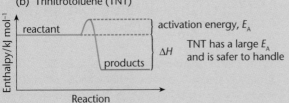

(b) Trinitrotoluene (TNT)

TNT has a large E_A and is safer to handle

Figure A2 *The activation energies for nitroglycerine and TNT*

TNT $(C_7H_5N_3O_6)$ is an effective explosive because it decomposes to give a large volume of gases:

$$2C_7H_5N_3O_6(l) \rightarrow 3N_2(g) + 7CO(g) + 5H_2O(g) + 7C(s)$$

From the equation, two moles of TNT decompose to give 15 moles of gases. To be an explosive, this must happen quickly and TNT decomposes very quickly – it has a very high rate of reaction.

Explosives also need to be chemically or energetically unstable. The best explosives contain their own oxygen. In other words, its molecules are made from atoms of oxygen as well as other atoms (look at the molecular formula for TNT). This means that the reaction does not have to rely on collisions between explosive molecules and oxygen molecules and so can proceed at a faster rate.

Questions

A1. How do the energy profiles for nitroglycerine and TNT show that the reactions are exothermic?

A2. What is activation energy?

A3. How does the energy profile for nitroglycerine show that it has a low activation energy?

A4. How do the E_A values for nitroglycerine and TNT affect the way that they can be handled?

A5. Which needs the larger imput of energy to start the reaction – nitroglycerine or TNT?

A6. Which three factors make TNT a good explosive?

A7. If nitroglycerine had a low rate of reaction, how would this affect

a. its explosive properties

b. its enthalpy profile?

KEY IDEAS

› Kinetics is the study of rates of reaction.

› The rate of a reaction is affected by the concentration of the reactants, the temperature, and the presence of a catalyst.

› Most collisions between reactants do not lead to a reaction.

› For a reaction to occur, reactants must have the correct orientation and collide with a minimum collision energy (the sum of the kinetic energies of the colliding particles) to break bonds in the reactants.

› The activation energy is the minimum energy that colliding molecules must have before they can react. Activation energy has the symbol E_A.

8.4 THE MAXWELL—BOLTZMANN DISTRIBUTION OF ENERGIES

Why do only some molecules have sufficient energy to react?

Temperature is an indication of the amount of kinetic energy stored in moving particles. The higher the temperature, the larger the quantity of kinetic energy stored in them and vice versa. The temperature of a gas sample tells us something about the average amount of energy that each molecule of the gas possesses.

At a given temperature, the molecules in a gas move at different speeds. They do not all have the same energy. At any instant in time, a tiny fraction of the molecules have very high kinetic energies. These move fastest. A tiny fraction have very low kinetic energies. These move very slowly. But most molecules have kinetic energies between the two extremes. There is a distribution of kinetic energies. Temperature reflects the average amount of kinetic energy stored in the molecules.

The distribution of kinetic energies was calculated statistically by the Scottish physicist James Clerk Maxwell in 1859, and was applied to gases by the Austrian physicist Ludwig Eduard Boltzmann in 1871. The result is the **Maxwell—Boltzmann distribution of molecular energies**. We can show this with a distribution curve, as in Figure 8, for molecular energies of molecules in a gas sample at a particular temperature, T_1. Note that all molecules have some energy. The temperature at which they have no kinetic energy is absolute zero, 0 K.

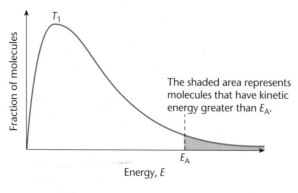

Figure 8 *Energy distribution in the molecules of a gas*

This enables us to describe the relationship between kinetic energy and activation energy graphically (Figure 8). We can use the graph to explain which

proportion of collisions are effective. Effective collisions happen when molecules have kinetic energy greater than the activation energy. The orange shaded part of the graph shows the proportion of molecules that have sufficient energy to react on collision. Providing they are correctly orientated, they will react.

QUESTIONS

Stretch and challenge

1. What causes some molecules in a gas at a set temperature to have much lower kinetic energies than the average kinetic energy?

8.5 THE EFFECT OF TEMPERATURE ON THE RATE OF A REACTION

As the temperature increases, more molecules move at higher speeds and have higher kinetic energies. Figure 9 shows two distribution curves for the energies of molecules. The curve labelled T_1 is at a lower temperature than the curve labelled T_2. For T_1, the proportion of molecules with sufficient energy to react is quite small. However, if the temperature is increased, the energy distribution is 'squashed' and moves to the right, as shown by the curve for T_2. The total area under the curves stays the same because the total number of molecules remains the same. However, at higher temperatures, a greater proportion of the molecules have energies equal to or greater than the activation energy. Therefore, more effective collisions happen that result in a reaction.

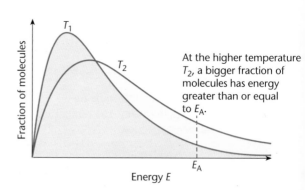

Figure 9 *The effect of increasing the temperature on kinetic energy distribution. The total area under curve T_1 will be the same as the total area under curve T_2.*

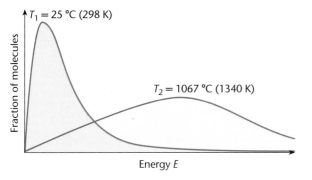

Figure 10 *Energy distribution for a large temperature difference*

Remember, however, that it is the sum of the kinetic energies of colliding particles (the collision energy) that must overcome the activation energy. So, strictly speaking, the Maxwell–Boltzmann distribution shows the proportion of particles that are likely to contribute sufficient energy for the collision energy to be equal to or greater than the activation energy.

The peaks in the graph represent the energy possessed by the highest number of molecules in the gas sample. The number is called the mode. The peak for the T_2 curve has an energy about 4.5 times greater than the peak for the T_1 curve (Figure 10). Also, at the higher temperature the molecules have a wider range of energies than at the lower temperature.

Figure 11 shows the energy distribution curves for gases near room temperature. There is a 10 K difference between the two temperatures. For an activation energy of 50 kJ mol^{-1}, the shaded area under the graph for the gases at 310 K is twice the shaded area for the gases at 300 K. This means that twice as many molecules at 310 K compared

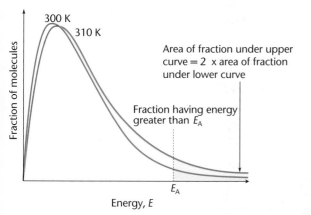

Figure 11 *Distribution curves showing the number of collisions that possess the minimum activation energy at 300 K and 310 K*

with molecules at 300 K have sufficient energy to overcome the activation energy and, therefore, react when they collide.

For reactions that happen in a range of about 300 to 400 K, the rate of reaction doubles with a 10 K temperature rise.

In the reaction between sodium thiosulfate solution and dilute hydrochloric acid, a yellow precipitate of sulfur is formed, as well as sodium chloride, sulfur dioxide and water.

$$Na_2S_2O_3(aq) + 2HCl(aq) \rightarrow$$
$$2NaCl(aq) + S(s) + SO_2(g) + H_2O(l)$$

The reaction is easily monitored because the time taken for the sulfur precipitate to form can be measured. The reaction happens at room temperature, but if the temperature of the reactants is increased, the reaction time is shorter and the rate of reaction increases.

QUESTIONS

2. Hydrogen reacts with both chlorine and iodine. The activation energies for these reactions are:
   $$H_2(g) + Cl_2(g) \rightarrow 2HCl(g) \; E_A = 25 \text{ kJ}$$
   $$H_2(g) + I_2(g) \rightarrow 2HI(g) \; E_A = 157 \text{ kJ}$$
   Predict which reaction will be faster at a particular temperature, and explain your answer.

3. Use the idea of activation energy to explain why food keeps fresher longer in a fridge or a freezer.

4. Explain why milk sours more quickly at room temperature than in a fridge.

8.6 THE EFFECT OF CONCENTRATION ON THE RATE OF A REACTION

Chemists define the rate of a chemical reaction as the rate at which a product is formed or the rate at which a reactant is used up. It is measured as a change in concentration of products or reactants over time. For reactions in solution, the units of rate are mol dm^{-3} s^{-1}, meaning change of concentration per cubic decimetre of solution per second.

Figure 12 shows how the concentration of a reactant decreases during the course of a typical reaction. If the rate of reaction were constant, the graph would be a straight line with a downward slope. Clearly, the rate changes with time, being fastest at the start of the reaction, when the concentration of reactants is greatest. At the end of the reaction we say that the rate is zero. At least one of the reactants must have been used up and, therefore, no more collisions between reactant particles can occur and lead to a reaction.

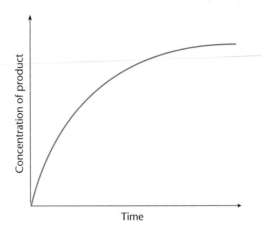

Figure 13 *The increase in the concentration of a product during a reaction*

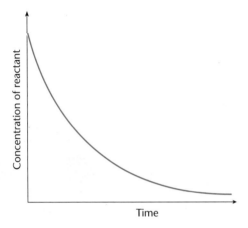

Figure 12 *The decrease in concentration of a reactant during a reaction*

Figure 13 is a graph for the increase in concentration of a product during a typical reaction. Again, the rate of formation of the product is fastest at the start of the reaction, slowing to zero at completion.

These graphs (Figures 12 and 13) are known as **rate curves**. The rate of reaction at any instant is given by the gradient (the slope) of the curve at that instant.

By drawing tangents to the curve, the rate of reaction at any time can be calculated (Figure 14). For example, to find the rate of reaction 50 s after the reaction began, we draw a tangent to the curve at 50 s. The gradient is found by making the tangent the hypotenuse of a triangle and reading off the number of moles and the time. Then:

rate of reaction at 50 s = gradient of tangent at 50 s

$$= \frac{(7.3 - 4.4) \text{ mol}}{(70 - 30) \text{ s}}$$

$$= 0.07 \text{ mol s}^{-1}$$

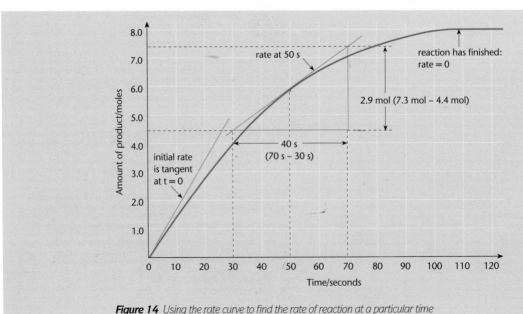

Figure 14 *Using the rate curve to find the rate of reaction at a particular time*

5. What do Figures 12 and 13 tell us about how the rate of reaction changes during the course of a reaction?

Explaining the effect of concentration

As discussed above, in the reaction between sodium thiosulfate solution and dilute hydrochloric acid, a precipitate of sulfur is formed, as well as sodium chloride, sulfur dioxide and water.

$$Na_2S_2O_3(aq) + 2HCl(aq) \rightarrow$$
$$2NaCl(aq) + S(s) + SO_2(g) + H_2O(l)$$

or

$$S_2O_3{}^{2-}(aq) + 2H^+(aq) \rightarrow S(s) + SO_2(g) + H_2O(l)$$

A range of concentrations of hydrochloric acid can be used. Similarly, the concentration of the sodium thiosulfate solution can be varied. If the temperature is kept constant, increasing the concentration of either or both reactants increases the frequency of collisions because there are more molecules in the same volume. Therefore, the number of collisions where the collision energy is equal to or greater than the activation energy increases. Increasing the concentration of reactants increases the rate of reaction.

Concentration in gaseous reactions

We can apply the same ideas about concentrations of solutions to reactions in gases. At a given temperature, increasing the pressure on a mixture of gases reduces the volume and increases the concentration of gas molecules. The molecules are forced closer together. This increases the rate of collisions and, therefore, there will be more effective collisions and the reaction rate increases. Collision theory explains why increasing the pressure of gases increases the rate of reaction.

6. In the Haber process for making ammonia, NH_3, nitrogen and hydrogen are reacted together. Increasing the pressure speeds up the rate of this reaction.

   a. Write a balanced equation for the reaction to make ammonia.

   b. Use the collision theory to explain why increasing the pressure speeds up the rate of this reaction.

Stretch and challenge

7. Use the ideal gas equation, $pV = nRT$, to explain why pressure and concentration are directly proportional to each other. Remember that n/V = concentration.

REQUIRED PRACTICAL ACTIVITY 3: APPARATUS AND TECHNIQUES

(PS 1.2, 2.1, 3.2 and 4.1; AT a, b, k)

Investigate how the rate of a reaction changes with temperature

Doing experiments to investigate rates of chemical changes gives you the opportunity to show that you can

▶ use appropriate apparatus to record volume of liquids, temperature and time

▶ use a water bath or other laboratory apparatus to heat reactants

▶ safely and carefully handle liquids, including irritant and toxic substances.

Apparatus

The rate of change of a reaction may be monitored in various ways.

Change	Examples of apparatus to measure change
increase or decrease in mass	balance of suitable range and precision
gas given off	displacement of water using, for example, an inverted measuring cylinder of burette; gas syringe (especially if gas produced is soluble in water)
colour change (or change in turbidity – cloudiness)	colorimeter; spectrophotometer (more accurate and precise than a colorimeter, but much more expensive; good for colour change, but not for turbidity)
'invisible' change in solution	sample withdrawn and analysed immediately before further reaction can take place; choice of apparatus depends on the quantitative analytical method used (titrations are often used)
pH	pH meter and electronic probe

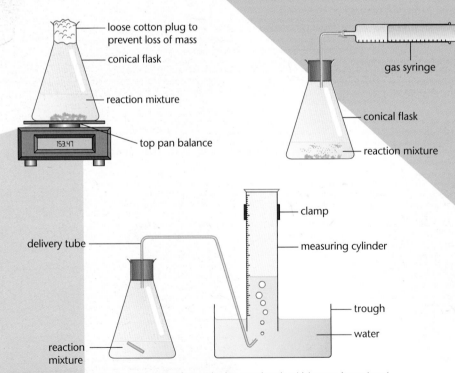

Figure P1 *Three experimental set-ups for monitoring reactions in which a gas is produced*

Figure P2 *A colorimeter may be used to monitor chemical reactions where there is a colour change. The reaction may happen in a colorimeter cuvette (sample tube) or samples may be withdrawn from a reaction mixture and put into a cuvette for measurment.*

As always in analytical work, it is important to use measurement apparatus that is suitable in terms of resolution, precision and range. Data logging allows data gathered electronically to be recorded and graphs of the variable against time to be generated by a computer. This is particularly useful for reactions that happen very quickly or very slowly.

Investigating the effect of temperature on the rate of reaction means that you need apparatus that will maintain the reaction mixture at a fixed temperature (not always easy if reactions are strongly exothermic or endothermic). Options include a simple water bath (beaker of water heated on a tripod and gauze or an electrical hotplate – only suitable if a tolerance of ± a few °C is acceptable) or a thermostated water bath (maintains the temperature at tolerances much less than a manual water bath).

Finally, you need a way of measuring time. Stopwatches and stop clocks are the most common way. Data loggers will do it automatically.

Figure P3 *Stopwatches and stop clocks are the most common way to measure time in the laboratory.*

Technique

The technique used will depend on the nature of the reaction and the analytical method used to measure and monitor changes. Here are some important considerations.

The apparatus needs to be assembled correctly.

Reactants should all be at the temperature being investigated. This may mean, for example, keeping two solutions in a water bath at the required temperature until they have both reached that temperature. At this stage they are ready to mix.

All other variables that may affect the rate of reaction, for example, quantities, concentration and particle size, must be kept constant for all temperatures investigated.

You need to decide how often to take samples and/ or measurements. Most of the time you will be given instructions, but if not then you may need to carry out some trial runs to get an idea of how quickly the reaction goes.

Unless a data logger is being used, you will need to carry out two operations simultaneously – reading the temperature and either taking an instrument reading, for example, colorimeter or balance, or withdrawing a sample.

QUESTIONS

P1. You are asked to investigate the effect of temperature on the reaction between calcium carbonate and hydrochloric acid.

   a. Write an equation for the reaction.

   b. Describe an experimental set-up you may use, explaining your choice.

Stretch and challenge

P2. You are investigating the reaction between magnesium and sulfuric acid. You decided to use 0.20 g of magnesium ribbon, an excess of 1 mol dm^{-3} sulfuric acid and to collect the hydrogen produced by downward displacement of water.

   a. Write an equation for the reaction.

   b. What volume of hydrogen do you calculate will be produced? A_r[Mg] = 24 and 1 mol of any gas at room temperature and pressure occupies about 24 dm^3.

   c. What piece of apparatus will you use to collect the hydrogen? Explain your choice.

P3. Why could the reaction between sulfur dioxide gas and oxygen gas be followed by measuring the changes in pressure with time?

KEY IDEAS

> The distribution of molecular energies in a gas is shown by the Maxwell–Boltzmann distribution curve.

> An increase in temperature increases the number of molecules with sufficient energy to overcome the activation energy and react when they collide. The rate of reaction increases.

> Small temperature increases lead to a large increase in rate. A 10 K temperature increase approximately doubles the rate for reactions near to room temperature.

> Increases in concentration (pressure for gases) increase the reaction rate since more reactants have sufficient energy to overcome the activation energy and react when they collide.

ASSIGNMENT 2: SUPER VEGETABLES

(MS 0.0, 0.1, 0.2, 3.1; PS 1.1, 1.2, 2.3, 3.1)

Photosynthesis is the endothermic reaction between carbon dioxide and water to make glucose and oxygen:

$$6CO_2(g) + 6H_2O(l) \rightarrow C_6H_{12}O_6(aq) + 6O_2(g)$$

$$\Delta H^{\theta} = +2802 \text{ kJ mol}^{-1}$$

In plants some of the glucose reacts with other molecules to produce proteins, fats and other compounds needed for plant growth. Plant growth, therefore, depends on the rate of photosynthesis. This is affected by the concentration of the reactants and by the temperature, as well as by the radiation from the Sun. Since the percentage of carbon dioxide in the air is 0.038, carbon dioxide concentration is often a limiting factor for the photosynthesis reaction.

Scientists have investigated the effects of carbon dioxide concentration on the growth of many plants. The table shows the results of growing spider lilies at two different concentrations of carbon dioxide.

All of the plants were grown for the same time and all other conditions were kept constant.

Just outside Rotterdam, waste carbon dioxide from Shell's oil refinery is now piped into 400 greenhouses used to produce vegetables and flowers. The carbon dioxide would otherwise end up in the atmosphere. The greenhouse operators would otherwise burn natural gas in their greenhouses to increase the carbon dioxide concentration, but most of this would end up in the atmosphere.

Figure A1 *Increasing carbon dioxide levels in a greenhouse can help to promote plant growth.*

Plant material	400 ppm carbon dioxide			700 ppm carbon dioxide		
	Experiment 1	Experiment 2	Mean	Experiment 1	Experiment 2	Mean
Mass of original plants/g	84.78	84.69	84.74	84.75	84.69	84.72
Mass of harvested plants/g	199.80	258.10	229.00	331.00	420.60	375.80

Table A1 *Results of growing spider lilies at two different concentrations (parts per million, ppm) of carbon dioxide*

The arrangement is estimated to prevent the emission of 170 000 tonnes of carbon dioxide into the atmosphere. By tripling the carbon dioxide concentration, productivity for the vegetables and flowers is up by 25%.

Questions

A1. Concentrations of gases present in large amounts in the atmosphere are usually expressed as percentages. Concentrations of gases present in small amounts in the atmosphere are expressed in parts per million (ppm).

   a. What percentage is 400 ppm?

   b. What percentage is 700 ppm?

A2. By what percentage did the original plants increase in mass in a carbon dioxide concentration of 400 ppm? Use mean values.

A3. By what percentage did the original plants increase in mass in a carbon dioxide concentration of 700 ppm?

A4. Explain the effect of increasing the carbon dioxide concentration in terms of collision theory.

Stretch and challenge

A5. Explain the term 'limiting factor'.

A6. Natural gas is methane (CH_4). Write an equation to show how burning natural gas increases carbon dioxide concentration.

A7. Plants stop growing at carbon dioxide concentrations less than 150 ppm. The rate of photosynthesis increases up to an optimum carbon dioxide concentration of 1500 ppm. Draw a sketch graph to show the effect of $CO_2(g)$ concentration on the rate of photosynthesis.

8.7 CATALYSTS

Figure 15 *Cross-section of a new catalytic converter (top) and broken pieces of used catalytic converters (bottom). The catalyst of the catalytic converter is chemically unchanged, but at a microscopic level, the surface of the used catalyst is rougher, showing its involvement in the reaction.*

A **catalyst** is a substance that alters the rate of a chemical reaction, without undergoing any permanent chemical change itself. It may change physically.

Catalysts play a crucial role in our lives. Protein catalysts called enzymes dramatically change the rate of chemical reactions in our bodies.

Catalysts are used in industry to produce many of the chemicals we rely on. The Haber process uses a finely divided iron catalyst to provide a surface for the nitrogen and hydrogen to react on to form ammonia. The catalytic converter in a petrol engine has a large metallic surface on which the carbon monoxide and nitrogen monoxide products from the engine react to form carbon dioxide and nitrogen gases.

A catalyst is not used up in a chemical reaction. You can always recover them after the reaction, chemically unchanged, but possibly changed physically. However, they are involved in the reaction. Only small amounts are needed because the catalyst can be used over and over again, although eventually most catalysts become contaminated and no longer effective.

A catalyst does not affect the amount of product obtained from a chemical reaction, but it does affect the rate at which the product is obtained. It often allows milder conditions to be used (lower temperatures and pressures). These are the reasons for using a catalyst.

How do catalysts alter the rate of a reaction?
We saw in Figure 6 that reactants interact to form a transition state and that energy is needed to reach this state (to overcome the activation energy). The profile shown in Figure 6 is a one-step process.

A catalyst provides an alternative route with at least one intermediate being formed. This intermediate forms between the catalyst and one or more of the reactants. This is a two-step process (Figure 16). Each step has a lower activation energy than the non-catalysed reaction. Therefore, more reactants have sufficient energy to react when they collide and so the rate of reaction increases. The profile of the two-step catalysed reaction has two peaks corresponding to the formation of two transition states, one for the change from reactants to reactive intermediate and the second for the change from reactive intermediate to products.

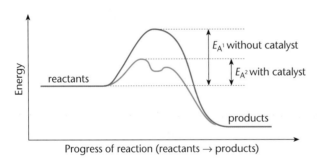

Figure 16 *Energy profiles for a catalysed and uncatalysed reaction*

In a two-step process it is the step with the highest activation energy that controls the rate of reaction. In Figure 16, it is the formation of the reactive intermediate (the first step). This is called the rate-determining step.

The effect of a catalyst can also show this information on the Maxwell–Boltzmann distribution curve, as in Figure 17. When a catalyst is used, more molecules have energy greater than the new activation energy. In terms of collisions, more molecules are going to collide with enough energy to react.

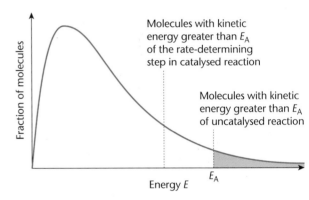

Figure 17 *Maxwell–Boltzmann distribution of energies with and without a catalyst.*

QUESTIONS

8. If leaded petrol is mistakenly used in an unleaded petrol engine, the emissions contain lead compounds. The catalyst becomes poisoned and no longer works. Explain how this happens.

Stretch and challenge

9. What is wrong with the statement: 'A catalyst works by lowering the activation energy'?

KEY IDEAS

- A catalyst alters the rate of reaction without being used up itself.

- A catalyst works by providing an alternative route with a lower activation energy.

ASSIGNMENT 3: NANOCATALYSTS

(MS 4.2, 4.3; PS 1.1, 1.2, 3.1)

There are two types of catalysts: heterogeneous catalysts and homogeneous catalysts. Heterogeneous catalysts are in a different state (or phase) as the reactant(s). The catalyst may be a solid and the reactants liquid or gas. Homogeneous catalysts are in the same state as the reactants. The catalyst and the reactants may both be liquid.

Heterogeneous catalysts

Catalytic converters fitted to car exhausts contain hetereogeneous catalysts. The catalyst is solid platinum, palladium or rhodium and the reactants are the exhaust gases. Hetereogeneous catalysts work by adsorbing the reactants on to the surface of the catalyst. This means that the catalyst forms a weak bond with the reactants. The reactants are held in place long enough for the reaction to occur. The products are then desorbed – they move away from the catalyst.

The larger the surface area of the catalyst, the more reactions between molecules it can catalyse in a given time and the faster the rate of reaction. Nanomaterials have particles no bigger than 100 nm. That is 1×10^{-7} m. (1 nm to a metre is about the same scale of size difference as a marble to Earth.) Nanocatalysts will have a much larger surface area than conventional catalysts. Research into nanocatalysts has developed explosively in the past decade, including research into nanocatalysts for use in catalytic converters in car exhausts.

Figure A1 *A piece of a car's used catalytic converter. The catalyst surface consists of nanoparticles of platinum, palladium and/or rhodium – these consist of just a few atoms – supported on a honeycombed structure made from an inert material.*

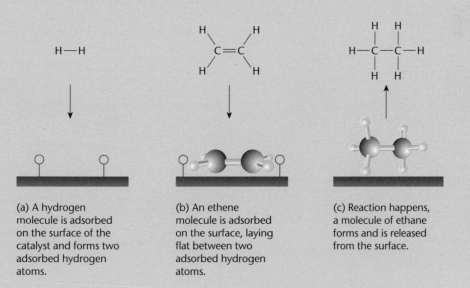

(a) A hydrogen molecule is adsorbed on the surface of the catalyst and forms two adsorbed hydrogen atoms.

(b) An ethene molecule is adsorbed on the surface, laying flat between two adsorbed hydrogen atoms.

(c) Reaction happens, a molecule of ethane forms and is released from the surface.

Figure A2 *The reaction between ethene and hydrogen on the surface of a heterogenous catalyst*

185

Questions

A1. What effect does a catalyst have on the activation energy and rate of reaction?

A2. Draw an energy profile to show how a catalyst works.

A3. Why are nanocatalysts mostly hetereogeneous catalysts?

A4. What are the likely benefits of nanocatalysts to the chemical industry?

A5. The principals of green chemistry include producing little waste, being safe and efficient and non-polluting. How can nanocatalysts help achieve these aims?

Stretch and challenge

A6. Explain how a nanocatalyst affects the rate of reaction.

PRACTICE QUESTIONS

1. Figure Q1 shows the Maxwell–Boltzmann distribution for a sample of gas at a fixed temperature.

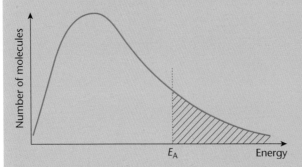

Figure Q1

E_A is the activation energy for the decomposition of this gas.

E_{mp} is the most probable value for the energy of the molecules.

a. On the appropriate axis of this diagram, mark the value of E_{mp} for this distribution. On this diagram, sketch a new distribution for the same sample of gas at a lower temperature.

b. With reference to the Maxwell–Boltzmann distribution, explain why a decrease in temperature decreases the rate of decomposition of this gas.

AQA Jul 2013 Unit 2 Question 3

2. A student carried out an experiment to determine the rate of decomposition of hydrogen peroxide into water and oxygen gas.

The student used 100 cm³ of a 1.0 mol dm⁻³ solution of hydrogen peroxide at 298 K and measured the volume of oxygen collected.

Curve R, in each of Figures Q2, Q3 and Q4, shows how the total volume of oxygen collected changed with time under these conditions.

a. Draw a curve on Figure Q2 to show how the total volume of oxygen collected will change with time if the experiment is repeated at 298 K using 100 cm³ of a 2.0 mol dm⁻³ solution of hydrogen peroxide.

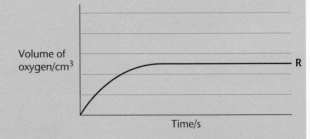

Figure Q2

b. Draw a curve on Figure Q3 to show how the total volume of oxygen collected will change with time if the experiment is repeated at 298 K using 100 cm³ of a 0.4 mol dm⁻³ solution of hydrogen peroxide.

(Continued)

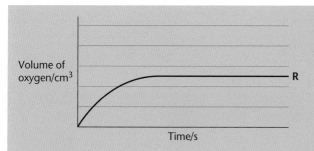

Figure Q3

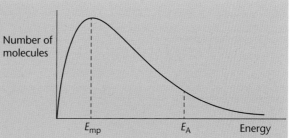

Figure Q5

c. Draw a curve on Figure Q4 to show how the total volume of oxygen collected will change with time if the original experiment is repeated at a temperature higher than 298 K. You should assume that the gas is collected at a temperature of 298 K.

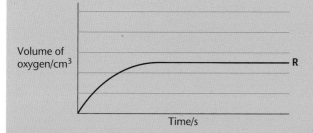

Figure Q4

d. Explain why the slope (gradient) of curve R decreases as time increases.

AQA Jun 2012 Unit 2 Question 1 a,b,c,d

3. The rate of a chemical reaction is influenced by the size of the activation energy. Catalysts are used to increase the rates of chemical reactions but are not used up in the reactions.

a. Give the meaning of the term activation energy.

b. Explain how a catalyst increases the rate of a reaction.

c. Figure Q5 shows the Maxwell–Boltzmann distribution of molecular energies, at a constant temperature, in a gas at the start of a reaction. On this diagram the most probable molecular energy at this temperature is shown by the symbol E_{mp}. The activation energy is shown by the symbol E_A.

To answer the questions c. i.–iv., you should use the words **increases**, **decreases** or **stays the same**. You may use each of these answers once, more than once or not at all.

  i. State how, if at all, the value of the most probable energy (E_{mp}) changes as the total number of molecules is increased at constant temperature.

 ii. State how, if at all, the number of molecules with the most probable energy (E_{mp}) changes as the temperature is decreased without changing the total number of molecules.

iii. State how, if at all, the number of molecules with energy greater than the activation energy (E_A) changes as the temperature is increased without changing the total number of molecules.

 iv. State how, if at all, the area under the molecular energy distribution curve changes as a catalyst is introduced without changing the temperature or the total number of molecules.

AQA Jun 2011 Unit 2 Question 1 a,b,c

4. Figure Q6 shows a Maxwell–Boltzmann distribution for a sample of gas at a fixed temperature. E_A is the activation energy for the decomposition of this gas.

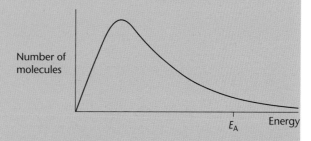

Figure Q6

(Continued)

a. i. On this diagram, sketch the distribution for the same sample of gas at a higher temperature.

ii. With reference to the Maxwell–Boltzmann distribution, explain why an increase in temperature increases the rate of a chemical reaction.

b. Dinitrogen oxide (N_2O) is used as a rocket fuel. The data in the table below show how the activation energy for the decomposition of dinitrogen oxide differs with different catalysts.

$$2N_2O(g) \rightarrow 2N_2(g) + O_2(g)$$

	E_A kJ mol^{-1}
Without a catalyst	245
With a gold catalyst	121
With an iron catalyst	116
With a platinum catalyst	136

i. Use the data in the table to deduce which is the most effective catalyst for this decomposition.

ii. Explain how a catalyst increases the rate of a reaction.

AQA Jan 2011 Unit 2 Question 2

5. The reaction between hydrogen gas and chlorine gas is very slow at room temperature.

$$H_2(g) + Cl_2(g) \rightarrow 2HCl(g)$$

a. Explain how the following factors affect the rate of the above reaction:

i. temperature

ii. pressure

iii. a catalyst.

b. i. Define the term *activation energy*.

ii. Give one reason why the reaction between hydrogen and chlorine is very slow at room temperature.

c. Suggest one reason why a solid catalyst for a gas-phase reaction is often in the form of a powder.

AQA Jan 2006 Unit 2 Question 2

6. a. Figure Q7 is a Maxwell–Boltzmann curve showing the distribution of molecular energies for a sample of gas at a temperature *T*.

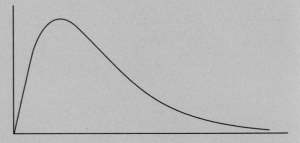

Figure Q7

i. Copy Figure Q7 and label the axes. Label a point on the *x*-axis to represent the activation energy.

ii. What does the area under the curve represent?

iii. State why this curve starts at the origin.

iv. Add a second curve to the diagram to represent the distribution of molecular energies when the temperature rises by 10 K. Label the curve T_1. Shade in the area that represents the number of molecules with sufficient energy to react when the temperature is raised 10 K.

b. The rate of a chemical reaction may be increased by an increase in reactant concentration, by an increase in temperature and by the addition of a catalyst. State which, if any, of these changes involves a different activation energy. Explain your answer.

AQA Jun 2002 Unit Question 5

7. a. Define the term *activation energy* for a reaction.

b. Give the meaning of the term *catalyst*.

c. Explain in general terms how a catalyst works.

d. In an experiment, two moles of gas W reacted completely with solid Y to form one mole of gas Z as shown in the equation below.

$$2W(g) + Y(s) \rightarrow Z(g)$$

(Continued)

The graph in Figure Q8 shows how the concentration of Z varied with time at constant temperature.

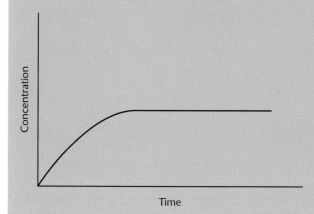

Concentration

Time

Figure Q8

i. Copy the axes from Figure Q8 and sketch a curve to show how the concentration of W would change with time in the same experiment. Label this curve W.

ii. Sketch a curve, on the Figure Q8 axes, to show how the concentration of Z would change with time if the reaction were to be repeated under the same conditions but in the presence of a catalyst. Label this curve Z.

iii. In terms of the behaviour of particles, explain why the rate of this reaction decreases with time.

AQA Jan 2003 Unit 2 Question 3

8. Catalytic converters in car exhausts consist of a honeycomb surface coated in platinum, palladium and rhodium. They convert carbon monoxide and nitrogen oxide (NO) in exhaust fumes to carbon dioxide and nitrogen.

a. Write an equation for the reaction.

b. Explain why the surface of the catalytic converter has a honeycomb structure.

c. Explain why the catalyst in a catalytic converter should last for the life of the car.

d. In terms of activation energy, explain how catalysts increase the rate of a reaction.

e. Draw and label an energy profile to show a reaction with and without a catalyst.

Practical skills question

9. Students are investigating the effect of temperature on the reaction between sodium thiosulfate solution and dilute hydrochloric acid. 40 cm^3 0.15 mol dm^{-3} sodium thiosulfate solution is placed in a conical flask and the solution heated to the required temperature in a water bath. 10 cm^3 of 1.0 mol dm^{-3} hydrochloric acid is placed in a separate conical flask and heated to the required temperature (see Figure Q9). The reactants are mixed and the time taken for a cross on a filter paper placed under the conical flask to disappear is measured. The experiment is repeated at different temperatures.

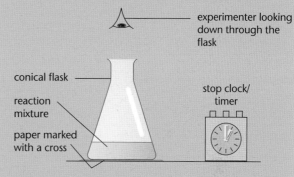

experimenter looking down through the flask

conical flask

reaction mixture

paper marked with a cross

stop clock/ timer

Figure Q9

a. Explain the advantage of using the following equipment.

i. A water bath instead of a Bunsen burner.

ii. A temperature probe instead of a thermometer.

b. Explain why it is necessary to use a separate measuring cylinder for sodium thiosulfate solution and for dilute hydrochloric acid.

c. Give three sources of error in this experiment.

d. Students had read that they could carry out the same experiment on a smaller scale using a porcelain spotting tile.

(Continued)

i. Describe a suitable procedure for using a spotting tile to investigate the effect of temperature on this reaction.

ii. Explain why using a spotting tile gives more accurate results for investigating the effect of concentration on rate than for investigating the effect of temperature on rate.

Stretch and challenge

10. Students are investigating the effect of concentration on the rate of reaction between magnesium ribbon and dilute hydrochloric acid. They are measuring the volume of hydrogen gas given off using the apparatus shown in Figure Q10.

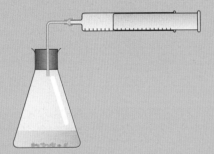

Figure Q10

a. The students were given 1.0 mol dm^{-3} hydrochloric acid. Describe a procedure to make the following dilutions from the given concentration of acid: 0.20 mol dm^{-3}, 0.40 mol dm^{-3}, 0.60 mol dm^{-3} and 0.80 mol dm^{-3}.

b. Describe how the hydrochloric acid dilutions and the 1.0 mol dm^{-3} acid are used to investigate the effect of concentration on the reaction with magnesium ribbon.

Multiple choice

11. What is activation energy?

    A. The maximum energy needed by colliding molecules before they can react

    B. The minimum energy needed by colliding molecules before they can react

    C. The energy given out when new bonds form

    D. The highest peak on the Maxwell–Boltzmann distribution of energies

12. Which of these usually increases the rate of a reaction?

    A. A lower temperature

    B. More dilute reactants

    C. Using a catalyst

    D. Using larger particles of solid reactants.

13. Which of these usually decreases a rate of reaction?

    A. A lower temperature

    B. Using a catalyst

    C. More concentrated reactants

    D. Using smaller particles of solid reactants.

14. How do catalysts alter the rate of a reaction?

    A. By increasing the rate of collisions

    B. By providing an alternative route with a higher activation energy

    C. By increasing the activation energy

    D. By providing an alternative route with a lower activation energy

9 EQUILIBRIA

PRIOR KNOWLEDGE

You may know about reversible reactions and that these reactions can proceed in both directions. You may also know that manufacturing ammonia by the Haber process is an example of a reversible reaction and how the conditions of the reaction are controlled to increase the yield of ammonia.

LEARNING OBJECTIVES

In this chapter, you will build on these ideas and find out how we can use ideas about reversible reactions to predict the effects of changing the conditions of a reaction. You will find out why this is important to processes in the chemical industry. You will learn how to describe a reversible reaction in terms of its equilibrium constant, K_c, and how to carry out calculations involving K_c.

(Specification 3.1.6)

The chemical industry in the UK is the largest and most diverse of all UK manufacturing industries, accounting for about 16% of all UK manufacturing sales. Its products supply diverse industries ranging from aerospace to pharmaceuticals, food and

Figure 1 *One of the active ingredients in some sunscreens is titanium dioxide. It is manufactured either by the sulfate process (starting material ilmenite, $FeTiO_3$) or the chloride process (starting material rutile, TiO_2).*

cosmetics, to name just a few. The UK is the fourth largest producer of chemicals in the EU.

Controlling and manipulating chemical reactions is vital in order for manufacturers to make a good profit. The problem with many chemical reactions the industry uses is that they are reversible. Instead of obtaining the desired product, a mixture of reactants and products is obtained. But, by changing the conditions of the chemical reactions, manufacturing chemists can increase the yield and make the process profitable.

9.1 THE DYNAMIC NATURE OF EQUILIBRIA

Many of the chemical reactions you have met so far seem to proceed in one direction only. If you burn magnesium in air you will see a brilliant white flame as the white powdery solid, magnesium oxide, is produced:

$$2Mg(s) + O_2(g) \rightarrow 2MgO(s)$$

The single-headed arrow in this equation indicates that all the magnesium forms magnesium oxide. The reaction goes to completion. For all intents and purposes the reaction is **irreversible**. However, some reactions that appear to be irreversible are not. They are reversible to a very small extent, but the quantities of reactants present at equilibrium are so low that you can ignore them.

Reversible change and equilibrium

Sparkling mineral water contains dissolved carbon dioxide, $CO_2(aq)$. When the cap is on the bottle, the air space above the mineral water contains carbon dioxide gas, $CO_2(g)$. If the bottle has been undisturbed for some time, the concentration of dissolved carbon dioxide and the concentration of carbon dioxide gas will be constant. An **equilibrium** has been established.

But although nothing appears to be changing, at the microscopic level carbon dioxide gas molecules are constantly moving in the space above the water. When they collide with the surface of the water some bounce back, but those with enough energy enter the water and dissolve to form $CO_2(aq)$. Likewise, the carbon dioxide molecules dissolved in the water are constantly moving. When they reach the surface, some have enough energy to escape and become gaseous, $CO_2(g)$.

A **reversible change** is one that can take place in both directions.

The reversible reaction is:

$$CO_2(g) \rightleftharpoons CO_2(aq)$$

When the overall concentrations of $CO_2(g)$ in the air space and $CO_2(aq)$ in water are not changing, the system is at equilibrium. Carbon dioxide molecules enter and leave the water, but the rate at which carbon dioxide molecules enter the water equals the rate at which they leave it.

This is a **dynamic equilibrium**.

When a reaction is in dynamic equilibrium, the forward and reverse reactions of reactants and products occur at the same rate and the concentrations are constant. The idea of reversible reactions in **dynamic equilibrium** is an important one.

Reversible chemical reactions

Ammonia manufacture
Ammonia is produced from hydrogen and nitrogen gases by the Haber process. The reaction is reversible. At the same time as ammonia is being produced, ammonia is also decomposing to produce nitrogen and hydrogen.

The forward reaction is: $3H_2(g) + N_2(g) \rightarrow 2NH_3(g)$
The reverse reaction is: $2NH_3(g) \rightarrow 3H_2(g) + N_2(g)$
You can write the reversible reaction as:

$$3H_2(g) + N_2(g) \rightleftharpoons 2NH_3(g)$$

Sparkling mineral water
In sparkling mineral water, carbon dioxide molecules are in solution, $CO_2(aq)$. They react with water molecules to produce hydrogen carbonate ions in solution, $HCO_3^-(aq)$, and hydrogen ions in solution, $H^+(aq)$. These ions react with one another to reform carbon dioxide molecules and water molecules.

The forward reaction is:
$$CO_2(aq) + H_2O(l) \rightarrow HCO_3^-(aq) + H^+(aq)$$
The reverse reaction is:
$$HCO_3^-(aq) + H^+(aq) \rightarrow CO_2(aq) + H_2O(l)$$
You can write the reversible reaction as:
$$CO_2(aq) + H_2O(l) \rightleftharpoons HCO_3^-(aq) + H^+(aq)$$

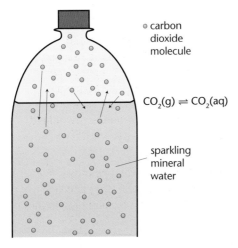

- carbon dioxide molecule

$CO_2(g) \rightleftharpoons CO_2(aq)$

sparkling mineral water

Figure 2 *The dynamic nature of the equilibrium in a bottle of sparkling mineral water.*

Haemoglobin

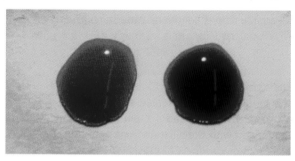

Figure 3 *Oxygenated blood (left) contains predominantly oxyhaemoglobin, whereas deoxygenated blood (right) contains predominantly haemoglobin.*

Many biochemical reactions are reversible. Haemoglobin in red blood cells reacts with oxygen to form oxyhaemoglobin (Figure 3). The reaction is reversible.

In the capillaries of your lungs, oxyhaemoglobin is produced:

haemoglobin + oxygen → oxyhaemoglobin

In your body cells, where conditions are different, the reverse reaction happens:

oxyhaemoglobin → haemoglobin + oxygen

The equation for the reversible reaction is written:

haemoglobin + oxygen ⇌ oxyhaemoglobin

QUESTIONS

1. Some nail polish removers contain propanone, $CH_3COCH_3(l)$. Write an equation for the dynamic equilibrium that exists in an undisturbed bottle of nail polish remover.

Monitoring forward and reverse reactions

Hydrogen and iodine react to produce hydrogen iodide in a reversible reaction:

$H_2(g) + I_2(g) \rightleftharpoons 2HI(g)$

Here is a description of two experiments, one starting with hydrogen iodide (experiment A) and the others starting with hydrogen and iodine (experiment B). In both cases:

❯ the gases are sealed in a glass bulb and kept at a constant temperature of 445 °C

❯ the reaction is stopped at regular intervals by cooling the bulb rapidly and the contents analysed.

The results are shown in Figure 4. The y-axis represents the amount of HI present in the reaction mixture and the x-axis is the time from mixing the two gases.

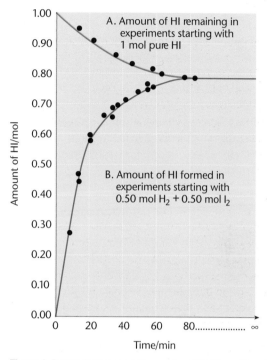

Figure 4 *Results for the reversible reaction of HI. The red line shows Experiment A, while the blue line shows Experiment B.*

Experiment A: Initially, 1 mol hydrogen iodide was put in the flask. After 84 minutes 0.78 mol hydrogen iodide was present. This number remained the same no matter how much longer the flask was left.

Experiment B: Initially, 0.5 mol hydrogen and 0.5 mol iodine were put in the flask. If the reaction went to completion, there would be 1 mol hydrogen iodide in the flask and zero moles of hydrogen and iodine. However, after 80 minutes there were 0.79 mol hydrogen iodide and this number remained the same no matter how much longer the flask was left.

So, regardless of whether the starting point was 1 mol HI(g) (Experiment A) or 0.5 mol $H_2(g)$ and 0.5 mol $I_2(g)$ (Experiment B) the same equilibrium mixture was produced:

0.78 mol HI(g)

0.11 mol $H_2(g)$

0.11 mol $I_2(g)$

193

QUESTIONS

2. **a.** Explain why rapid cooling apparently stops the reactions in Experiments A and B.

   b. Imagine that, instead of being reversible, the reaction goes to completion. If the starting mixture contains two moles of H_2 and one mole of I_2, how many moles of HI would be formed?

   c. Describe the shapes of the curves in Figure 4 in terms of reaction rates. (See Chapter 8 for information about rates of reaction.)

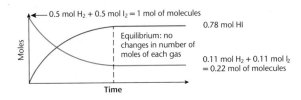

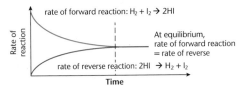

Figure 5 *Changes in amounts of reactants and products and rates of forward and reverse reactions for $H_2(g) + I_2(g) \rightleftharpoons 2HI(g)$. The vertical dotted lines show where equilibrium is reached.*

Figure 5a shows how the amount of hydrogen, iodine and hydrogen iodide change during Experiment B. The equilibrium mixture contains 0.78 mol HI, 0.11 mol H_2 and 0.11 mol I_2, all in the gaseous state.

Figure 5b shows how the rates of forward and reverse reactions change. Initially the rate of the forward reaction ($H_2 + I_2 \rightarrow 2HI$) is at its fastest because the reactants are at their highest concentrations. The forward rate slows down as reactant concentrations decrease. As the concentration of HI increases, the backward rate increases. At equilibrium the rates of forward and reverse reactions are equal.

Once equilibrium is established, the concentrations of the gases do not change. Both the forward and reverse reactions continue at the same rate. A dynamic equilibrium is established.

The importance of a closed system

The glass bulbs used for Experiments A and B in the previous section were sealed so that no gas could escape. This is a **closed system**. No substances are exchanged with the surroundings and, therefore, no reactants or products are lost.

An equilibrium can be maintained only when the equilibrium concentrations of reactants and products are kept constant. If a reactant is allowed to escape, its concentration decreases and the rate of the forward reaction decreases. If a product is allowed to escape, its concentration decreases and the rate of the reverse reaction decreases. In both cases, the original state of equilibrium no longer exists.

The equilibrium mixture of hydrogen, iodine and hydrogen iodide gases is an example of a **homogeneous** equilibrium. This means that all the substances are in the same phase, in this case the gas phase. Equilibria between substances in different phases are **heterogeneous equilibria.**

The thermal decomposition of calcium carbonate may appear to be irreversible. This is because heating is usually carried out with an open system and the carbon dioxide escapes. In fact, the reaction is reversible. However, equilibrium is only established if the system is closed and the carbon dioxide does not escape.

$$CaCO_3(s) \rightleftharpoons CaO(s) + CO_2(g)$$

QUESTIONS

3. What type of equilibrium is established when calcium carbonate is heated in a closed system?

4. State in which of the following there is a dynamic equilibrium:

   a. An unopened bottle of fizzy lemonade.

   b. A bottle of fizzy lemonade with the top left off.

   c. A sealed bag containing wet washing.

   d. Wet washing hanging on a washing line on a windy day.

9.2 THE EQUILIBRIUM CONSTANT, K_c

Dynamic equilibrium can be established in homogeneous reactions and heterogeneous reactions. For the moment we will just look at homogeneous equilibria.

If a reversible reaction is allowed to reach equilibrium and the concentration of all the substances measured, these values can be combined into an expression to give an **equilibrium constant**, or K_c. The subscipt 'c' is to indicate concentration.

In the reversible reaction

$$aA + bB \rightleftharpoons cC + dD$$

a, b, c and d are the number of moles of reactants (A and B) and products (C and D), as shown in the equation.

The equation for the equilibrium constant is:

$$K_c = \frac{[C]^c [D]^d}{[A]^a [B]^b}$$

Note that the right-hand side of the equation goes on the top line, and the left-hand side goes on the bottom line. The square brackets indicate concentrations, in $mol\ dm^{-3}$, in the equilibrium mixture.

The value of K_c for a chemical reaction depends on temperature, but for a given temperature it is a constant. It is important to give the temperature for each K_c value.

For example, sulfur dioxide gas reacts with oxygen gas to produce sulfur trioxide gas. All substances are gases. The reaction is:
$$2SO_2(g) + O_2(g) \rightleftharpoons 2SO_3(g)$$

The equation for the equilibrium constant is:

$$K_c = \frac{[SO_3]^2}{[SO_2]^2 [O_2]}$$

The reaction to make ammonia is:

$$3H_2(g) + N_2(g) \rightleftharpoons 2NH_3(g)$$

The equation for the equilibrium constant is:

$$K_c = \frac{[NH_3]^2}{[H_2]^3 [N_2]}$$

Although both examples involve gases, the concentration is still measured in $mol\ dm^{-3}$.

There is another constant, K_p, that uses partial pressures rather than concentrations and is used for gaseous reactions.

The units of K_c are found by substituting the concentration units in the equation for K_c. In the example to produce sulfur trioxide, the units of K_c are:

$$\frac{(mol\ dm^{-3})^2}{(mol\ dm^{-3})^2 (mol\ dm^{-3})}$$

After cancelling, the units for K_c for the equilibrium reaction are:

$$\frac{1}{(mol\ dm^{-3})}$$

or $mol^{-1}\ dm^3$.

Different equilibrium reactions have different units and each needs to be calculated.

Calculations involving K_c

You may be asked to calculate a value for K_c from concentrations at equilibrium, or to calculate concentrations at equilibrium given the initial concentrations of the reactants and K_c.

Worked example 1

(MS 2.1, 2.3 and 2.4)

Calculate K_c for the reaction between ethanoic acid and ethanol.

Step 1 The reaction is: $CH_3COOH(l) + CH_3CH_2OH(l) \rightleftharpoons CH_3COOCH_2CH_3(l) + H_2O(l)$

$$K_c = \frac{[CH_3COOCH_2CH_3][H_2O]}{[CH_3COOH][CH_3CH_2OH]}$$

Step 2 The concentration of each reactant and product at equilibrium at 298 K is shown in Table 1.

Reagent	Concentration at equilibrium/mol dm^{-3}
CH_3COOH	0.33
C_2H_5OH	0.33
$CH_3COOC_2H_5$	0.67
H_2O	0.67

Table 1

$$K_c = \frac{0.67 \times 0.67}{0.33 \times 0.33} = 4.1$$

For this reaction, K_c has no units because they cancel each other out.

Worked example 2

(MS 2.1, 2.3 and 2.4)

Calculate the concentration of each reagent at equilibrium in the reaction:

$CH_3COOH(l) + CH_3CH_2OH(l) \rightleftharpoons$
$\qquad CH_3COOCH_2CH_3(l) + H_2O(l)$

if $K_c = 4.1$ and the concentration of each reactant at the start of the reaction is 1.0 mol dm^{-3}.

Step 1 From the equation, equal moles of $CH_3COOCH_2CH_3$ and H_2O are present at equilibrium. We use x to represent this. Then, the concentrations at equilibrium are as shown in Table 2.

Since $K_c = \dfrac{[CH_3COOCH_2CH_3][H_2O]}{[CH_3COOH][CH_3CH_2OH]}$

$4.1 = \dfrac{(x)(x)}{(1.0 - x)(1.0 - x)} = \dfrac{x^2}{(1.0 - x)^2}$

Step 2 Taking the square root of each side:

$2.0 = \dfrac{x}{(1.0 - x)}$

$2.0 - 2.0x = x$

$2.0 = 3x$, so $x = 0.67$ mol dm^{-3}

	CH_3COOH	C_2H_5OH	$CH_3COOC_2H_5$	H_2O
Concentration at the start of the reaction / mol dm^{-3}	1.0	1.0	0	0
Concentration at equilibrium / mol dm^{-3}	1.0 − x	1.0 − x	x	x

Table 2

What does K_c tell us?

If the value of K_c is 1, or near to 1, the concentrations of reactants and products are similar and the position of the equilibrium lies near the centre. If K_c is greater than 1, the position of the equilibrium lies towards the products. Similarly, if K_c is less than 1, the position of the equilibrium lies towards the reactants.

A useful rule of thumb is:

> if K_c is greater than 10^5, the reaction almost goes to completion

> if K_c is less than 10^{-5}, the reaction hardly happens.

Stretch and challenge

b. When PCl_5 is heated in a closed container, an equilibrium is established: $PCl_5(g) \rightleftharpoons PCl_3(g) + Cl_2(g)$.

The following concentrations in mol dm^{-3} were measured at equilibrium:

PCl_5	0.077
PCl_3	0.123
Cl_2	0.123

Calculate the value of K_c.

QUESTIONS

5. Write expressions for the equilibrium constants for:

   a. the reversible reaction between hydrogen and iodine gas to produce hydrogen iodide

   b. the reaction shown in the equation:
   $2NO(g) + O_2(g) \rightleftharpoons 2NO_2(g)$

6. a. Calculate the value of K_c for the reaction $H_2(g) + I_2(g) \rightleftharpoons 2HI(g)$ if, at equilibrium, there is 0.220 mol hydrogen gas, 0.220 mol iodine gas and 0.780 mol hydrogen iodide gas.

KEY IDEAS

> Many chemical reactions are reversible.

> In a closed system, an equilibrium is established for a reversible reaction.

> Equilibria are dynamic because the forward and reverse reactions continue to proceed at the same rate.

> When a reaction is at equilibrium, the concentrations of reactants and products are constant.

> In a homogeneous equilibrium, the concentration of the substances can be combined to give an expression for the equilibrium constant.

9.3 CHANGES THAT AFFECT A SYSTEM IN A HOMOGENEOUS EQUILIBRIUM

The **position of equilibrium** in a reaction tells you the proportion of products to reactants. If product concentrations at equilibrium are high and reactant concentrations are low, we say that the position of the equilibrium lies to the right. If reactant concentrations are high and product concentrations are low at equilibrium, we say the position of equilibrium lies to the left. These descriptions must refer to the ratio of quantities indicated in a chemical equation.

The position of equilibrium can be changed by changes in temperature, pressure and concentration. It is not affected by catalysts.

Le Chatelier's principle

In 1884, the French chemist Henri Le Chatelier formulated his famous principle, which can be simply stated as:

The position of the equilibrium of a system moves to minimise the effect of any imposed change in conditions.

The term 'imposed change' means any change in temperature or pressure or concentration made to the system. Le Chatelier's principle is a useful tool that can be applied to any reaction that reaches equilibrium, and is used to predict the effect of changes to conditions.

The process to manufacture ammonia from nitrogen and hydrogen was developed by Fritz Haber. Haber applied Le Chatelier's principle to the process.

Figure 6 *Ammonia produced in the Haber process is used to make artificial fertilisers. Without these, we could only feed 50% of the world's population.*

The effect of temperature on equilibria

When nitrogen gas and hydrogen gas are heated to 1000 °C at atmospheric pressure, the position of equilibrium lies towards the reactants, that is, to the left. The reaction mixture at equilibrium contains less than 1% of ammonia. The equation is:

$$N_2(g) + 3H_2(g) \rightleftharpoons 2NH_3(g)$$
$$\Delta H^\ominus = -92 \text{ kJ mol}^{-1}$$

You can predict the effect of increasing the temperature on the position of the equilibrium for this reaction. Le Chatelier's principle tells you that, in any reaction, if you raise the temperature, and so transfer energy to the reaction mixture, the equilibrium shifts to absorb energy. The negative ΔH above tells you that the reaction is exothermic in the forward direction. Therefore, raising the temperature shifts the equilibrium in the direction of the endothermic change so that the system can absorb the energy. This means the equilibrium position moves to the left, increasing the concentrations of nitrogen and hydrogen gas, and reducing the concentration of ammonia.

Since the concentrations of the reactants increase and the concentrations of the products decrease, the value of K_c changes. For the Haber process,

$$K_c = \frac{[NH_3]^2}{[H_2]^3 [N_2]}$$

Increasing the temperature means that the value of $[H_2]^3[N_2]$ is now larger, and the value of $[NH_3]^2$ smaller, so the value of K_c decreases. Decreasing the temperature has the opposite effect and K_c increases.

For reversible reactions at equilibrium that are endothermic in the forward direction, increasing the temperature increases the value of K_c.

But for the Haber process, raising the temperature reduces the equilibrium yield of ammonia (Figure 7). However, Haber needed a high temperature to increase the rate of a reaction and bring it to equilibrium faster, even if the yield was less good.

Here is an example that may be studied easily in the laboratory. A solution of cobalt chloride in water is pink due to $Co(H_2O)_6^{2+}$ ions. If concentrated hydrochloric acid is added, a violet-coloured solution forms. Add more concentrated hydrochloric acid and the solution becomes blue because it contains mainly $CoCl_4^{2-}$ ions.

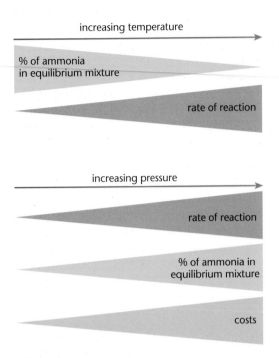

Figure 7 *Effects of changing conditions of the Haber process*

The violet-coloured solution contains a mixture of $[Co(H_2O)_6]^{2+}$ ions and $CoCl_4^{2-}$ ions in equilibrium:

$$[Co(H_2O)_6]^{2+}(aq) + 4Cl^-(aq) \rightleftharpoons CoCl_4^{2-}(aq) + 6H_2O(l)$$

pink   blue

The forward reaction is endothermic and the reverse reaction is exothermic.

Figure 8 *The effect of temperature on the position of equilibrium of $[Co(H_2O)_6]^{2+}(aq) + 4Cl^-(aq) \rightleftharpoons CoCl_4^{2-}(aq) + 6H_2O(l)$. A blue solution forms at high temperatures and a pink solution forms at low temperatures.*

If a test tube containing the violet-coloured solution is put into:

- boiling water, the solution turns blue
- crushed ice, the solution turns pink.

Taking it out and allowing the solution to get back to room temperature produces a violet-coloured solution in each case.

Increasing the temperature favours the forward (endothermic) reaction and moves the position of equilibrium to the right. Decreasing the temperature favours the reverse (exothermic) reaction and moves the position of equilibrium to the left.

QUESTIONS

7. For these equilibria, predict the effect of increasing the temperature on the equilibrium position and on the value of K_c, and give your reasons:

   a. $N_2(g) + O_2(g) \rightleftharpoons 2NO(g)$

   $\Delta H^\ominus = +90 \text{ kJ mol}^{-1}$

   b. $2SO_2(g) + O_2(g) \rightleftharpoons 2SO_3(g)$

   $\Delta H^\ominus = -98 \text{ kJ mol}^{-1}$

   c. $C_2H_4(g) + H_2O(g) \rightleftharpoons C_2H_5OH(g)$

   $\Delta H^\ominus = -46 \text{ kJ mol}^{-1}$

The effect of pressure on equilibria

Haber used Le Chatelier's principle to predict that, if he increased the pressure at a fixed temperature, the equilibrium would shift to the right and give a better yield of ammonia. He wrote:

'It was clear that a change to using maximum pressure would be an advantage. It would improve the position of equilibrium and probably the rate of reaction as well.'

Why did Haber think pressure would move the equilibrium to the right?

The equation shows that there are fewer moles of gaseous product on the right of the equation than moles of gaseous reactant on the left. In a gas, pressure depends on the number of molecules colliding with the walls of the gas's container and exerting a force on them. When Haber increased the pressure, the equilibrium shifted to the right, as the reaction opposed the change imposed on it. More ammonia was formed, while the pressure remained constant (Figure 7).

By comparison, in the reversible reaction

$$H_2(g) + I_2(g) \rightleftharpoons 2HI(g)$$

the number of molecules is the same on both sides, so an increase in pressure has no effect on the equilibrium position.

QUESTIONS

8. Use Le Chatelier's principle to predict the effect of raising the pressure on each of these reactions at equilibrium:

   a. $CH_4(g) + H_2O(g) \rightleftharpoons CO(g) + 3H_2(g)$

   b. $CO(g) + H_2O(g) \rightleftharpoons CO_2(g) + H_2(g)$

   c. $2NO_2(g) \rightleftharpoons N_2O_4(g)$

   d. $2NO(g) + O_2(g) \rightleftharpoons 2NO_2(g)$

   e. $C_2H_6(g) \rightleftharpoons C_2H_4(g) + H_2(g)$

The effect of concentration on equilibria

The position of the equilibrium for reactions in the liquid phase is not affected by pressure, but it is affected by changes in concentration.

Ethyl ethanoate, $CH_3COOC_2H_5(l)$, is used as nail polish remover for acrylic nails. It is an ester made in a reversible reaction from ethanol, $C_2H_5OH(l)$, and ethanoic acid, $CH_3COOH(l)$:

$$C_2H_5OH(l) + CH_3COOH(l) \rightleftharpoons CH_3COOC_2H_5(l) + H_2O(l)$$

If the concentration of ethanol is increased, then Le Chatelier's principle states that the position of the equilibrium will shift to minimise the imposed change. The position of the equilibrium moves to the right to minimise the imposed change and re-establish the equilibrium. The amount of product made increases. This does not change the value of K_c.

$$K_c = \frac{[CH_3COOCH_2CH_3][H_2O]}{[CH_3COOH][CH_3CH_2OH]}$$

There is a larger value for the concentration of ethanol on the lower line of the expression, but also a larger value for the concentration of the products on the top line, and K_c does not change. When the concentration of a reagent is changed, the concentrations of the others adjust to maintain the value of K_c.

Similarly, if ethyl ethanoate is added to the system, then the position of the equilibrium shifts to the left to decrease the concentration of ethyl ethanoate, but the value of K_c does not change.

The effect of a catalyst

A catalyst increases the rate of reaction by providing an alternative route with a lower activation energy. It does not alter the position of the equilibrium. The concentrations of the products and reactants remain exactly the same and the value of K_c does not change, but equilibrium is reached more quickly. The effect of adding a catalyst is to increase the rates of both forward and backward reactions equally, so that it takes a shorter time to reach equilibrium. Fritz Haber carried out about 6500 experiments to find the best catalyst for the Haber process. Today, finely divided iron is used.

ASSIGNMENT 1: FRITZ HABER – HERO AND VILLAIN

(PS 1.1, 1.2, 2.3, 3.1)

Today we are concerned about dwindling oil reserves and developing alternative energy resources. 100 years ago, however, the concern was a shortage of nitrogen compounds, which were needed for fertilisers and to make explosives. The sources of these were natural and the main ones were bird droppings from Peru (where they were metres deep) and sodium nitrate, found in the Chilean deserts. These had to be transported by boat and

the sources were running out fast at a time when agriculture was expanding in northern Europe and the demands for nitroglycerine and trinitrotoluene (TNT) were high.

At the start of the 20th century, Germany was preparing for war. Britain controlled the seas and was likely to threaten the import of Chilean nitrate. In order to meet its demand for nitrogen compounds, Germany had to find a way of making them from nitrogen gas in the air, but nitrogen gas is very unreactive. Many scientists had previously tried to convert nitrogen gas into a form that could be used for nitrate production, but had failed.

Figure A1 *This is the Bahia Blanca factory in Argentina. It manufactures ammonia using the Haber process. The raw materials are nitrogen from the air and hydrogen, made by reacting water with methane (natural gas) in steam reforming.*

Fritz Haber made the breakthrough in 1909. He synthesised 100 g of ammonia from nitrogen and hydrogen gases, using the reaction:

$$N_2(g) + 3H_2(g) \rightleftharpoons 2NH_3(g)$$

Fritz Haber was a hero to the agricultural world.

In contrast, Haber contributed to the development of chlorine gas and other lethal gases as chemical weapons. He personally oversaw their deployment on the battlefield. Being of Jewish origin, Haber was forced to flee Nazi Germany in 1933, despite his contributions. Many of his family were killed in the gas chambers of the Nazi concentration camps by the very gases he had developed.

Questions

A1. What is the effect on yeild of increasing the pressure?

A2. What is the effect on yield of increasing the temperature?

A3. The standard enthalpy change of the reaction is -92 kJ mol^{-1}. How are the experimental results for the effects of temperature in agreement with Le Chatelier's principle?

A4. How are the experimental results for the effects of pressure in agreement with Le Chatelier's principle?

A5. Modern Haber process plants tend to operate at pressures between 5000 and 20 000 kPa. Why are higher pressures not used?

A6. The temperature usually used is 700 K. Why is a lower temperature not used?

A7. What effect does the finely divided iron catalyst have on the process?

A8. The boiling points of the gases involved in the Haber process are shown in Table A1. How can ammonia be separated from the unreacted nitrogen and hydrogen?

Gas	Boiling point/K
N_2	77
H_2	20
NH_3	240

Table A1 *Boiling points of nitrogen, hydrogen and ammonia*

A9. Explain why the reaction conditions of 15 000 kPa and 700 K are chosen.

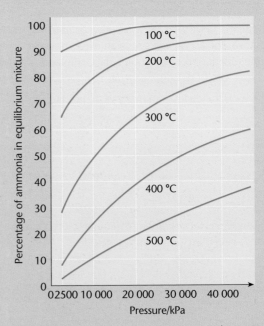

Figure A2 *The relationship between pressure and percentage ammonia in the equilibrium mixture*

ASSIGNMENT 2: A LUNAR SOLUTION

(PS 1.1, 1.2, 2.3)

Figure A1 *Impression of a lunar base.*

A permanently manned lunar base will need a power supply. Transporting any material to the moon will be expensive. The use of solar energy is one solution, but a lunar day lasts for 14 Earth days and a lunar night is equally long. So the challenge is to store enough solar energy to last 14 Earth days. Analysis of lunar soil samples has shown a calcium oxide content between 9% and 16%. One suggestion is to use the reaction of calcium oxide with water to store solar energy.

$$CaO(s) + H_2O(l) \rightleftharpoons Ca(OH)_2(s)$$
$$\Delta H^\ominus = -94.6 \text{ kJ mol}^{-1}$$

Calcium hydroxide decomposes at 823 K. Solar energy could be used to heat the calcium hydroxide during the day.

Questions

A1. What effect will heating calcium hydroxide have on

   a. the position of the equilibrium

   b. the concentrations of calcium oxide and water at equilibrium?

A2. During the lunar night, the temperature falls.

   a. What effect will this have on the position of equilibrium?

   b. What effect will a lower temperature have on the concentration of calcium hydroxide at equilibrium?

   c. Why will the reaction release energy when the temperature falls?

ASSIGNMENT 3: AN EQUILIBRIUM CASE STUDY

(MS 4.1, 4.2; PS 1.1, 1.2, 2.3)

Figure A1 *Water hyacinths are considered to be an invasive species outside their natural habitat of the Amazon basin. They are one of the fastest growing plants known. Herbicides containing copper(II) sulfate are one way to control this invasive plant.*

Figure A2 *Copper sulfate solution is blue (left). Adding a few drops of concentrated hydrochloric acid to it changes the colour to green. Adding water to the green solution changes it back to blue.*

Copper(II) sulfate crystals, $CuSO_4.5H_2O$, dissolve in water to give a blue solution containing $[Cu(H_2O)_6]^{2+}$ ions and SO_4^{2-} ions. In a $[Cu(H_2O)_6]^{2+}$ ion, six water molecules form coordinate bonds with a Cu^{2+} ion, using one of their lone pair of electrons.

If chloride ions are added, the following equilibrium is set up:

$$[Cu(H_2O)_6]^{2+}(aq) + 4Cl^-(aq) \rightleftharpoons CuCl_4^{2-}(aq) + 6H_2O(l)$$

pale blue   pale yellow

The six water molecules can be replaced by four chloride ions in a reversible reaction.

The equilibrium may be investigated using a simple test tube experiment.

About 5 cm^3 of copper sulfate solution are put into a test tube. A few drops of concentrated hydrochloric acid are added and the solution becomes green. Now water is added to the contents of the test tube. The blue colour is restored, though a little paler than before.

Questions

A1. Explain the colour change when concentrated hydrochloric acid is added to copper sulfate solution.

A2. Explain the colour change when water is then added to the solution of copper sulfate with concentrated hydrochloric acid.

A3. Is this a heterogeneous or homogeneous equilibrium?

A4. How do these experiments support Le Chatelier's principle?

A5. Explain why the copper sulfate solution does not become yellow.

Stretch and challenge

A6. What are the shapes of these ions?

    a. $Cu(H_2O)_6^{2+}$

    b. $CuCl_4^{2-}$

KEY IDEAS

› Le Chatelier's principle states that the position of the equilibrium of a system changes to minimise the effect of any imposed change in conditions.

› Increasing the temperature shifts the position of the equilibrium in the direction of the endothermic reaction and changes the value of K_c.

› Decreasing the temperature shifts the position of the equilibrium in the direction of the exothermic reaction and changes the value of K_c.

› For a gaseous system, increasing the pressure shifts the position of the equilibrium in the direction with fewer moles of gas. This reduces the pressure.

› For a gaseous system, decreasing the pressure shifts the position of the equilibrium in the direction with more moles of gas. This increases the pressure.

› Increasing the concentration of a reactant shifts the position of the equilibrium to the products side, but does not change the value of K_c.

› Increasing the concentration of a product shifts the position of the equilibrium to the reactant side, but does not change the value of K_c.

› A catalyst does not affect the position of the equilibrium or the value of K_c.

› A catalyst increases the rate of the forward and reverse reactions equally, so that it takes a shorter time to reach equilibrium.

9.4 INDUSTRIAL PROCESSES AND EQUILIBRIA

Chemical industries must be economically viable – they need to make a good profit. As you have read, if the production involves reversible chemical reactions, conditions can be manipulated to increase the yield. Le Chatelier's principle predicts that a low temperature and a high pressure will give the best yield in the Haber process. The problem is that the lower the temperature, the slower the rate of reaction. A higher yield may be obtained, but it would take too long to obtain it. It is more economically advantagous to make little often and a **compromise temperature** of 700 K is used.

High pressures are expensive to maintain and technologically difficult. The pressure used for the Haber process, which is between 5000 and 20000 kPa, is not that high and is a **compromise pressure**.

Manufacturing methanol, CH_3OH

The manufacture of methanol is an example of an industrial process that involves a chemical reaction in dynamic equilibrium.

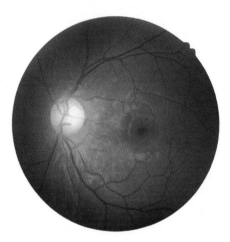

Figure 9 *This photograph of the retina was taken during a routine eye examination. The bright area with converging blood vessels is where the optic nerve joins the back of the eye. Methanol is toxic and ingesting as little as 10 cm³ will break down the optic nerve and cause blindness – 30 cm³ is fatal.*

Methanol is manufactured from carbon monoxide gas and hydrogen gas. The reversible reaction is:

$$CO(g) + 2H_2(g) \rightleftharpoons CH_3OH(g)$$
$$\Delta H^\ominus = -90.7 \text{ kJ mol}^{-1}$$

Since the forward reaction is exothermic, increasing the temperature shifts the position of the equilibrium in the endothermic direction towards the reactants. A low temperature increases the yield, but decreases the rate of reaction. It would take a long time to reach equilibrium. A compromise temperature of 523 K is used.

Since three moles of reactants produce one mole of products, increasing the pressure moves the position of the equilibrium towards the products. But, high pressures are expensive to maintain and a compromise pressure of 5000 to 10000 kPa is used.

The reaction is carried out using a catalyst of copper, zinc oxide and aluminium oxide. It has no effect on the position of the equilibrium, but it does decrease the time taken to reach it.

QUESTIONS

9. Much of the hydrogen used in the Haber process is manufactured from methane in the reaction:

$$CH_4(g) + H_2O(g) \rightleftharpoons CO(g) + 3H_2(g)$$
$$\Delta H^\ominus = +210 \text{kJ mol}^{-1}$$

a. Use Le Chatelier's principle to predict
   i. which temperature will give the best yield and how this will effect the rate of reaction
   ii. which pressure will give the best yield and how it will effect the rate of reaction

b. Explain why a catalyst is used.

ASSIGNMENT 4: SULFURIC ACID

(PS 1.1, 1.2, 2.3, 2.4)

Figure A1 *Sulfuric acid is made naturally in Venus' upper atmosphere by the action of UV light on carbon dioxide, sulfur dioxide and water. Here it exists as droplets of liquid, making sulfuric acid clouds. When it rains, it rains concentrated sulfuric acid.*

Sulfuric acid is a very important chemical and it is claimed that a country's sulfuric acid production is a good indicator of its industrial strength. Sulfuric acid is manufactured in the Contact process. The first stage is to burn sulfur in oxygen to make sulfur dioxide. The sulfur dioxide is then reacted with oxygen to make sulfur trioxide:

$$2SO_2(g) + O_2(g) \rightleftharpoons 2SO_3$$
$$\Delta H^\ominus = -196\text{kJ mol}^{-1}$$

The reaction is reversible and industrial chemists manipulate the conditions to give the highest yield in the shortest time.

Finally, the sulfur trioxide is reacted indirectly with water to produce sulfuric acid.

In practice, a pressure of 100–200 kPa is used because this gives a 98% yield. It does not make economic sense to use a higher pressure.

Questions

A1. In the reaction to produce sulfur trioxide, what happens to the position of equilibrium and the value of K_c if the temperature is increased?

A2. Will a higher or lower temperature give a higher yield of sulfur trioxide?

A3. Explain why a compromise temperature of 450 °C is used.

A4. What is the effect on the position of equilibrium if the pressure is increased?

A5. What is the effect on the concentration of sulfur trioxide at equilibrium and the value of K_c if the pressure is increased?

A6. The Contact process uses a vanadium pentoxide catalyst. What effect does the catalyst have on:

a. the position of the equilibrium

b. the rate of reaction

c. the value of K_c?

KEY IDEAS

> Many chemical reactions used in industrial processes are reversible.

> Temperature, pressure, concentration and a catalyst are employed to give the highest yield in the fastest time.

> A reaction may favour a low temperature, but this lowers the rate of reaction and a compromise temperature is used.

> A reaction may favour a high pressure, but high pressures are expensive and difficult to maintain. A compromise pressure may be used.

PRACTICE QUESTIONS

1. Ammonia is manufactured by the Haber process in which the following equilibrium is established.

$$N_2(g) + 3H_2(g) \rightleftharpoons 2NH_3(g)$$

   a. Give two features of a reaction at equilibrium.

   b. Explain why a catalyst has no effect on the position of an equilibrium.

   c. Figure Q1 shows how the equilibrium yield of ammonia varies with changes in pressure and temperature.

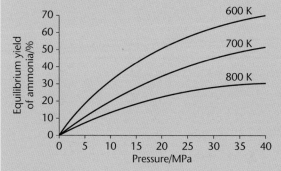

Figure Q1

   i. State the effect of an increase in pressure at constant temperature on the yield of ammonia. Use Figure Q1 to help you. Use Le Chatelier's principle to explain this effect.

   ii. State the effect of an increase in temperature at constant pressure on the yield of ammonia. Use Figure Q1 to help you. Use Le Chatelier's principle to explain this effect.

   d. At equilibrium, with a pressure of 35 MPa and a temperature of 600 K, the yield of ammonia is 65%.

   i. State why industry uses a temperature higher than 600 K.

   ii. State why industry uses a pressure lower than 35 MPa.
   (Do not include references to safety.)

   AQA Jan 2012 Unit 2 Question 2

2. a. The following equilibrium is established between colourless dinitrogen tetraoxide gas (N_2O_4) and dark brown nitrogen dioxide gas.

$$N_2O_4(g) \rightleftharpoons 2NO_2(g) \quad \Delta H = +58 \text{ kJ mol}^{-1}$$

   i. Give two features of a reaction at equilibrium.

   ii. Use Le Chatelier's principle to explain why the mixture of gases becomes darker in colour when the mixture is heated at constant pressure.

   iii. Use Le Chatelier's principle to explain why the amount of NO_2 decreases when the pressure is increased at constant temperature.

   AQA Jun 2011 Unit 2 Question 5b,c,d

3. An equilibrium is established between oxygen and ozone molecules as shown:

$$3O_2(g) \rightleftharpoons 2O_3(g) \quad \Delta H = +284 \text{ kJ mol}^{-1}$$

   a. State Le Chatelier's principle.

   b. Use Le Chatelier's principle to explain how an increase in temperature causes an increase in the equilibrium yield of ozone.

   AQA Jan 2011 Unit 2 Question 1c

4. Colourless solutions of X(aq) and Y(aq) react to form an orange solution of Z(aq) according to the following equation.

$$X(aq) + Y(aq) \rightleftharpoons Z(aq) \quad \Delta H = -20 \text{ kJ mol}^{-1}$$

   a. A student added a solution containing 0.50 mol of X(aq) to a solution containing 0.50 mol of Y(aq) and shook the mixture. After 30 seconds, there was no further change in colour. The amount of Z(aq) at equilibrium was 0.20 mol. Deduce the amount of X(aq) at equilibrium.

   b. Sketch a graph on the axes in Figure Q2 to show how the amount of Z(aq) changed from the time of initial mixing until 60 seconds had elapsed.

(continued)

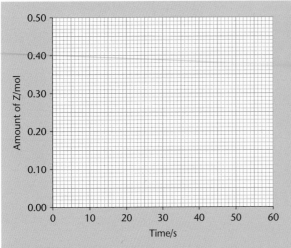

Figure Q2

c. The student prepared another equilibrium mixture in which the equilibrium concentrations were: $X(aq) = 0.40$ mol dm^{-3}, $Y(aq) = 0.30$ mol dm^{-3} and $Z(aq) = 0.35$ mol dm^{-3}.

  i. Write an expression for the equilibrium constant, Kc, for this reaction.

  ii. Calculate the value of the equilibrium constant, Kc.

  iii. The student added a few drops of $Y(aq)$ to a portion of the equilibrium mixture of $X(aq)$, $Y(aq)$ and $Z(aq)$. Suggest how the colour of the mixture changed. Give a reason for your answer.

  iv. The student warmed a second portion of the equilibrium mixture. Predict the colour change, if any, when the equilibrium mixture was warmed.

AQA 2014 specimen paper

5. This equation represents a reaction in equilibrium:

$X(g) + 2Y(g) \rightleftharpoons 2Z(g)$

a. Explain what is meant by a *reaction in equilibrium*.

b. State and explain the effect on the yield of Z if the overall pressure is increased.

c. Write an expression for the equilibrium constant, K, for this reaction.

d. An increase in temperature causes a decrease in the yield of Z.

  i. State and explain what can be deduced about the enthalpy change for the forward reaction.

  ii. What happens to the value of K_c when the temperature is raised?

AQA Jan 2007 Unit 2 Question 4

6. In the Haber Process for the manufacture of ammonia, nitrogen and hydrogen react as shown in this equation:

$N_2(g) + 3H_2(g) \rightleftharpoons 2NH_3(g)$

$\Delta H^{\ominus} = -92 \text{ kJ mol}^{-1}$

Table Q1 shows the percentage yield of ammonia, under different conditions of pressure and temperature, when the reaction has reached dynamic equilibrium.

Temperature/K	600	800	1000
% yield of ammonia at 10 MPa	50	10	2
% yield of ammonia at 20 MPa	60	16	4
% yield of ammonia at 50 MPa	75	25	7

Table Q1

a. Explain the meaning of the term *dynamic equilibrium*.

b. Use Le Chatelier's principle to explain why, at a given temperature, the percentage yield of ammonia increases with an increase in overall pressure.

c. Give a reason why a high pressure of 50 MPa is not normally used in the Haber Process.

d. Many industrial ammonia plants operate at a compromise temperature of about 700 K.

  i. State and explain, by using Le Chatelier's principle, one advantage, other than cost, of using a temperature lower than 700 K.

  ii. State the major advantage of using a temperature higher than 700 K.

  iii. Hence explain why 700 K is referred to as a *compromise temperature*.

e. i. Write an expression for the equilibrium constant, K_c, for the reaction to produce ammonia.

(continued)

ii. Predict and explain which temperature shown in Table Q1 will have the lowest value for K_c.

AQA June 2006 Unit 2 Question 3

7. Hydrogen is produced on an industrial scale from methane as shown by this equation:

$$CH_4(g) + H_2O(g) \rightleftharpoons CO(g) + 3H_2(g)$$
$$\Delta H^\ominus = +205 \text{ kJ mol}^{-1}$$

a. State Le Chatelier's principle.

b. The following changes are made to this reaction at equilibrium. In each case, predict what would happen to the yield of hydrogen from a given amount of methane. Use Le Chatelier's principle to explain your answers.

   i. The overall pressure is increased.

   ii. The concentration of steam in the reaction mixture is increased.

   iii. The temperature is increased.

c. i. Write an expression for K_c for the reaction.

   ii. Explain why this reaction is an example of a homogeneous equilibriaum.

AQA January 2004 Unit 2 Question 1

8. Sulfur dioxide gas reacts with oxygen gas to produce sulfur trioxide. The equation is:
$$2 SO_2(g) + O_2(g) \rightleftharpoons 2 SO_3(g).$$
K_c for the reaction is written as:

$$K_c = \frac{[SO_3]^2}{[SO_2]^2[O_2]}$$

a. i. What does K_c mean?

   ii. State what the squared brackets mean in the expression for K_c.

   iii. State what the super script numbers mean in the expression for K_c.

   iv. What happens to the value of K_c if the position of equilibrium moves to the right?

b. i. Write a balanced equation for the reaction between nitrogen gas and hydrogen gas to produce ammonia.

   ii. Write an expession for K_c for the reaction.

9. Ethanol and ethanoic acid react to produce ethyl ethanoate and water. The reaction is reversible and the equation is:
$$C_2H_5OH(l) + CH_3COOH(l) \rightleftharpoons$$
$$CH_3COOC_2H_5(l) + H_2O(l)$$
$$\Delta H = -2.0 \text{ kJ mol}^{-1}$$

1.0 mol ethanol was added to 1.0 mol ethanoic acid and the mixture was allowed to reach equilibrium. The following concentrations were measured at equilibrium.

Compound	Concentration at equilibrium/mol dm^{-3}
C_2H_5OH	0.33
CH_3COOH	0.33
$CH_3COOC_2H_5$	0.67
H_2O	0.67

a. i. Write an expression for K_c for the reaction.

   ii. Calculate the value of K_c for the reaction.

b. The units for K_c values depend on the reaction. They can be worked out by including units for concentration in the expression for K_c and cancelling where appropriate. Explain why the value for K_c has no units in this example.

c. Predict and explain the effect on K_c when the equilibrium mixture is at a higher temperature.

d. Predict and explain the effect on K_c when ethanol is added to the equilibrium mixture at constant temperature.

Stretch and challenge

10. When nitrogen dioxide (NO_2) is heated in a sealed container, the following reaction takes place:
$$2NO_2(g) \rightleftharpoons 2NO(g) + O_2(g)$$

21.3 g of nitrogen dioxide were heated to a constant temperature in a sealed container with a volume of 10.0 dm^3. At equilibrium, the mixture contained 7.04 g oxygen.

a. Calculate the number of moles of oxygen and hence nitrogen monoxide (NO) at equilibrium.

(continued)

b. Calculate the number of moles of nitrogen dioxide originally used and hence the number of moles of nitrogen dioxide present at equilibrium.

c. Write an expression for K_c for the equilibrium and calculate its value for this reaction.

d. Explain why it is important to keep the temperature constant.

Multiple choice

11. In which of these reactions does increasing the pressure have no effect on the position of the equilibrium?

   A. $CO_2(g) + 3H_2(g) \rightleftharpoons CH_3OH(g) + H_2O(g)$
   B. $CO_2(g) + 2H_2(g) \rightleftharpoons CH_3OH(g)$
   C. $CH_4(g) + H_2O(g) \rightleftharpoons CO(g) + 3H_2(g)$
   D. $CO(g) + H_2O(g) \rightleftharpoons CO_2(g) + H_2(g)$

12. What is happening in a dynamic equilibrium?

   A. The forward and reverse reactions occur at the same rate

   B. The rate of the forward reaction is higher than the rate of the reverse reaction

   C. The rate of the forward reaction is slower than the rate of the reverse reaction

   D. The rate of both forward and reverse reactions is zero

13. How does a catalyst affect a reversible reaction?

   A. It shortens the time taken to reach equilibrium

   B. It moves the position of the equilibrium to the right

   C. It moves the position of the equilibrium to the left

   D. It has no effect on the position of equilibrium or the reaction rate

14. When is a compromise temperature used in an industrial process involving a reversible reaction?

   A. When the reaction to produce the product is endothermic and the reaction rate is slow

   B. When the reaction to produce the product is exothermic and the reaction rate is slow

   C. When there are less moles of reactants than products in the balanced equation

   D. When there are more moles of reactants than products in the balanced equation

10 REDOX REACTIONS

PRIOR KNOWLEDGE

You may know about oxidation and reduction reactions and have studied some examples. You may have learnt that electrons are lost when a substance is oxidised and that they are gained when a substance is reduced. You may also know that oxidation and reduction occur in displacement reactions and at the electrodes during electrolysis.

LEARNING OBJECTIVES

(Specification 3.1.7)

Figure 1 *Blueberries and other fruits and vegetables are good natural sources of antioxidants. Antioxidants can be natural or man-made. Vitamin C supplements (most are processed from corn) are thought to work as antioxidants.*

Your body's trillion or so cells are under constant threat. Releasing energy from your food in cells during respiration is an oxidation reaction. Inevitably it produces some free radicals. Free radicals contain an unpaired electron and are highly reactive. Left to their own devices, they scavenge around the body grabbing or donating electrons and causing damage. Health experts think that free radical reactions are responsible for the development of diseases such as cancer, heart disease, stroke and many other serious illnesses.

But the body has its own defences. Foods such as fruit and vegetables contain chemicals collectively known as antioxidants (Figure 1). The government advises us to eat at least five portions of these a day. In your body cells, antioxidants readily donate electrons to the free radicals and prevent them from doing damage. The free radicals are reduced and the antioxidant is oxidised.

10.1 OXIDATION AND REDUCTION

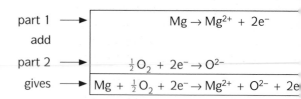

part 1 ⟶

add

part 2 ⟶

gives ⟶

$$Mg \rightarrow Mg^{2+} + 2e^-$$

$$\tfrac{1}{2}O_2 + 2e^- \rightarrow O^{2-}$$

$$Mg + \tfrac{1}{2}O_2 + 2e^- \rightarrow Mg^{2+} + O^{2-} + 2e$$

The $2e^-$ on either side of the equation cancel one another out, leaving the ions in the MgO lattice:

$$Mg + \tfrac{1}{2}O_2 \rightarrow Mg^{2+} + O^{2-}$$

or

$$Mg + \tfrac{1}{2}O_2 \rightarrow MgO$$

or, including state symbols, $Mg(s) + \tfrac{1}{2}O_2(g) \rightarrow MgO(s$

Because magnesium has lost electrons, we say it is oxidised. Oxidation is electron loss. Part 1 is the half-equation for oxidation. Note that the charges either side of the half-equation balance.

Oxygen gains electrons and is reduced. Reduction is electron gain. Part 2 is the half-equation for reduction Again, note that the charges either side of the half-equation balance.

Figure 3 summarises the loss and gain of electrons in oxidation and reduction reactions.

Figure 2 *Fireworks – sparks, bright light and colours. All are due to oxidation reactions.*

Nearly every element known reacts to form compounds. When elements react, electrons in the outer shells of their atoms are rearranged. They are either transferred from one atom to another to form an ionic bond or shared to form covalent bonds, some of which may be polar.

The brilliant white light from the firework in Figure 2 is the result of magnesium powder reacting with oxygen. The same reaction can be carried out in the laboratory using magnesium ribbon. Its equation is:

$$Mg(s) + \tfrac{1}{2}O_2(g) \rightarrow MgO(s)$$

Magnesium oxide, MgO, has a giant structure consisting of magnesium ions, Mg^{2+}, and oxide ions, O^{2-}. The formation of ions involves the transfer of two electrons from the outermost shell of the magnesium atom, which gives it a stable noble gas configuration to form Mg^{2+}. The oxygen atom gains the two electrons to form O^{2-}, which also has a stable noble gas configuration.

The equation for the reaction can be described of as being made up of two parts:

Part 1: $Mg \rightarrow Mg^{2+} + 2e^-$
Part 2: $\tfrac{1}{2}O_2 + 2e^- \rightarrow O^{2-}$

Equations that show each half of the overall reaction are called **half-equations**. Adding them together gives the full equation.

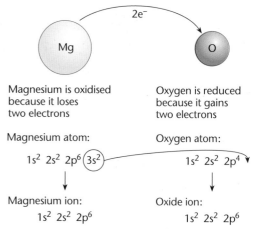

Figure 3 *A redox reaction occurs because electrons are transferred when magnesium reacts with oxygen.*

Oxidation and reduction reactions go hand in hand. In order for something to transfer electrons, something has to gain them, so if one reactant is oxidised, another must be reduced. Reduction–oxidation reactions are referred to as **redox reactions**.

A useful aid for remembering which word refers to the loss or gain of electrons is OIL RIG:

OXIDATION **IS** **L**OSS

REDUCTION **IS** **G**AIN

idising and reducing agents

e oxidising agent is the reactant that does the dising. When magnesium burns in oxygen, gnesium is oxidised and oxygen is the oxidising ent. Oxidising agents are electron acceptors.

e reducing agent is the reactant that does the ucing. So, when magnesium burns in oxygen, gen is reduced and magnesium is the reducing ent. Reducing agents are electron donors.

another example, sodium metal burns in chlorine to produce sodium chloride:

$(s) + \frac{1}{2}Cl_2(g) \rightarrow NaCl(s)$

e sodium atom transfers one electron to a chlorine m. The half-equations are:

$\rightarrow Na^+ + e^-$

$l_2 + e^- \rightarrow Cl^-$

dium is oxidised because it loses one electron. orine is the oxidising agent because it gains an ctron. Chlorine is reduced because it gains one ctron. Sodium is the reducing agent because it nsfers an electron.

idising agents are electron acceptors and are emselves reduced.

ducing agents are electron donors and are emselves oxidised.

les for writing half-equations

portant rules to follow when writing any f-equation for an oxidation or a reduction ction are:

The equation must balance for atoms.

The equation must balance for charge.

th rules apply also to full equations.

QUESTIONS

1. For the following reactions, i. write the full equation, ii. write the half-equations, iii. state which reactant is being oxidised and which is being reduced, iv. state which reactant is the oxidising agent and which is the reducing agent.

   a. Sodium reacting with bromine to form sodium bromide, NaBr.

   b. Sodium burning in oxygen to form sodium oxide, Na_2O.

   c. Calcium reacting with iodine to form calcium iodide, CaI_2.

   d. Magnesium reacting with nitrogen to form magnesium nitride, Mg_3N_2.

2. Potassium is a Group 1 element. It readily reacts with oxygen to form K_2O.

   a. Write a full balanced equation for the reaction.

   b. Write the two half-equations for the reaction.

   c. Explain why potassium is a good reducing agent.

3. Calcium is in Group 2 and reacts with cold water.

   a. Write a full balanced equation for the reaction with water.

   b. Write the two half-equations for the reaction.

   c. Explain why calcium is a good reducing agent.

KEY IDEAS

› Oxidation is the process of electron loss.

› Reduction is the process of electron gain.

› If a reaction involves oxidation, then reduction must also occur.

› Reduction-oxidation reactions are called redox reactions.

› Oxidising agents are electron acceptors.

› Reducing agents are electron donors.

10.2 OXIDATION STATES

The concept of oxidation states makes it easy to work out what has been oxidised and what has been reduced in a reaction. The reactions you have looked at so far involved ionic compounds.

Sodium chloride consists of Na^+ ions and Cl^- ions. Sodium is assigned the oxidation state +1 and chlorine is assigned the oxidation state -1. You may find the term oxidation number used instead of oxidation state in some text books and websites, but it means the same thing.

Magnesium oxide consists of Mg^{2+} ions and O^{2-} ions. The oxidation state of oxygen is -2. For simple ions, oxidation state is the same as the charge on the ion.

The reactions between hydrogen and oxygen and between magnesium and oxygen have similar equations. In both, oxygen acts as an oxidising agent.

$$H_2(g) + \tfrac{1}{2}O_2(g) \rightarrow H_2O(l)$$

$$Mg(s) + \tfrac{1}{2}O_2(g) \rightarrow MgO(s)$$

However, in the hydrogen/oxygen reaction all the reactants and products are molecules with covalent bonding. Therefore, we cannot assign oxidation states to hydrogen and oxygen in a water molecule from the charges on the ions – there aren't any.

IUPAC has published rules for assigning oxidation states.

IUPAC rules for assigning oxidation states (2014)

Rule 1: The oxidation state of a 'free' (uncombined) element is always zero.

Rule 2: For a simple monoatomic ion (formed from one type of atom only) the oxidation state is the charge on the ion.

Rule 3: Hydrogen has an oxidation state of $+1$ and oxygen has an oxidation state of -2 in most compounds. The exceptions are hydrogen which has an oxidation state of -1 in hydrides of active metals (for example, LiH) and oxygen which has an oxidation state of -1 in peroxides (for example, H_2O_2).

Rule 4: The sum of the oxidation states of all the atoms in a neutral molecule must be zero. For ions, the sum of the oxidation states of the constituent atoms must equal the charge on the ion.

Variable oxidation states

Some elements have more than one oxidation state. To avoid confusion, the oxidation state of an element in a particular compound is shown by Roman numerals in brackets after the element in the chemical name. For example, iron has more than one oxidation state, but $+2$ and $+3$ are the commonest. Iron has an oxidation state of $+2$ in iron(II) sulfate and $+3$ in iron(III) sulfate.

Some examples are given in Tables 1 and 2. Note that there is no space between the bracketed Roman

numerals and the element being described. An **oxyanion** is made from atoms of two elements, one which is oxygen.

Formula	Old name	Name using oxidation states
$FeSO_4$	ferrous sulfate	iron(II) sulfate
$Fe_2(SO_4)_3$	ferric sulfate	iron(III) sulfate
Cu_2O	cuprous oxide	copper(I) oxide
CuO	cupric oxide	copper(II) oxide

Table 1 *Naming compounds containing metals with various oxidation states*

Formula	Old name	Name using oxidation states
ClO^-	hypochlorite ion	chlorate(I)
ClO_2^-	chlorite ion	chlorate(III)
ClO_3^-	chlorate	chlorate(V)
ClO_4^-	perchlorate	chlorate(VII)

Table 2 *Naming oxyanions containing chlorine*

Unfortunately the use of oxidation states for naming compounds is inconsistent. Where compounds contain an element with variable oxidation state, it is common to give the oxidation state in the name. For example, copper(II) sulfate and manganese(IV) oxide. Use of oxidation states is much less consistent with non-metal elements. For example:

> CO is always called carbon monoxide rather than carbon(II) oxide and CO_2 is carbon dioixide rather than carbon(IV) oxide

> H_2SO_4 is nearly always called sulfuric acid rather than sulfuric(VI) acid and H_2SO_3 is sulfurous acid rather than sulfuric(III) acid.

In your A-level course you will probably only use oxidation states when naming oxyanions containing chlorine and when naming compounds of transition metals. However, you may see them used in other books, on some websites and even on the labels of containers of some chemical compounds.

Assigning oxidation states – some examples

Lithium in lithium oxide
Lithium is a Group 1 metal and forms Li^+ ions. Since the charge on this ion is $+1$, it is assigned an oxidation state of $+1$ (rule 2).

fur in sulfuric acid

e formula of sulfuric acid is H_2SO_4. Oxygen has
oxidation state of -2 (rule 3) and hydrogen of $+1$
e 3). This just leaves the oxidation state of sulfur
determine. Since the oxidation states of all atoms
ions in a compound must add up to zero (rule 4),
ur has an oxidation state of $+6$:

$\times$ H) + S + (4 $\times$ O) = 0

$\times$ +1) + S + (4 $\times$ -2) = 0

$+6$

orine in the chlorate(V) ion, ClO_3^-

e charge on the ClO_3^- ion is -1. According to rule
the oxidation states in an ion must add up to its
arge. The oxidation state of Cl is:

+ (3 $\times$ O) = -1

+ (3 $\times$ -2) = -1

= $+5$

QUESTIONS

4. What is the oxidation state of each
element in:

a. Ag^+ g. NaCl

b. NO_3^- h. HNO_3

c. PCl_5 i. MnO_4^-

d. Al_2O_3 j. K_2SO_4

e. SO_4^{2-} k. H_3PO_4

f. PO_4^{3-} l. $Cr_2O_7^{2-}$?

5. The same element can have different
oxidation states in different compounds.
What is the oxidation state of iron in:

a. FeO

b. Fe_2O_3

c. Fe_3O_4?

Stretch and challenge

6. What is the oxidation state of chlorine in:

a. Cl_2 e. $HClO_2$

b. ClF   f. $HClO_3$

c. Cl^- g. $HClO_4$?

d. HClO

7. A 70 kg human being contains about
3.5 kg of the compound $Ca_5(PO_4)_3OH$,
a major constituent of bones and teeth.
What is the oxidation state of phosphorus
in this compound?

8. Nitrogen forms three oxides: NO, NO_2
and N_2O_4. What is the oxidation state of
nitrogen in these compounds?

Oxidation state and redox reactions

Oxidation states can be used to define oxidation
and reduction.

Oxidation occurs when the oxidation state of an
element in a reaction increases.

Reduction occurs when the oxidation state of an
element in a reaction decreases.

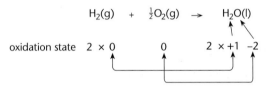

$$H_2(g) \quad + \quad \tfrac{1}{2}O_2(g) \quad \rightarrow \quad H_2O(l)$$

oxidation state   2 $\times$ 0   0   2 $\times$ +1   -2

Figure 4 *Changes in oxidation states in the reaction of H_2 and O_2*

The oxidation state of all elements is zero, regardless
of their structure and bonding. Metallic elements
have giant structures and metallic bonding. Many
non-metal elements have molecular structures with
covalent bonding, for example O_2, P_4 and S_8. Some
non-metal elements have giant structures with
covalent bonding, for example carbon and silicon. But
in all cases the oxidation state of the element is zero.

In the reaction between hydrogen and oxygen to form
water, hydrogen is said to be oxidised because its
oxidation state increases from 0 to +1, while oxygen
is reduced because its oxidation state decreases from
0 to -2 (Figure 4). To make the equation balanced,
for one oxygen being reduced from 0 to -2, two
hydrogens must be oxidised from 0 to +1.

Figure 5 *Vanadium is a transition metal. Like all transition metals it has several oxidation states. In the $VO_2^+(aq)$ ion it is $+5$ and its solution is yellow. In the $VO^{2+}(aq)$ ion it is $+4$ and its solution is blue. $V^{3+}(aq)$ is green and $V^{2+}(aq)$ is purple. These ions are not equally stable in aqueous solution in air. For example, $V^{2+}(aq)$ oxidises rapidly to ions in which vanadium has a higher oxidation state than $+2$.*

Figure 6 *Malachite contains basic copper(II) carbonate. It has been ground and used as eye shadow since about 5000BCE. Cleopatra probably wore green eye shadow made from malachite to protect her eyes from the Sun, as well as for cosmetic reasons. Copper(II) carbonate contains Cu^{2+} ions with an oxidation state of $+2$.*

Examples of redox reactions

Magnesium and nitrogen

The reaction is:

	$3Mg(s)$	$+$ $N_2(g)$	$\rightarrow$	$Mg_3N_2(s)$
Oxidation states	$Mg = 0$			$Mg = +2$
		$N = 0$		$N = -3$

In this reaction, magnesium is oxidised because its oxidation state has increased from 0 to +2. Nitrogen is reduced because its oxidation state has decreased from 0 to −3.

Nitrogen and hydrogen

When nitrogen reacts with hydrogen in the Haber process, the full equation is:

	$N_2(g)$	$+$ $3H_2(g)$	$\rightarrow$	$2NH_3(g)$
Oxidation states	$N = 0$			$N = -3$
		$H = 0$		$H = +1$

The nitrogen has again been reduced because its oxidation state has decreased from 0 to −3. Hydrogen has been oxidised because its oxidation state has increased from 0 to +1. Oxidation states can be used to work out whether a reaction is a redox reaction.

Lithium and water

When lithium reacts with water, hydrogen is given off and lithium hydroxide forms. If we assign oxidation states to each element, we can determine what has been oxidised and what has been reduced.

	$2Li(s)$	$+$ $2H_2O(l)$	$\rightarrow$	$2LiOH(aq)$	$+$ $H_2(g)$
Oxidation states	$Li = 0$			$Li = +1$	
		$H = +1$		$H = +1$	$H = 0$
		$O = -2$		$O = -2$	

The oxidation state of lithium has increased; lithium has been oxidised. The oxidation state of one of the hydrogens in water has decreased from +1 in water to 0 in hydrogen gas. This is reduction. The other hydrogen from water remains in the +1 oxidation state in lithium hydroxide; it is not reduced.

Sodium halides and concentrated sulfuric acid

The reaction between sodium chloride and concentrated sulfuric acid is:

	$NaCl(s)$	$+$ $H_2SO_4(aq)$	$\rightarrow$	$NaHSO_4(aq)$	$+$ HCl
Oxidation states	$Na = +1$			$Na = +1$	
	$Cl = -1$				$Cl =$
		$H = +1$		$H = +1$	$H =$
		$S = +6$		$S = +6$	
		$O = -2$		$O = -2$	

The oxidation states for all the elements in the reaction have not changed. This is not a redox reaction.

However, the reaction between sodium iodide and concentrated sulfuric acid is:

	$8NaI(s)$ + $5H_2SO_4(aq)$ → $4Na_2SO_4(aq)$ + $4I_2(s)$ + $H_2S(g)$ + $4H_2O(l)$
Oxidation states	Na = +1 Na = +1
	I = −1 I = 0
	H = +1 H = +1 H = +1
	S = +6 S = +6 S = −2
	O = −2 O = −2 O = −2

This is a redox reaction because the oxidation state of some of the sulfur has decreased from +6 to −2, so has been reduced. The oxidation state of iodine has increased from −1 to 0 and been oxidised.

QUESTIONS

9. Use oxidation states to identify what has been oxidised and what has been reduced in these reactions:

   a. $2Li(s) + Cl_2(g) → 2LiCl(s)$

   b. $N_2(g) + O_2(g) → 2NO(g)$

   c. $2NO(g) + O_2(g) → 2NO_2(g)$

10. Use oxidation states to decide which of these reactions are redox reactions:

   a. $SO_2(aq) + H_2O(l) \rightleftharpoons H_2SO_3(aq)$

   b. $K_2SO_3(s) + 2HNO_3(aq) →$
   $2KNO_3(aq) + SO_2(g) + H_2O(l)$

   c. $Na_2S_2O_3(aq) + 2HCl(aq) →$
   $SO_2(g) + S(s) + 2NaCl(aq) + H_2O(l)$

11. Use oxidation states to identify what has been oxidised and what has been reduced in the reaction between sodium hydroxide solution and chlorine gas to produce sodium chloride, sodium chlorate (NaClO) and water.

ASSIGNMENT 1: REVERSING NATURE

(MS 02; PS 1.1, 1.2, 3.2)

Figure A1 *The water in the oceans and most of of the rocks in the Earth's crust are products of oxidaton reactions.*

Earth's atmosphere contains 21% oxygen by volume (23% by mass for dry air). Oxygen is an excellent oxidising agent. The ten most abundant compounds in Earth's crust are oxides. As a result, the Earth's crust is 46.6% by mass oxygen. Scientists think that the elements that make up the Earth were originally made in the stars. Most of the metals and non-metals in the Earth's crust have been oxidised at some time in their geological history. The oceans are water − an oxide of hydrogen.

Only the very unreactive metals are found uncombined in nature. They are called native metals. The rest are present chemically combined in ores. For example, iron ores include Fe_2O_3 and Fe_3O_4, aluminium occurs as bauxite, Al_2O_3, and tungsten occurs as wolframite, WO_3. Extracting metals from their ores involves reversing nature's oxidation process and reducing the metal ore.

Ore	Formula	Colour
cuprite	Cu_2O	red
chalcocite	Cu_2S	dark grey
malachite	$CuCO_3.Cu(OH)_2$	bright green
azurite	$2CuCO_3.Cu(OH)_2$	blue

Table A1 *Some copper minerals found in Earth's crust*

Questions

A1. What are the oxidation states of iron in Fe_2O_3?

A2. Fe_2O_3 is reduced in the blast furnace using carbon or carbon monoxide as the reducing agent.

    a. Write and equation for the reduction of Fe_2O_3 with carbon monoxide.

    b. What are the changes in oxidation state for the reaction?

    c. Write a half-equation for the reduction of iron in Fe_2O_3.

A3. Left unprotected, iron metal reacts with oxygen and water in the atmosphere to form rust. If you take the formula of rust to be $Fe_2O_3.nH_2O$, explain, using oxidation states, why this is an oxidation reaction.

A4. Aluminium is extracted from Al_2O_3 by electrolysis. Aluminium ions move to the cathode, and oxide ions move to the anode and lose electrons.

    a. Write a half-equation for the reaction occuring at the cathode.

    b. Write a half-equation for the reaction occurring at the anode.

    c. Where do oxidation and reduction processes occur during electrolysis?

A5. Tungsten is extracted by reduction with hydrogen gas.

    a. What is the oxidation state of tungsten in WO_3?

    b. Write a half-equation to show the reduction of tungston ore.

Stretch and challenge

A6. Look at the list of copper minerals given in Table A1.

    a. From the examples of copper compounds that you have learned about in chemistry, what is the most common oxidation state for copper in its compounds? Give the names and formulae of three copper compounds in which copper is in this oxidation state.

    b. What is the oxidation state of copper in each mineral in Table A1?

    c. All the minerals in Table A1 contain copper ions. What colour do you associate with copper(II) ions, Cu^{2+}?

    d. You are given 1 kg of each ore. What is the maxiumum mass of copper that can be obtained from each?

    (A_r: H 1, C 12, O 6, S 32, Cu 63.5)

    e. The formulae of malachite and azurite makes it appear that they are mixtures, but they are not. Find out more about the structure and bonding in these two minerals.

ASSIGNMENT 2: GOLD CHEMISTRY

(PS 1.1, 1.2)

Figure A1 *Tutankhamun's death mask is more than 3000 years old. Gold's lack of chemistry has preserved it.*

Gold used to be known as the *royal metal*. It was the metal of kings and queens. Its chemistry is limited, but it does react very slowly with chlorine gas to produce auric chloride, $AuCl_3$. (Auric means gold ions with three positive charges.)

Early chemists discovered that a mixture of nitric acid and hydrochloric acid was able to dissolve gold. They called their potent mixture *aqua Regis* or *aqua regia*, meaning *royal water*. Aqua regia is a concentrated mixture of three moles of hydrochloric acid to one mole of nitric acid and is most effective when used hot. Neither of the acids will dissolve gold alone. The nitric acid dissolves a very small amount of gold to form Au^{3+} ions. The Au^{3+} ions then react with the Cl^- ions in hydrochloric acid in a reversible reaction to form $AuCl_4^-$ ions. The position of the equilibrium lies far to the right and the Au^{3+} ions are removed. More gold can now be dissolved by the nitric acid.

Questions

A1. **a.** Write a full equation for the reaction between solid gold and chlorine gas.

    b. Use oxidation states to identify what has been oxidised and what has been reduced.

    c. Write half-equations to show the oxidation and reduction reactions.

A2. **a.** What is the oxidation state of gold in $AuCl_4^-$ ions?

    b. Write an equation for the reaction between Au^{3+} ions and Cl^- ions.

    c. Use oxidation states to decide whether this is a redox reaction.

    d. When the reaction to produce $AuCl_4^-$ ions is at equilibrium, which components are in greater concentration, the reactants or the products?

    e. If more Au^{3+} ions are introduced to the system, what happens to the position of the equilibrium and why?

A3. Gold also forms a compound with rubidium, RbAu. What is the oxidation state of gold in RbAu and why is this unusual?

KEY IDEAS

❯ IUPAC rules are used to assign oxidation states.

❯ Oxidation occurs when the oxidation state of an element increases.

❯ Reduction occurs when the oxidation state of an element decreases.

❯ Oxidation states can be used to identify redox reactions.

❯ Oxidation stated can be used to name compounds.

10.3 REDOX EQUATIONS

Redox equations, sometimes called ionic equations, show what is oxidised and what is reduced in a redox reaction. They do not include ions that are unchanged and present before and after the reaction.

For some simple reactions involving ions, we can work out the redox equation from the full balanced equation. For most redox reactions, we can work out the overall redox equation by combining the oxidation and reduction half-equations.

The photograph in Figure 7 shows the redox reaction between zinc metal and copper(II) sulfate solution.

Figure 7 *When zinc metal is dipped into copper(II) sulfate solution, a redox reaction occurs.*

The overall equation for this reaction is:

$$Zn(s) + CuSO_4(aq) \rightarrow ZnSO_4(aq) + Cu(s)$$

Since the reaction involves ions, the equation can be expanded to list the ions separately.

$$Zn(s) + Cu^{2+}(aq) + SO_4^{2-}(aq) \rightarrow$$
$$Zn^{2+}(aq) + SO_4^{2-}(aq) + Cu(s)$$

The sulfate ions are the same on both sides of the equation and do not change. They do not take part in the reaction and are called **spectator ions**. Cancelling out the sulfate ions (removing them from each side of the equation) leaves:

$$Zn(s) + Cu^{2+}(aq) \rightarrow Zn^{2+}(aq) + Cu(s)$$

This is the overall redox equation. The charges, as well as the atoms, on both sides of the equation must balance.

From the overall redox equation, the two half-equations for the oxidation of zinc and reduction of copper(II) ions, respectively, are:

$$Zn(s) \rightarrow Zn^{2+}(aq) + 2e^-$$ oxidation

$$Cu^{2+}(aq) + 2e^- \rightarrow Cu(s)$$ reduction

QUESTIONS

12. For the reaction between zinc and aqueous copper(II) sulfate, explain in terms of loss and gain of electrons why zinc is oxidised and copper is reduced.

13. Expand the following equations to show all the ions separately, and write a redox equation for each reaction:

   a. $Cu(s) + 2AgNO_3(aq) \rightarrow$
      $\quad\quad\quad\quad Cu(NO_3)_2(aq) + 2Ag(s)$

   b. $Mg(s) + 2HCl(aq) \rightarrow MgCl_2(aq) + H_2(g)$

   c. $Mg(s) + 2AgNO_3(aq) \rightarrow$
      $\quad\quad\quad\quad Mg(NO_3)_2(aq) + 2Ag(s)$

Combining half-equations

Redox reaction for magnesium and hydrochloric acid

When magnesium reacts with dilute hydrochloric acid, the magnesium is oxidised and the hydrogen ions are reduced. You may need to use the full equation and oxidation states to work out what has been oxidised and what has been reduced.

The half-equation for the oxidation of magnesium is:

$$Mg(s) \rightarrow Mg^{2+}(aq) + 2e^-$$

The half-equation for the reduction of hydrogen is:

$$2H^+(aq) + 2e^- \rightarrow H_2(g)$$

Before these can be added together by combining all the reactants and all the products, the number of electrons transferred must balance. The half-equations show magnesium losing two electrons and hydrogen gaining two electrons.

	$Mg(s) \rightarrow$	$Mg^{2+}(aq) + 2e^-$
$2H^+(aq) + 2e^- \rightarrow$		$H_2(g)$
$Mg(s) + 2H^+(aq) + 2e^- \rightarrow$		$Mg^{2+}(aq) + H_2(g) + 2e^-$

Since $2e^-$ is on both sides of the equation, they cancel each other out.

The redox equation is:

$$Mg(s) + 2H^+(aq) \rightarrow Mg^{2+}(aq) + H_2(g)$$

Note that the chloride ions do appear appear in the redox equation. These are spectator ions.

Some redox reactions are more complicated, as the following examples show.

Redox reaction for sulfur dioxide and chlorine

First, decide what has been oxidised and what has been reduced. The full equation for the reaction between sulfur, chlorine and water is:

	$SO_2(g)$	+	$Cl_2(aq)$	+	$2H_2O(l)$	→	$H_2SO_4(aq)$	+	$2HCl(aq)$
Oxidation states	$S = +4$						$S = +6$		
	$O = -2$				$O = -2$				
			$Cl = 0$						$Cl = -1$
					$H = +1$		$H = +1$		$H = +1$

The oxidation state of chlorine changes from 0 to −1 and chlorine is reduced. The half-equation is:

$Cl_2(g) + 2e^- → 2Cl^-(aq)$

The oxidation state of sulfur changes from $+4$ to $+6$. This means that sulfur transfers two electrons and is oxidised:

$SO_2(g) → SO_4{}^{2-}(aq) + 2e^-$

But this half-equation does not balance, either in terms of atoms or charges. This is where the water shown in the balanced reaction equation comes in:

$SO_2(g) + 2H_2O(l) → SO_4{}^{2-}(aq) + 4H^+(aq) + 2e^-$

Both the numbers of atoms and the charges now balance. You will often see the involvement of water and H^+ ions to balance half-equations.

You can add the two half-equations together. The electrons cancel out because there are $2e^-$ on both sides of the equation.

$SO_2(g) + 2H_2O(l) → SO_4{}^{2-}(aq) + 4H^+(aq) + 2e^-$
$Cl_2(g) + 2e^- → 2Cl^-(aq)$
$SO_2(g) + Cl_2(aq) + 2H_2O(l) → SO_4{}^{2-}(aq) + 4H^+(aq) + 2Cl^-(aq)$

Redox reaction for nitric acid and copper

Copper reacts with cold, dilute nitric acid to produce nitrogen monoxide and copper(II) nitrate.

The full balanced equation is:

	$3Cu(s)$	+	$8HNO_3(aq)$	→	$3Cu(NO_3)_2(aq)$	+	$2NO(g)$	+	$4H_2O(l)$
Oxidation states	$Cu = 0$				$Cu = +2$				
			$H = +1$						$H = +1$
			$N = +5$		$N = +5$		$N = +2$		
			$O = -2$		$O = -2$		$O = -2$		$O = -2$

The oxidation state of copper has increased from 0 to $+2$ and copper has been oxidised. The half-equation is:

$Cu(s) → Cu^{2+}(aq) + 2e^-$

The oxidation state of nitrogen has decreased from $+5$ to $+2$ and nitrogen has been reduced. The difference in the oxidation states show that nitrogen has gained three electrons and the half-equation is:

$NO_3{}^-(aq) + 3e^- → NO(g)$

This is unbalanced for charges and atoms. As in the reaction between sulfur dioxide and chlorine above, water and H^+ ions must be involved to balance the half-equation:

$NO_3{}^-(aq) + 4H^+(aq) + 3e^- → NO(g) + 2H_2O(l)$

Adding the two half-equations together gives the redox equation, but the numbers of electrons transferred needs to balance first. The oxidation half-equation for copper is multiplied by three, and the reduction half-equation by two:

$$3Cu(s) \rightarrow 3Cu^{2+}(aq) + 6e^-$$
$$2NO_3^-(aq) + 8H^+(aq) + 6e^- \rightarrow 2NO(g) + 4H_2O(l)$$
$$3Cu(s) + 2NO_3^-(aq) + 8H^+(aq) \rightarrow 3Cu^{2+}(aq) + 2NO(g) + 4H_2O(l)$$

The electrons now cancel out.

Redox reaction for potassium manganate(VII) and iron(II)

Potassium manganate(VII) is purple. When it reacts with acidified iron(II) sulfate, the manganate(VII) ions are reduced to manganese(II) ions, which are nearly colourless in solution. The iron(II) ions are oxidised to iron(III) ions. The colour change is used when titrating potassium manganate(VII) with iron(II) ions to determine the concentration of the iron(II) ions (see Assignment 3).

The overall redox equation can be deduced from the half-equations.

The iron(II) is oxidised to iron(III). The half-equation is:

$$Fe^{2+}(aq) \rightarrow Fe^{3+}(aq) + e^-$$

The manganese ion has gained five electrons and the reduction half-equation is:

$$MnO_4^- + 5e^- \rightarrow Mn^{2+}$$

But the half-equation for manganese is not balanced for oxygen. Oxygen combines with hydrogen ions in the acidic solution to form water and the balanced half-equation is:

$$MnO_4^-(aq) + 8H^+(aq) + 5e^- \rightarrow Mn^{2+}(aq) + 4H_2O(l)$$

The numbers of atoms are equal on both sides, and the total charge on the left equals the total charge on the right.

While the manganate(VII) is reduced, the iron(II) is oxidised to iron(III). The half-equation is:

$$Fe^{2+}(aq) \rightarrow Fe^{3+}(aq) + e^-$$

Before combining the half-equations, the oxidation half-equation must be multiplied by five so that the same number of electrons are transferred:

$$MnO_4^-(aq) + 8H^+(aq) + 5e^- \rightarrow Mn^{2+}(aq) + 4H_2O(l)$$
$$5Fe^{2+}(aq) \rightarrow 5Fe^{3+}(aq) + 5e^-$$
$$MnO_4^-(aq) + 5Fe^{2+}(aq) + 8H^+(aq) \rightarrow Mn^{2+}(aq) + 5Fe^{3+} + 4H_2O(l)$$

QUESTIONS

Stretch and challenge

14. Identify the spectator ions in the reaction between potassium manganate(VII) solution and acidified iron(II) sulfate solution.

15. The dichromate(VI) ion is used in volumetric analysis as an oxidising agent.

It oxidises iron(II) ions to iron(III) ions. The half-equations are:

$$Fe^{2+}(aq) \rightarrow Fe^{3+}(aq) + e^-$$
$$Cr_2O_7^{2-}(aq) + 14H^+(aq) + 6e^- \rightarrow 2Cr^{3+}(aq) + 7H_2O(l)$$

Write the overall redox equation.

KEY IDEAS

› Half-equations can be written for the oxidation and reduction reactions which occur in a redox reaction.

› Half-equations can be combined to give an overall redox equation.

ASSIGNMENT 3: IRON IN THE BLOOD

(MS 1.1, 1.2, 1.3; PS 1.1, 1.2, 2.3, 3.2, 3.3, 4.1)

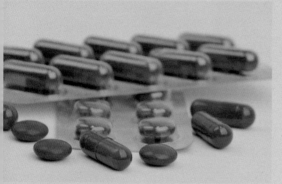

Figure A1 *Iron is essential in the diet for good health. People at risk of iron deficiency can supplement their intake with iron supplements.*

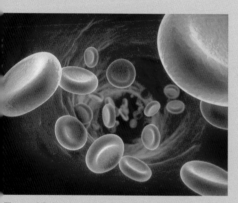

Figure A2 *Haemoglobin in red blood cells contains iron. If insufficient iron is ingested in the diet, insufficient haemoglobin will be present and the ability to transport oxygen around the body will be reduced. The person is anaemic.*

The European Union recommended daily allowance (RDA) for iron intake is 14 mg. In the USA, the RDA is 10–18 mg (30 mg for pregnant women).

An iron supplement claims to contain the RDA for iron (14 mg). The iron is present as iron(II) sulfate. You can check this claim using the redox reaction between iron(II) ions and manganate (VII) ions.

A student crushes five tablets using a pestle and mortar and transfers the crushed tablets to a 100 cm³ beaker. A minimum volume of 1 mol dm⁻³ sulfuric acid is added to dissolve the tablets and the solution is transferred to a 100 cm³ volumetric flask. The apparatus used to dissolve the tablets

is washed with the 1 mol dm⁻³ sulfuric acid and the washings added to the flask. The solution is made up to the graduation mark with 1 mol dm⁻³ sulfuric acid.

25 cm³ portions of this are titrated against 0.010 mol dm⁻³ potassium manganate(VII) solution. The results are shown in Table A1.

		Titration		
	Rough	1	2	3
Final burette reading/cm³	7.50	13.80	20.05	26.25
Initial burette reading/cm³	1.05	7.50	13.80	20.05
Titre/cm³	6.45	6.30	6.25	6.20

Table A1 *Results of experiment*

Questions

A1. What is the average titre?

A2. How many moles of MnO_4^- ions are contained in this titre?

A3. Write the overall equation for the redox reaction.

A4. How many moles of Fe^{2+} ions react with one mole of MnO_4^- ions?

A5. Calculate the number of moles of Fe^{2+} ions in the 25 cm³ of iron tablet solution.

A6. What is the number of moles of Fe^{2+} ions in the 100 cm³ solution?

A7. What mass of iron is contained in five tablets?

A8. What mass of iron is contained in one tablet?

A9. Are the manufacturer's claims correct?

A10. Errors on the apparatus used are: burette, ±0.05 cm³; 100 cm³ volumetric flask, ±0.05 cm³; 25 cm³ pipette ±0.06 cm³. Calculate the percentage (or precision) error on each measurement and the overall percentage (or precision) error.

PRACTICE QUESTIONS

1. The price of copper is increasing as supplies of high-grade ores start to run out. The mineral covellite (CuS), found in low-grade ores, is a possible future source of copper.

   a. When copper is extracted from covellite, a reaction occurs between copper(II) sulfide and nitric acid to form a dilute solution of copper(II) sulfate.

      i. Balance the equation for this reaction.

      $3CuS(s) +$ _____$HNO_3(aq)$ _____ $CuSO_4(aq) +$ _____$NO(g) +$ _____ $H_2O(l)$

      ii. Give the oxidation state of nitrogen in each of the following.

      HNO_3 _____

      NO _____

      iii. Deduce the redox half-equation for the reduction of the nitrate ion in acidified solution to form nitrogen monoxide and water.

      iv. Deduce the redox half-equation for the oxidation of the sulfide ion in aqueous solution to form the sulfate ion and $H^+(aq)$ ions.

   AQA Jul 2013 Unit 2 Question 4a

2. The manufacture of food grade phosphoric acid for use in cola drinks begins with the production of pure white phosphorus from the mineral fluoroapatite, $Ca_5F(PO_4)_3$

   a. Complete the following equation for the manufacture of phosphorus.

      _____ $Ca_5 F(PO_4)_3 + 9SiO_2 +$ _____$C \rightarrow$ $9CaSiO_3 + CaF_2 +$ _____$CO +$ _____P

   b. As the phosphorus cools, it forms white phosphorus, P_4. Give the oxidation state of phosphorus in each of the following.

      P_4 _____

      $H_3 PO_4$ _____

   AQA Jun 2012 Unit 2 Question 3a, b

3. Iodine reacts with concentrated nitric acid to produce nitrogen dioxide (NO_2).

   a. i. Give the oxidation state of iodine in each of the following.

      I_2 _____

      HIO_3 _____

      ii. Complete the balancing of the following equation.

      $I_2 + 10HNO_3 \rightarrow$ _____$HIO_3 +$ _____$NO_2 +$ _____H_2O

   b. In industry, iodine is produced from the $NaIO_3$ that remains after sodium nitrate has been crystallised from the mineral Chile saltpetre. The final stage involves the reaction between $NaIO_3$ and NaI in acidic solution. Half-equations for the redox processes are:

      $IO_3^- + 5e^- + 6H^+ \rightarrow 3H_2O + \frac{1}{2}I_2$

      $I^- \rightarrow \frac{1}{2}I_2 + e^-$

      Use these half-equations to deduce an overall ionic equation for the production of iodine by this process. Identify the oxidising agent.

   AQA Jan 2012 Unit 2 Question 5

4. Reactions that involve oxidation and reduction are used in a number of important industrial processes.

   a. Iodine can be extracted from seaweed by the oxidation of iodide ions. In this extraction, seaweed is heated with MnO_2 and concentrated sulfuric acid.

      i. Give the oxidation state of manganese in MnO_2

      ii. Write a half-equation for the reaction of MnO_2 in acid to form Mn^{2+} ions and water as the only products.

      iii. In terms of electrons, state what happens to the iodide ions when they are oxidised.

   b. Chlorine is used in water treatment. When chlorine is added to cold water it reacts to form the acids HCl and HClO. The following equilibrium is established.

      $Cl_2 (aq) + H_2O(l) \rightleftharpoons$ $H^+ (aq) + Cl^- (aq) + HClO(aq)$

(Continue

Give the oxidation state of chlorine in Cl_2 and in HClO.

AQA Jan 2011 Unit 2 Question 10 a, bi

. **a.** Define an *oxidising agent* in terms of electrons.

b. For the copper in copper oxide, explain the meaning of the term *oxidation state*.

c. Copy and complete Table Q1 by deducing the oxidation state of each of the stated elements in the given ion or compound.

	Oxidation state
Carbon in CO_3^{2-}	
Phosphorus in PCl_4^+	
Nitrogen in Mg_3N_2	

Table Q1

d. In acidified aqueous solution, nitrate ions, NO^{-3}, react with copper metal forming nitrogen monoxide, NO, and copper(II) ions.

  i. Write a half-equation for the oxidation of copper to copper(II) ions.

  ii. Write a half-equation for the reduction, in an acidified solution, of nitrate ions to nitrogen monoxide.

  iii. Write an overall equation for this reaction.

AQA Jan 2005 Unit 2 Question 2

5. Chlorine and bromine are both oxidising agents. In aqueous solution, bromine oxidises sulfur dioxide, SO_2, to sulfate ions, SO_4^{2-}

a. Deduce the oxidation state of sulfur in SO_2 and in SO_4^{2-}

b. Deduce a half-equation for the reduction of bromine in aqueous solution.

c. Deduce a half-equation for the oxidation of SO_2 in aqueous solution forming SO_4^{2-} and H^+ ions.

d. Use these two half-equations to construct an overall equation for the reaction between aqueous bromine and sulfur dioxide.

AQA Jun 2004 Unit 2 Question 4

7. **a.** In terms of electrons, what happens to an oxidising agent during a redox reaction?

b. Consider the following redox reaction.

$$SO_2(aq) + 2H_2O(l) + 2Ag^+(aq) \rightarrow 2Ag(s) + SO_4^{2-}(aq) + 4H^+(aq)$$

  i. Identify the oxidising agent and the reducing agent in this reaction.

  ii. Write a half-equation to show how sulfur dioxide is converted into sulfate ions in aqueous solution.

c. Fe^{2+} ions are oxidised to Fe^{3+} ions by ClO_3^- ions in acidic conditions. The ClO_3^- ions are reduced to Cl^- ions.

  i. Write a half-equation for the oxidation of Fe^{2+} ions in this reaction.

  ii. Deduce the oxidation state of chlorine in ClO_3^- ions.

  iii. Write a half-equation for the reduction of ClO_3^- ions to Cl^- ions in acidic conditions.

  iv. Hence, write an overall equation for the reaction.

AQA Jun 2003 Unit 2 Question 2

8. Nitrogen can have various oxidation states from -3 to $+5$ when it forms molecules and ions.

a. State the oxidation state of nitrogen in the following species:

  i. N_2

  ii. NH_3

  iii. NH_4Cl

  iv. HNO_3

  v. HNO_2

  vi. NO

  vii. NO_2

  viii. NO_3^-

b. i. When copper reacts with warm concentrated nitric acid, NO_2 is formed. Write the half-equation for the conversion of NO_3^- to NO_2 in the presence of hydrogen ions.

  ii. Write the overall equation for the reaction between warm concentrated nitric acid and copper.

(Continued)

9. Water and oxygen are required when iron rusts. The rusting process involves several steps. In the first step, Fe is converted to Fe^{2+}.

   a. i. Write a half-equation to show the conversion of Fe to Fe^{2+}.

      ii. Give the oxidation states of the reactant and the product.

      iii. Identify this as oxidation or reduction.

   b. Write a half-equation for the reaction of oxygen in the presence of water to give OH^- ions.

   c. Use the two half-equations to write an overall ionic equation to show the first stage of rusting.

   d. The Fe^{2+} and OH^- ions formed react to give a precipitate of iron(II) hydroxide. Write an ionic equation to show this reaction.

   e. The second step of the rusting process involves the oxidation of Fe^{2+} in $Fe(OH)_2$ to Fe^{3+} in $Fe(OH)_3$. Write a half equation to show this oxidation reaction.

Stretch and challenge

10. Hydrogen fuel cells can be used in vehicles to provide electricity to power the vehicle. Hydrogen fuel is oxidised at the anode and the oxidising agent is reduced at the cathode. The electrons transferred are diverted to produce an electric current.

   a. i. Write a half-equation for the oxidation of hydrogen fuel at the anode.

      ii. Write a half equation for the reduction of the oxidising agent at the cathode.

      iii. Combine the two half-equations to give an overall redox equation.

      iv. In terms of the number of electrons transferred, explain why twice as much hydrogen as oxygen needs to be supplied to a fuel cell.

   b. Methanol, CH_3OH, can be used as an alternative fuel in fuel cells. Methanol and water react at the anode to produce carbon dioxide. At the cathode, hydrogen ions and oxygen combine to produce water.

      i. Write half-equations for oxidation and reduction reactions.

      ii. Combine the half-equations to give an overall redox equation.

Multiple choice

11. Which vanadium species has an oxidation state of +5?

   A. V^{2+}

   B. V^{3+}

   C. VO^{2+}

   D. VO_2^+

12. Which statement is correct?

   A. Oxidation is the process of electron gain

   B. Oxidation is the process of electron gain and oxidising agents are electron acceptors

   C. Reduction is the process of electron gain and reducing agents are electron acceptors

   D. Reduction is the process of electron gain and reducing agents are electron donors

13. Which is the correct half-equation for the oxidation of magnesium to produce magnesium ions?

   A. $Mg \rightarrow Mg^{2+} + 2e^-$

   B. $Mg + 2e^- \rightarrow Mg^{2+}$

   C. $Mg \rightarrow Mg^{2+} + e^-$

   D. $Mg^{2+} + 2e^- \rightarrow Mg$

14. Which is the correct half-equation for the reduction of oxide ions to oxygen gas?

   A. $O^{2-} + 2e^- \rightarrow O_2$

   B. $2O^{2-} \rightarrow O_2 + 4e^-$

   C. $2O_2 + 2e^- \rightarrow O^{2-}$

   D. $O^{2-} \rightarrow O_2 + 2e^-$

11 GROUP 2, THE ALKALINE EARTH METALS

PRIOR KNOWLEDGE

You may have learnt about calcium compounds in your GCSE course. You may have heated calcium carbonate to produce calcium oxide and added water to produce calcium hydroxide. You may also know some industrial uses of these calcium compounds. In Chapter 1, you read about trends in ionisation energies in Group 2 and in Chapter 6, you found out how calcium oxide and calcium carbonate are used to remove sulfur dioxide from flue gases.

LEARNING OBJECTIVES

In this chapter, you will find out more about the trends in properties of the Group 2 metals and of some of their compounds, and how these determine their uses in agriculture and medicine. You will also find out how magnesium is used in the extraction if titanium.

(Specification 3.2.2)

Harry Potter and the Philosopher's Stone was JK Rowling's first book in the Harry Potter series. In it, Harry discovers his magical heritage and learns of the existence of the Philosopher's Stone, a magical elixir of life reputed to impart immortality.

Early chemists, who were called alchemists, also searched for the Philosopher's Stone; a legendary substance said to turn metals such as lead and copper into gold and silver. The magic was never realised, however, although many claimed to have discovered the Philosopher's Stone. In the 1600s, an Italian shoemaker discovered a silvery white mineral which emitted a red glow when heated with flour. If the combusted material was exposed to sunlight, it would then glow in the dark for up to an hour. Excitedly, he thought he had discovered the Philosopher's Stone, but sadly his fortune never materialised and his discovery became a curiosity.

Today we know the silvery white mineral as barytes (sometimes spelt barites). It is an impure form of barium sulfate and the red glow when heated was probably due to copper impurities. Reduction with carbon produces barium sulfide which is phosphorescent – in sunlight, it absorbs radiation and then emits it later, or glows in the dark.

Figure 1 *Barytes minerals may be tinted yellow, brown, blue or green because of the impurities present.*

Barium is a Group 2 metal, also known as an alkaline earth metal. Beryllium, magnesium, calcium, strontium and radium are the other Group 2 metals.

11.1 TRENDS IN PHYSICAL PROPERTIES

Group 2 elements are all metals with a shiny silvery white appearance (see Figure 2). Their compounds are all found in the Earth's crust and are widely distributed in rock structures. Calcium and magnesium are the most common Group 2 metals, while strontium and barium are less common and beryllium and radium are rare. Table 1 shows the percentage by mass of each Group 2 metal in the Earth's crust. None are found uncombined.

Group 2 metal	Percentage by mass in the Earth's crust
beryllium	1.9×10^{-4}
magnesium	2.9
calcium	5.0
strontium	3.6×10^{-2}
barium	3.4×10^{-2}
radium	9.9×10^{-12}

Table 1 *Abundance of Group 2 metals in Earth's crust*

Figure 2 *The Group 2 elements. From the left: Beryllium, magnesium, calcium, strontium, barium and radium. Radium is extremely reactive. It is also radioactive. In an inert atmosphere, radium is silvery white and its reactivity makes it luminescent.*

The alternative name, alkaline earth metals, comes from their metal oxides. 'Earth' was used by early chemists to describe all non-metallic substances that are insoluble in water and did not decompose on heating. The discovery that the 'Earths' were not elements but metal oxides came later, in 1789 when Antoine Lavoisier predicted that they might be metal oxides. Sir Humphrey Davies later confirmed the prediction by extracting some of the metals by electrolysis.

Figure 3 *Beryllium is rare and occurs as 2ppm in the Earth's crust. Emeralds are one of the naturally occurring compounds of beryllium; beryllium aluminium silicate, $Be_3Al_2(SiO_3)_6$.*

Atomic number	Element	Outer electron configuration	Metallic radius/ nm	Ionic radius/ nm	First ionisation energy/ kJ mol^{-1}	Electronegativity (scale 0 to 4)	Melting point/ °C
4	Be	$2s^2$	0.112	0.027	900	1.5	1278
12	Mg	$3s^2$	0.160	0.072	736	1.2	650
20	Ca	$4s^2$	0.197	0.100	590	1.0	850
38	Sr	$5s^2$	0.215	0.113	548	1.0	768
56	Ba	$6s^2$	0.221	0.136	502	0.9	714
88	Ra	$7s^2$	0.220	0.140	510	0.9	700

Table 2 *Some properties of Group 2 elements*

Electronic configurations

Group 2 elements have two electrons in their outer s sub-shell. They have metallic bonding where each metal atom transfers two electrons to create the sea of delocalised electrons (Figure 4).

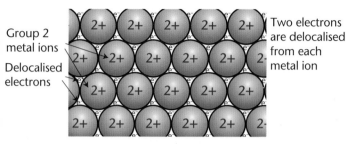

Group 2 metal ions

Delocalised electrons

Two electrons are delocalised from each metal ion

Figure 4 *Metallic bonding in Group 2 metals (this illustrates the bonding, but does not show the structure).*

Moving down Group 2, each element has an extra shell of electrons. The electron configuration for beryllium is $1s^2 2s^2$ and that for magnesium is $1s^2 2s^2 2p^6 3s^2$.

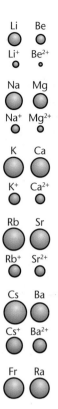

Figure 5 *The atomic and ionic radii of Group 1 and 2 elements*

QUESTIONS

1. Write the full electron configuration for:
   a. calcium
   b. strontium.

Trends in atomic radius

The atomic radius is the distance from the centre of an atom to its outermost electrons. It is usually measured as half the distance between the nuclei of two atoms (see Chapter 4). Since Group 2 elements are metals, the metallic radii values are used.

Moving down Group 2, the atomic radii increase. Each successive element has an extra shell of electrons and a larger nuclear charge. The additional shell of electrons shields the outer 2s electrons from the nuclear charge. The outer electrons are attracted less strongly and they move further away from the nucleus. The atomic radius increases. Figure 5 illustrates the trends in ionic and atomic radii of Group 1 and 2 elements.

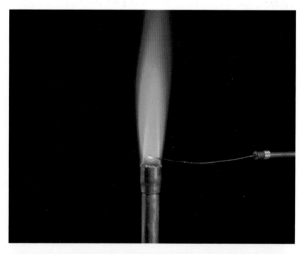

Figure 6 *Group 2 metals can be identified from their emission spectra. A spectroscope splits the brick red light emitted during the calcium flame test into its individual wavelengths, producing a characteristic line spectrum. The lines correspond to the energy given out when excited electrons fall back to their ground state.*

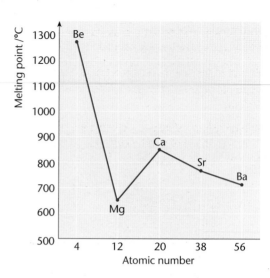

Figure 8 *Trend in melting points of Group 2 elements*

Trend in first ionisation energy

The first ionisation energy is the energy needed to remove one mole of electrons from one mole of atoms in the gaseous state (see Chapter 1). It is measured in $kJ\,mol^{-1}$. First ionisation energies decrease down Group 2 as extra shells of electrons are added. These successfully shield the outer 2s electrons from the nuclear charge and it is easier to remove an electron. So, the first ionisation energy decreases down Group 2, as shown in Figure 7.

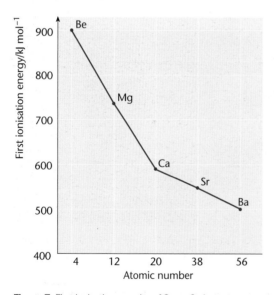

Figure 7 *First ionisation energies of Group 2 elements*

Trend in melting points

The trend in melting points for Group 2 elements is not so clear cut, but does generally decrease as you go down Group 2. Figure 8 shows the trend.

The melting point of a metal depends on the strength of the bonds formed when the element's atoms join together to make giant structures with metallic bonding. When metals melt, the metal ions gain sufficient kinetic energy to break free from their fixed positions in the surrounding sea of delocalised electrons. The melting point is an indication of the energy that must be transferred to the metal for it to melt.

Group 2 elements have giant structures and metallic bonding, with the same number (two) of delocalised electrons per metal ion. The attraction between the positive metal ion and the delocalised electrons holds the metal together. The stronger this attraction, the higher the melting point of the metal because more energy is needed to overcome the stronger bonds.

Moving down Group 2, successive shells of electrons shield the metal ion nucleus from the delocalised electrons, reducing the attraction. It becomes easier to break the metallic bonds and the melting point is lower.

Looking again at Figure 8, you will see that the melting point of magnesium is lower than might be predicted from the trend in the melting points of the other Group 2 metals. This is because the strength of the metallic bond depends not only on the charges and radii of the metal ions, but also on the pattern in which the metals ions pack together.

Beryllium and magnesium have hexagonal close packed structures.

Calcium, strontium and barium have cubic structures. These are not as closely packed as hexagonal structures. Calcium and strontium are face centred cubic and barium is body centred cubic.

This difference in packing patterns accounts for the 'jump' in calcium compared to the 'fall' from beryllium to magnesium.

However, magnesium's unusually low melting point of 650 °C makes it a useful reducing agent in the extraction of titanium metal.

11.2 EXTRACTING TITANIUM

Titanium has a high tensile strength, similar to steel. Steel is relatively cheap and readily available, but titanium has a significant advantage for many applications – its density is half that of steel. It does not corrode, even under extreme conditions, because it forms a layer of protective oxide on its surface when exposed to air.

Titanium makes up 0.66% by mass of the Earth's crust and is potentially a very useful metal (see Figure 9). The problem is that it is difficult and expensive to extract.

Titanium is extracted from rutile, TiO_2, and ilmenite, $FeTiO_3$. Unfortunately, reduction with carbon produces titanium carbide, which makes the metal brittle and useless. Reduction of the oxide with more reactive metals also results in impurity problems and it is difficult and expensive to eliminate the reducing agent from the titanium. But, titanium(IV) chloride (Figure 10) is more easily reduced.

The process involves two steps. First titanium(IV) oxide is converted to titanium(IV) chloride, $TiCl_4$, by heating with carbon in a stream of chlorine gas. The reaction is:

$$TiO_2(s) + 2Cl_2(g) + 2C(s) \rightarrow TiCl_4(g) + 2CO(g)$$

Titanium(IV) chloride is then reduced using magnesium metal as the reducing agent. The process is called the

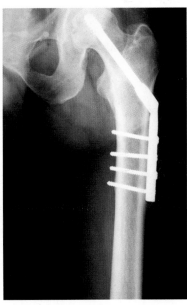

Figure 9 *The expense of extraction limits the uses of titanium metal to the aero industry and artificial joints. The fan of this aeroplane engine (left) is made from titanium. The X-ray (right) shows an artificial titanium hip joint.*

Figure 10 *Titanium(IV) chloride is a liquid consisting of $TiCl_4$ molecules at room temperature. It can be purified by fractional distillation.*

Kroll process and Figure 11 shows the furnace used. Magnesium is oxidised, from oxidation state 0 in the metal to oxidation state +2 in magnesium chloride.

Titanium(IV) chloride and liquid magnesium are heated to about 1200 °C in an inert atmosphere of helium or argon gas.

$$TiCl_4(l) + 2Mg(l) \rightarrow Ti(l) + 2MgCl_2(l)$$
$$\Delta H^{\ominus} = -504 \text{ kJ mol}^{-1}$$

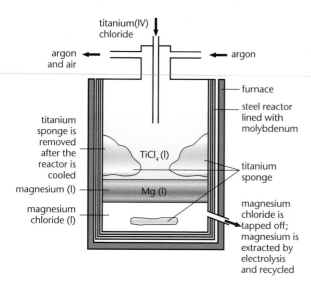

Figure 11 *Extracting titanium by the Kroll process*

The process is lengthy and takes 36 to 50 hours, plus four days to cool. It is called a batch process because the furnace does not operate continuously. It is cooled after each operation, the contents emptied and the process started again.

QUESTIONS

4. a. What is the percentage atom economy when titanium(IV) chloride is reduced with magnesium?

    b. Electrolysis is used to extract magnesium from the magnesium chloride produced. The magnesium is then recycled. What is the percentage atom economy when magnesium is recycled?

Stretch and challenge

5. Industrial processes are either continuous processes or batch processes. The blast furnace used to extract iron from iron ore runs 365 days a year, 24 hours a day. Reactants are continuously added and products removed. It is an example of a continuous process. In contrast, titanium is extracted in a batch process. What are the advantages and disadvantages of using:

    a. a batch process

    b. a continuous process?

KEY IDEAS

› Atomic radii increase down Group 2 as extra shells of electrons shield the nuclear charge from the outer electrons and they move further away.

› First ionisation energies decrease down Group 2 as successive electron shells shield the nuclear charge and it becomes easier to remove an electron.

› Melting points generally decrease down Group 2 as metallic bonds become weaker.

› Magnesium is used to reduce titanium(IV) chloride in the Kroll process.

11.3 TRENDS IN CHEMICAL PROPERTIES

The reactions of Group 2 elements with water

Group 2 metals have two electrons in their outer shells which are readily transferred to other elements and compounds. This makes Group 2 metals good reducing agents. They can reduce, for example, water to hydrogen. The general reaction is:

$$M(s) + 2H_2O(l) \rightarrow M(OH)_2(s \text{ or } aq) + H_2(g)$$

where M represents a Group 2 metal.

The full redox equation is:

$$M(s) + 2H_2O(l) \rightarrow M^{2+}(aq) + 2OH^-(aq) + H_2(g)$$

You can write this as half-equations. The oxidation reaction is:

$$M \rightarrow M^{2+} + 2e^-$$

The reduction reaction is:

$$2H_2O + 2e^- \rightarrow 2OH^- + H_2$$

Adding these two half-equations and cancelling out the number of electrons on each side gives the overall equation for the reaction.

The ease with which the metal can transfer electrons depends on the first and second ionisation energies.

First ionisation energy: $M(g) \rightarrow M^+(g) + e^-$

Second ionisation energy: $M^+(g) \rightarrow M^{2+}(g) + e^-$

You have seen that the first ionisation energy of the elements decreases down Group 2. The second isation energy follows a similar pattern. So, the ﹖er an element is in Group 2, the easier it is to ﹖nsfer two electrons.

ce the transfer of electrons away from an atom ﹖xidation, the energy needed to oxidise Group 2 ﹖tals is less as you go down the group.

﹖gnesium reacts only very slowly with cold water ﹖roduce magnesium hydroxide and hydrogen gas. ﹖akes several days to produce noticeable amounts ﹖ydrogen gas. This is because the magnesium ﹖droxide formed is almost insoluble and forms a ﹖er on the magnesium metal. If hot water is used, ﹖ rate of reaction increases.

$$\text{(s)} + 2H_2O(l) \rightarrow Mg(OH)_2(s) + H_2(g)$$

﹖t, if steam is passed over heated magnesium, ﹖exothermic reaction occurs (Figure 12) and the ﹖gnesium hydroxide initially formed is decomposed ﹖ magnesium oxide and hydrogen gas:

$$\text{(s)} + H_2O(l) \rightarrow MgO(s) + H_2(g)$$

QUESTIONS

5. What is the oxidation state of Group 2 elements in their compounds?

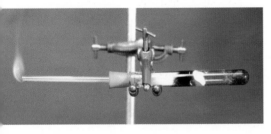

﹖ure 12 *Magnesium reacts exothermically with steam.*

﹖ure 13 *Calcium reacts with cold water.*

The first and second ionisation energies of calcium are lower than those of magnesium, and so calcium transfers its two outer electrons more easily. Calcium reacts with cold water to form calcium hydroxide and hydrogen gas (Figure 13). The resulting solution is commonly called limewater and is used to test for carbon dioxide gas (Figure 14).

$$Ca(s) + 2H_2O(l) \rightarrow Ca(OH)_2(aq) + H_2(g)$$

Figure 14 *Limewater is calcium hydroxide solution. When carbon dioxide is bubbled through a test tube containing limewater, a white precipitate of calcium carbonate forms. This is a common test for carbon dioxide.*

Strontium and barium react with increasing vigour with water to produce metal hydroxide and hydrogen.

QUESTIONS

7. a. Write a full equation for the reaction of strontium with cold water.
   b. Write half-equations to show the oxidation and reduction reactions.

The solubilities of Group 2 hydroxides

A **solute** is a substance that is present in a solution. A **solution** is made by dissolving the substance in a **solvent**. It is a homogeneous mixture of a solute and solvent. The most common solvent used to dissolve substances is water, but water does not dissolve all substances.

Different compounds have different solubilities in water. The solubility of a substance measures the maximum concentration of the solution possible. The same is true of other solvents.

231

Chemists often group substances into three main categories of solubility, using these terms:

> insoluble: the substance does not dissolve or, if it does, only in tiny quantities

> sparingly soluble: only very small quantities of the substance dissolve

> soluble: the substance dissolves.

These terms are useful when comparing a number of substances under the same conditions (volume and temperature of solvent). However, they are not quantitative.

When magnesium hydroxide, a white powder, is added to water and shaken, it appears insoluble. However, testing the pH of the water shows that it has become slightly alkaline (pH just more than 7). The increase in pH shows that there must be hydroxide ions in solution and, therefore, that a small amount of magnesium hydroxide dissolved.

$$Mg(OH)_2(s) + aq \rightarrow Mg^{2+}(aq) + 2OH^-(aq)$$

We describe magnesium hydroxide as sparingly soluble.

As you go down Group 2, the solubility of the hydroxides increases (Figure 15). Calcium hydroxide is described as soluble. A solution of calcium hydroxide in water is commonly called limewater. Barium hydroxide is soluble.

The general equation for these dissolution processes is

$$M(OH)_2(s) + aq \rightarrow M^{2+}(aq) + 2OH^-(aq)$$

where M represents a Group 2 metal.

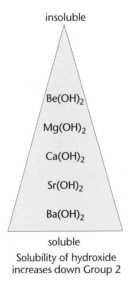

insoluble

Be(OH)$_2$

Mg(OH)$_2$

Ca(OH)$_2$

Sr(OH)$_2$

Ba(OH)$_2$

soluble

Solubility of hydroxide
increases down Group 2

Figure 15 *Solubilities of hydroxides of Group 2 elements in water*

Solubility is the maximum amount of solute that will dissolve in a solvent at a stated temperature (solubility varies with temperature). It may be measured in grams per 100 g of solvent, grams per 100 cm^3 of solvent, grams per 1000 cm^3 of solvent or moles per dm^3 of solution.

For example, the solubility of calcium hydroxide is often given as 1 g in 1 dm^3 of water. Since the relative formula mass of calcium hydroxide is 74, this is approximately 0.014 mol dm^{-3} of solution. The solubility of barium hydroxide is about 0.1 mol dm$^-$ of solution.

QUESTIONS

8. When Group 2 metals react with water, which hydroxides form insoluble solids, and which are in solution?

9. Write an equation to represent the dissolution of solid barium hydroxide in water.

11.4 USES OF GROUP 2 HYDROXIDES

Magnesium hydroxide is used in medicines for digestive upsets and as a laxative. An 8% aqueous suspension is known as Milk of Magnesia. Since magnesium hydroxide is only sparingly soluble, a suspension of particles is formed, which settles on standing. Magnesium hydroxide is also available in tablet form (for example, Milk of Magnesia tablets) and is present in many indigestion remedies.

Stomach acid is hydrochloric acid (about 0.1 mol dm^{-3}) and excess stomach acid causes pain and discomfort. Also, some acid leaks up into the stomach and oesophagus, a process called acid reflux, and causes irritation. Magnesium hydroxide is used to neutralise excess acid in the oesophagus. The reaction is:

$$Mg(OH)_2(s) + 2HCl(aq) \rightarrow MgCl_2(aq) + 2H_2O(l)$$

The resulting magnesium chloride solution is neutral (pH 7).

Tooth decay is caused when bacteria in the mouth feed on sugary foods and produce acidic toxins. Neutralising this acidity is an important feature of toothpastes. Many have magnesium hydroxide added for this purpose.

cium hydroxide is commonly known as slaked
e. The name comes from the highly exothermic
ction that happens when water is added to
cium oxide (lime).

)(s) + H$_2$O(l) → Ca(OH)$_2$(aq)

s used in agriculture to neutralise soil acidity.
hough it is called soil acidity, it is the acidity
water that surrounds the soil particles that is
asured (Figure 16). The healthy growth of plants
bend on soil pH and the best growth for each
cies is only obtained in a narrow pH range. For
st plants, this is between pH 6 and 7. If soil pH
ues are too high or too low, absorption of plant
rients is interfered with and crops demonstrate
or growth. To the agriculture industry, this is a low
ld and a low profit. Adjusting soil pH is important
mprove agricultural yields.

ny of the uses of calcium hydroxide and
gnesium hydroxide involve neutralisation reactions.
ese may be summarised by the equation

(aq) + OH$^-$(aq) → H$_2$O(l)

Figure 16 *Clay soils, soils with a high organic material content and soils with high aluminium content are often acidic. pH indicators are used in soil testing kits to measure acidity.*

ASSIGNMENT 1: PRECISION FARMING

PS 1.1, 1.2, 2.1, 4.1)

he increasing global population means there is
n increasing demand for food, and so improving
agricultural management worldwide is essential to
meet this demand.

n the past, farmers have treated their fields
uniformly and not taken into account variations
within a field. So chemicals such as calcium hydroxide
slaked lime) would be spread on the whole field,
egardless of whether it was needed all over or not.
A better system is to identify parts of the field that
are deficient in a particular additive and only apply
he right amount where it is needed. This also
applies to fertilisers, pesticides, herbicides and other
additives used.

Global positioning systems (GPS), along with
echnologies that enable the farmer to vary the rate
at which additives are spread (called variable rate
echnologies), have enabled this location-specific
approach to soil management. It is called precision
arming. Tractors are equipped with a GPS that can
pinpoint their position to within one metre.

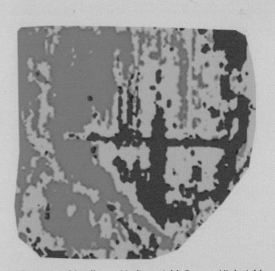

Red = Low yield; Yellow = Medium yield; Green = High yield

Figure A1 *This map shows the yields of crops obtained from a field. The red areas and yellow areas give low and medium yields, respectively, and are the areas to be treated. If the soil pH falls below the optimum range for the crop, these areas can be treated with calcium hydroxide. Money is not wasted treating the green areas.*

Sensors on board the tractor make measurements of soil properties, including pH, and software generates maps showing variations in, for example, soil pH, water content and crop health. This information is relayed by satellite to a control centre where the data is compared with the ideal conditions for plant growth. Instantly, any adjustments needed are sent back to the tractor where an onboard computer regulates the addition of, for example, calcium hydroxide to adjust soil pH and fertilisers to regulate any nutrient deficiencies. All this happens before the tractor has had time to move further, so soil treatment happens at the location given by the GPS.

Figure A2 *Using GPS systems to add the right amount of calcium hydroxide or other additive only where it is needed benefits the crops, soil and ground water.*

Questions

Stretch and challenge

A1. What type of sensor could be used to measure the acidity of soil?

A2. The best yields of wheat are obtained from soil with pH values from 6.0 to 7.0. What data would be gathered by the tractor to trigger the addition of lime to the soil?

A3. The hydroxide ions in calcium hydroxide react with hydrogen ions in soil to produce water. Write an equation to show this.

A4. What is the advantage of using calcium hydroxide as a fine powder rather than in pellet form?

A5. Farmers also use other calcium compounds to increase soil pH. The term 'liming' is generally used for any calcium compound used and can get confusing because 'lime' is the common name for calcium oxide. In theory, any Group 2 hydroxide could be used. Strontium hydroxide and barium hydroxide dissolve to give solutions with higher pH values than calcium hydroxide, but this makes them highly corrosive.

   a. Write an ionic equation to show how powdered limestone (an impure form of calcium carbonate) can decrease soil acidity.

   b. Two solutions are made, one by adding calcium hydroxide to water until no more will dissolve, and the other by adding barium hydroxide to water until no more will dissolve. Explain why the pH of barium hydroxide solution is higher than the pH of calcium hydroxide solution.

   c. Suggest a reason why strontium and barium hydroxides are not used to raise soil pH.

A6. a. What type of bonding is present in solid calcium hydroxide?

   b. What type of structure does calcium hydroxide have?

Stretch and challenge

A7. Dissolution of an ionic solid requires bonds between the ions to be broken. The freed ions become surrounded by and bonded to water molecules. Hydrated ions are formed.

   a. Why is less energy required to break up the barium hydroxide lattice than to break up the calcium hydroxide lattice?

   b. Explain why breaking up a lattice is an endothermic process and forming hydrated ions is an exothermic process.

ASSIGNMENT 2: ELEMENT-120

(PS 1.1, 1.2, 2.3, 3.2)

The known Group 2 elements occur naturally, but experiments have been carried out to synthesis the element with atomic number 120. This is likely to be the next member of Group 2.

Figure A1 *The Flerov Laboratory of Nuclear Reactions was founded in 1957.*

The first attempt was carried out in March 2007 at the Flerov laboratory of Nuclear Reactions in Dubna, Russia. The reaction carried out to try to produce the next member of Group 2 was a nuclear reaction. Plutonium-244 ions were bombarded with iron-58 ions in a particle accelerator. However, no

atoms with atomic number 120 were identified. The second attempt was carried out in April 2007 in Germany when uranium-238 ions were bombarded with nickel-64 ions. Again, no atoms with atomic number 120 were identified.

Element-120's properties are predicted to be closer to calcium and strontium than to barium and radium. Scientists expect element 120 to have a first ionisation energy higher than that of barium and radium.

Questions

A1. Why is element-120 likely to be in Group 2?

A2. How many protons and electrons will an atom of element-120 have?

A3. In theory, why might plutonium-244 ions and iron-58 ions react to produce element-120?

A4. In terms of electrons, protons and neutrons, how do nuclear reactions differ from ordinary chemical reactions?

A5. What is the trend in first ionisation energies down Group 2?

A6. Use Figure 7 to suggest a value for the first ionisation energy for element 120.

Stretch and challenge

A7. Attempts to synthesis element-120 have also been made using curium ions and using californium ions. Which elements would they need to be bombarded with, theoretically, to produce element-120?

11.5 THE RELATIVE SOLUBILITIES OF THE GROUP 2 SULFATES

The solubilities of the Group 2 sulfates decrease down Group 2 (Figure 17). The trend is the opposite to that for the solubilities of the Group 2 hydroxides. Magnesium sulfate is very soluble in water, calcium sulfate is sparingly soluble, and strontium sulfate and barium sulfate are insoluble.

Soluble compounds of barium are very poisonous. However, barium sulfate is insoluble and can pass

through the body without dissolving in body fluids and being absorbed. It is used as the 'barium meal' given to patients when X-ray images of their digestive system are needed. Normally, X-rays are only absorbed by hard bone material in the body, which is why only your skeleton shows up on an X-ray. You cannot see the soft tissues that make up your organs. But barium sulfate does absorb X-rays, so as the barium meal fills the digestive tract, an X-ray image shows any abnormalities (see Figure 18).

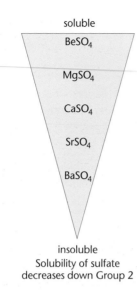

soluble

$BeSO_4$

$MgSO_4$

$CaSO_4$

$SrSO_4$

$BaSO_4$

insoluble
Solubility of sulfate
decreases down Group 2

Figure 17 *The solubilities of the sulfates of Group 2 elements in water*

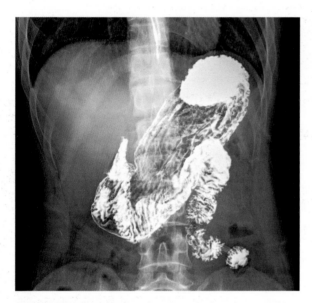

Figure 18 *Barium sulfate is commonly used to investigate conditions of the large intestine using X-rays.*

Identifying sulfates

When two solutions of ionic compounds are mixed, a mixture of ions is produced. If any of the ions in solution combine to produce an insoluble compound, a solid forms. It is called a **precipitate**.

Precipitates are useful for identifying anions (negatively charged ions). Barium sulfate is insoluble. If two solutions are mixed, one containing barium ions and the other containing sulfate ions, a white precipitate of barium sulfate forms (Figure 19). The equation for the reaction that occurs when barium

chloride solution and sodium sulfate solution are mixed is:

$$BaCl_2(aq) + Na_2SO_4(aq) \rightarrow BaSO_4(s) + 2NaCl(aq)$$

The ionic equation for the formation of the precipitate is

$$Ba^{2+}(aq) + SO_4^{2-}(aq) \rightarrow BaSO_4(s)$$

Note that the state symbols show the formation of the solid precipitate.

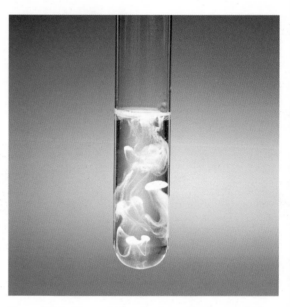

Figure 19 *A white precipitate of barium sulfate forms as barium chloride solution is added to a solution containing sulfate ions.*

The reaction may be used to identify sulfate ions in solution. Barium chloride is soluble, so a solution of barium chloride provides the barium ions.

$$BaCl_2(s) + aq \rightarrow Ba^{2+}(aq) + 2Cl^-(aq)$$

However, if the solution being tested contains carbonate ions, a white precipitate of barium carbonate forms and the results can be misleading.

$$Ba^{2+}(aq) + CO_3^{2-}(aq) \rightarrow BaCO_3(s)$$

Therefore, dilute hydrochloric acid or nitric acid is added to the barium chloride solution. The acid reacts with any carbonate ion present producing carbon dioxide and water and removing them from the solution.

$$CO_3^{2-}(aq) + 2H^+(aq) \rightarrow CO_2(g) + H_2O(l)$$

A white precipitate of barium sulfate when an acidified solution of barium chloride is added confirms the presence of sulfate ions.

$$Ba^{2+}(aq) + SO_4^{2-}(aq) \rightarrow BaSO_4(s)$$

QUESTIONS

10.a. Name the spectator ions in the reaction between barium chloride solution and sodium sulfate solution.

b. What happens to the spectator ions when the precipitate of barium sulfate forms?

c. Why is sulfuric acid not used to acidify barium chloride solution?

Sparingly soluble calcium sulfate

Another application of sulfate solubility is the use of calcium oxide and calcium carbonate to remove sulfur dioxide from waste gases or flue gases. You read about this in Chapter 6.

Sulfur dioxide acidifies rain and needs to be removed from waste gases. The reactions with calcium oxide and calcium carbonate are:

$SO_2(g) + CaCO_3(s) \rightarrow CaSO_3(aq) + CO_2(g)$; then

$2CaSO_3(s) + O_2(g) \rightarrow 2CaSO_4(s)$

and

$SO_2(g) + CaO(s) \rightarrow CaSO_3(s)$; then

$2CaSO_3(s) + O_2(g) \rightarrow 2CaSO_4(s)$

Calcium sulfate is sparingly soluble. Only a very small amount dissolves in any steam present. Most forms as solid calcium sulfate which is removed and used in the building industry.

KEY IDEAS

> The solubility of Group 2 hydroxides increases down the group.

> Magnesium hydroxide is used in medicine to neutralise stomach acid.

> Calcium hydroxide is used in agriculture to neutralise soil acidity.

> The solubility of Group 2 sulfates decreases down the group.

> Barium sulfate is insoluble and is used in barium meals to X-ray the digestive system.

> Acidified barium chloride solution is used to identify sulfate ions. A white precipitate of barium sulfate forms.

> Calcium oxide and calcium carbonate are used to remove sulfur dioxide from flue gases.

ASSIGNMENT 3: ANALYSIS BY PRECIPITATION

(MS 2.3; PS 1.1, 1.2, 2.3, 3.2)

If you mix sodium nitrate and copper(II) sulfate solutions, you simply get a mixture of $Na^+(aq)$, $NO_3^-(aq)$, $Cu^{2+}(aq)$ and $SO_4^{2-}(aq)$ ions in solution. Sodium nitrate solution is colourless, and copper(II) sulfate solution is blue. The mixture is pale blue.

If you mix sodium carbonate solution (colourless solution) and copper(II) nitrate solution (blue solution), a green precipitate forms:

$2Na^+(aq) + CO_3^{2-}(aq) + Cu^{2+}(aq) + 2NO_3^-(aq) \rightarrow$
$\qquad CuCO_3(s) + 2Na^+(aq) + 2NO_3^-(aq)$

The $Na^+(aq)$ and $NO_3^-(aq)$ stay in solution, but when the copper ions collide with the carbonate ions, a precipitate of copper(II) carbonate forms. The $Na^+(aq)$ and $NO_3^-(aq)$ ions are spectator ions. Cancelling these out, the ionic equation becomes:

$CO_3^{2-}(aq) + Cu^{2+}(aq) \rightarrow CuCO_3(s)$

The information in Table A1 is useful in deciding whether or not a precipitate forms.

Soluble	Insoluble
all nitrates	none
all sulfates, except Pb^{2+}, Ba^{2+}, Hg^{2+}, Ca^{2+}	$PbSO_4$, $BaSO_4$, $HgSO_4$, $CaSO_4$
all chlorides, except Ag^+, Hg^{2+}, Pb^{2+}	$AgCl$, $HgCl_2$, $PbCl_2$
bromides and iodides follow a similar pattern to chlorides	
Na_2CO_3, K_2CO_3, $(NH_4)_2CO_3$	all carbonates, CO_3^{2-}, except Na^+, K^+, NH_4^+

Table A1 Soluble and insoluble ions

Students were given a sample of seawater and asked to carry out analyses of it.

Qualitative analysis

The qualitative tests that they carried out and the results are shown in Table A2.

Test		Result
1	Addition of a few drops of acidified barium chloride solution to 2 cm depth of sea water in a test tube	A white precipitate formed
2	Addition of a few drops of silver nitrate solution to 2 cm depth of sea water in a test tube	A white precipitate formed that darkened in the light

Table A2 *Analysing sea water*

Quantitative analysis

Then the students carried out a quantitative analysis. They took 100 cm^3 of sea water and added barium chloride solution until no more precipitate formed. The barium sulfate precipitate was filtered, washed, dried and weighed. They obtained 1.51 g of barium sulfate.

Questions

A1. Write ionic equations for the reactions that produce a precipitate:

    a. $Pb(NO_3)_2(aq) + CuSO_4(aq)$

    b. $Na_2CO_3(aq) + Cu(NO_3)_2(aq)$

    c. $AgNO_3(aq) + FeCl_2(aq)$

    d. $MgSO_4(aq) + CuCl_2(aq)$

    e. $MgCl_2(aq) + Pb(NO_3)_2(aq)$

    f. $CuCl_2(aq) + Mg(NO_3)_2(aq)$

    g. $(NH_4)_2CO_3(aq) + CaCl_2(aq)$

    h. $BaCl_2(aq) + CaSO_4(aq)$

A2. a. What does the result of qualitative test 1 show?

    b. Carry out an internet search to find out what qualitative test 2 shows.

    c. Write ionic equations to show the reactions in tests 1 and 2.

A3. In the quantitative analysis:

    a. how many moles of barium sulfate did the students obtain?

    b. How many moles of sulfate ions in sea water were needed to produce that many moles of barium sulfate?

    c. What is the concentration of the sulfate ions in sea water in mol dm^{-3}?

Stretch and challenge

A4. Use the solubility table to plan how you could test the solubility of Group 2 hydroxides by mixing solutions of soluble Group 2 salts with sodium hydroxide solution. What would you expect to observe?

PRACTICE QUESTIONS

1. Barium metal reacts very quickly with dilute hydrochloric acid, but it reacts more slowly with water.

    a. Write an equation for the reaction of barium with water.

    b. A solution containing barium ions can be used to show the presence of sulfate ions in an aqueous solution of sodium sulfate. Write the simplest ionic equation for the reaction that occurs and state what is observed.

    c. State one use of barium sulfate in medicine. Explain why this use is possible, given that solutions containing barium ions are poisonous.

        AQA Jan 2013 Question 1c

2. Group 2 elements and their compounds have a wide range of uses.

    a. For parts a.i. to a.iii., draw a ring around the correct answer to complete each sentence.

(Continued)

i. From $Mg(OH)_2$ to $Ba(OH)_2$, the solubility in water:
decreases/increases/stays the same

ii. From Mg to Ba, the first ionisation energy:
decreases/increases/stays the same

iii. From Mg to Ba, the atomic radius:
decreases/increases/stays the same

b. Explain why calcium has a higher melting point than strontium.

c. Acidified barium chloride solution is used as a reagent to test for sulfate ions.

i. State why sulfuric acid should not be used to acidify the barium chloride.

ii. Write the simplest ionic equation for the reaction that occurs when acidified barium chloride solution is added to a solution containing sulfate ions.

AQA Jan 2012 Unit 2 Question 7

3. Group 2 metals and their compounds are used commercially in a variety of processes.

a. Strontium is extracted from strontium oxide (SrO) by heating a mixture of powdered strontium oxide and powdered aluminium. Consider the standard enthalpies of formation in Table Q1.

$$3SrO(s) + 2Al(s) \rightarrow 3Sr(s) + Al_2O_3(s)$$

	SrO(s)	Al_2O_3(s)
$\Delta_fH^{\ominus}$/ kJ mol^{-1}	-590	-1669

Table Q1 *Some standard enthalpies of formation*

i. Use these data and the equation to calculate the standard enthalpy change for this extraction of strontium.

The use of powdered strontium oxide and powdered aluminium increases the surface area of the reactants.

ii. Suggest one reason why this increases the reaction rate.

iii. Suggest one major reason why this method of extracting strontium is expensive.

b. Explain why calcium has a higher melting point than strontium.

c. Magnesium is used in fireworks. It reacts rapidly with oxygen, burning with a bright white light. Magnesium reacts slowly with cold water.

i. Write an equation for the reaction of magnesium with oxygen.

ii. Write an equation for the reaction of magnesium with cold water.

iii. Give a medical use for the magnesium compound formed in the reaction of magnesium with cold water.

AQA Jun 2013 Unit 2 Question 11

4. The reactivity of the elements in Group 2 is related to their ionisation energies.

a. Write an equation, including state symbols, to show the first ionisation energy for magnesium.

b. Explain the trend in first ionisation energies down Group 2 from magnesium to barium.

c. Magnesium reacts very slowly with cold water. Write an equation for the reaction.

d. How does the rate of reaction between magnesium and water compare with that of the reaction between calcium and water?

e. Calcium hydroxide is soluble in water. The solution is commonly called lime water and is used to test for carbon dioxide. State the trend in solubilities of the hydroxides down Group 2.

f. Calcium hydroxide can be used to raise the pH of soil. Write a full balanced equation for the reaction between calcium hydroxide and hydrochloric acid. Include state symbols.

5. Group 2 metals have many uses in medicine. Magnesium hydroxide is found in many digestive remedies where indigestion is caused by excess stomach acid.

a. Magnesium hydroxide medicine is sold as Milk of Magnesia tablets.

i. Is it possible to dissolve a Milk of Magnesia tablet in cold water? Explain your answer, referring to the solubilities of the Group 2 hydroxides.

(Continued)

ii. What type of chemical reaction takes place between Milk of Magnesia tablets and stomach acid?

iii. Stomach acid is dilute hydrochloric acid. Write an ionic equation for its reaction with Milk of Magnesia tablets.

b. Magnesium hydroxide can also be used as a laxative. Here it is the Mg^{2+} ions that produce the effect. Other magnesium compounds, such as magnesium sulfate, can have similar effects.

i. State the observations when magnesium sulfate is added to water.

ii. Explain your answer with reference to the solubilities of the Group 2 sulfates.

c. Barium salts are very toxic. Why can barium sulfate be used internally in the large intestine in medicine without causing toxic effects?

6. 7.7% of all the ions present in sea water are sulfate ions, SO_4^{2-}.

a. i. Describe a chemical test that chemists could use to confirm the presence of sulfate ions in sea water.

ii. Describe the change that happens if sulfate ions are present.

iii. Write an ionic equation for the reaction.

b. i. Describe the trend in solubilities of the Group 2 sulfates?

ii. Describe the trend in solubilities of the Group 2 hydroxides?

c. Sea water also contains calcium ions (1.2% of all ions in sea water) and magnesium ions (3.7% of all ion in sea water).

i. Name a calcium compound used in agriculture and describe its use.

ii. Name a magnesium compound used in medicine and describe its use.

iii. Explain why magnesium instead of carbon is used to extract titanium from its ore.

7. Table Q2 shows the metallic radii of some Group 2 metals.

Group 2 metal	Metallic radius/nm
magnesium	0.160
calcium	0.197
strontium	0.215
barium	0.221

Table Q2

a. Explain the trend in metallic radius as you go down Group 2.

b. i. Write an equation to show the first ionisation of barium.

ii. Explain why this value is lower than the first ionisation of magnesium.

c. Group 2 elements react with water.

i. Write a full equation for the reaction between calcium and water.

ii. State the trend down the group of the reaction of Group 2 metals with water?

iii. Explain this trend.

8. Group 2 elements are metals. Many of their physical properties can be explained by considering the bonding.

a. What type of bonding is present in a solid Group 2 metal?

b. The melting points of four metals are given in Table Q3.

Group1 metal	Melting point/K	Group 2 metal	Melting point/K
sodium	371	magnesium	922
potassium	336	calcium	1112

Table Q3

i. Explain why the Group 2 metals in the table have higher melting points than the Group 1 metals.

ii. The Group 2 metals in the table are not typical of the trend in melting points in this group. What is the overall trend?

iii. Explain this trend.

(Continued)

c. Apart from magnesium, metals with lower melting points in Group 2 also have lower first ionisation energies than those in Group 1. Explain why.

9. Chemists frequently use chemical tests to show the presence of anions such as the sulfate ion.

   a. What is the trend in solubilities of the Group 2 sulfates?

   b. i. What chemical is used to test for the presence of sulfate ions?

      ii. State the observations when this chemical is added to a solution of sodium sulfate.

      iii. Write an ionic equation to show the reaction in ii.

      iv. Explain why magnesium chloride solution cannot be used to test for the presence of sulfate ions.

Stretch and challenge

10. A data book gives the solubilities of Group 2 hydroxides shown in Table Q4:

Group 2 hydroxide	Solubility/ mol/100 g saturated solution at 20 °C
$Mg(OH)_2$	2.00×10^{-5}
$Ca(OH)_2$	1.53×10^{-3}
$Sr(OH)_2$	3.37×10^{-3}
$Ba(OH)_2$	1.50×10^{-2}

Table Q4

   a. Calculate the concentration of the saturated solutions for each hydroxide in $mol\,dm^{-3}$. Assume 100 g solution has a mass of 100 g.

   b. How many grams of strontium hydroxide are present in $1\,dm^3$ saturated solution at 20 °C?

c. i. Describe how titrating a standard solution of hydrochloric acid against a saturated solution of calcium hydroxide can be used to confirm the solubility of calcium hydroxide in mol/100 g saturated solution.

   ii. Name the major source of error in this experiment.

Multiple choice

11. Which statement explains why titanium cannot be extracted from its ore by reduction with carbon?

    A. Titanium carbide forms instead of titanium metal

    B. The process is too expensive

    C. Titanium carbonate forms

    D. Reduction with magnesium is a cheaper process

12. What is the electron configuration of the outer electron shell in group two Group 2 elements?

    A. $2p^2$

    B. $2d^2$

    C. $2s^1$

    D. $2s^2$

13. Which pair of ores contain titanium?

    A. Rutile and galena

    B. Rutile and ilmenite

    C. Ilmenite and galena

    D. Galena and bauxite

14. Which chemical is commonly used to remove sulfur dioxide from flue gas?

    A. Calcium carbonate

    B. Calcium sulfate

    C. Barium sulfate

    D. Barium carbonate

12 GROUP 7(17), THE HALOGENS

PRIOR KNOWLEDGE

You may have learnt about the physical properties of Group 7(17) elements and seen chlorine, bromine and iodine in the laboratory. You may already know that these halogens become less reactive as you go down Group 7(17). You may even have carried out an electrolysis experiment with sodium chloride solution as the electrolyte.

LEARNING OBJECTIVES

In this chapter, you will find out more about the trends in physical properties of the halogens and apply some of the concepts, such as electronegativity, that you met earlier in this book. You will learn about some chemical reactions of the halogens, their compounds, their uses.

(Specification 3.2.3.1, 3.2.3.2)

Sterilising swimming pool water is essential to safeguard the health of the users and prevent the spread of infectious disease, regardless of whether the pool is a public swimming pool, spa pool or private pool. Traditionally, the two most common methods of sterilisation are to use chlorine, Cl_2, or ozone, O_3. Chlorine gas dissolves in the water to give hydrochloric acid, HCl(aq), and chloric(I) acid, HClO(aq). It is chloric acid that sterilises the water. Ozone sterilises the water by oxidising any microbes in the water. Both methods are expensive to use.

Since the 1980s, a method based on the electrolysis of salt water has been used increasingly to sterilise swimming pools. Rather than adding chlorine gas, the electrolysis of salt water is used to produce chlorine, which sterilises the water. Salt (sodium chloride) is added to the water to give a 3000 ppm (0.3% by mass) sodium chloride solution. This is very dilute; human taste does not detect salt below 3500 ppm. A small electrolytic cell is placed in the pumping system and as the salt solution is pumped through, chlorine gas is given off at the anode. The chlorine sterilises the water. Salt is cheaper than chlorine gas or ozone and the system is easy and cheaper to maintain. The main disadvantage is the initial cost of installation.

Figure 1 *Disinfection is one process in the treatment of waste water so that it can be returned to into the environment and used later for drinking and bathing. Chlorination is one method of disinfection.*

12.1 TRENDS IN PHYSICAL PROPERTIES

Group 7(17) elements are called the halogens. Their atoms have seven electrons in their outer shells and the halogens are the most reactive group of non-metals. None of them is found in nature as the uncombined element. Many halogens form compounds with metals such as those in Groups 1 and 2. The compounds are called metal halides.

Chlorine occurs most commonly chemically combined as sodium chloride in sea water, underground deposits (rock salt) and salt flats. Fluorine occurs chemically combined in the minerals fluorspar, CaF_2, and cryolite, Na_3AlF_6. Naturally occurring compounds of bromine and iodine are less common in the Earth's crust, but small quantities are found in sea water. Bromine can be extracted from sea water, where it is present at from 65 to 70 ppm. Astatine is an artificially produced radioactive element. Its most stable isotope has a half life of 8.5 hours. Table 1 describes the appearances of the halogens.

The physical properties of the halogens are shown in Table 2.

Atomic number	Symbol	State and colour at room temperature
9	F	Fluorine is a pale yellow gas consisting of flourine molecules, F_2
17	Cl	Chlorine is a yellow-green gas consisting of chlorine molecules, Cl_2
35	Br	Bromine is a dark red liquid consisting of bromine molecules, Br_2
53	I	Iodine is a shiny grey-black crystalline solid consisting of iodine molecules, I_2
85	As	Astatine is very radioactive and very rare

Table 1 *The halogens*

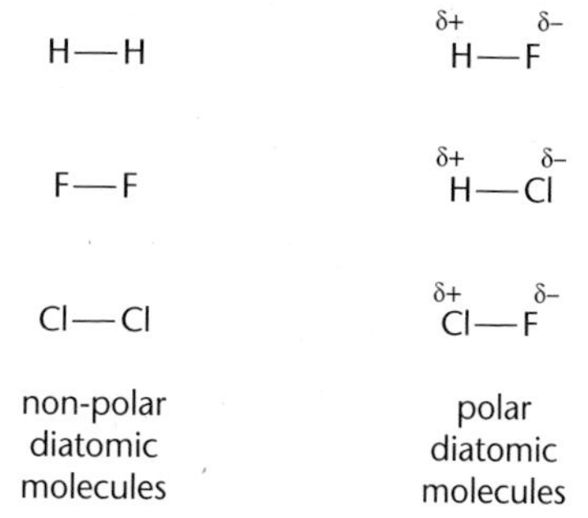

$$H—H$$

$$F—F$$

$$Cl—Cl$$

non-polar
diatomic
molecules

$$\overset{\delta+}{H}—\overset{\delta-}{F}$$

$$\overset{\delta+}{H}—\overset{\delta-}{Cl}$$

$$\overset{\delta+}{Cl}—\overset{\delta-}{F}$$

polar
diatomic
molecules

Figure 2 *Bonds between like atoms are non-polar. Bonds between unlike atoms are polar – how polar the bonds are depends on the electronegativity difference between the two atoms.*

Electronegativity

The electronegativity of an atom is a measure of its power to attract the bonding pair of electrons in a covalent bond. The electronegativity of all the halogens is high, but it decreases down Group 7(17), as seen in Table 2.

Fluorine atoms are small. When one forms a covalent bond it gets close to the nucleus of the other atom. With just one inner electron shell to shield the other atom from the fluorine nucleus, the attractive force for shared electrons is strong. Fluorine is the most electronegative element.

Chlorine is a larger atom. When it forms a covalent bond, its nucleus cannot get as close to the other atom as the nucleus of a fluorine atom can. Also, a chlorine atom has two inner shells of electrons that shield the charge of the nucleus. Chlorine has a lower electronegativity than fluorine. This is why the molecule ClF is polarised (Figure 2).

As the atomic radius of the atom increases, so does the number of shells of electrons that shield the charge on the nucleus. Both size and number of shells reduce the power of the nucleus to attract the bonding pair of electrons. So bromine is less electronegative than chlorine, and iodine is less electronegative than bromine:

F > Cl > Br > I

decreasing electronegativity

Halogen	Electron configuration	Atomic radius/nm	Ionic radius/nm	Melting point/°C	Boiling point/°C	Electro-negativity
fluorine	$1s^2\,2s^2\,2p^5$	0.071	0.133	– 220	– 188	4.0
chlorine	$1s^2\,2s^2\,2p^6\,3s^2\,3p^5$	0.099	0.180	– 101	– 35	3.0
bromine	$[Ar]\,3d^{10}\,4s^2\,4p^5$	0.114	0.195	– 7	59	2.8
iodine	$[Kr]\,4d^{10}\,5s^2\,5p^5$	0.133	0.215	114	184	2.5
astatine	$[Xe]\,5d^{10}\,6s^2\,6p^5$	0.140	–	?	?	–

Table 2 *Physical properties of the halogens*

Boiling point

All halogen elements exist naturally as diatomic molecules made from two halogen atoms covalently bonded and sharing one electron from each atom. This may be shown using dot-and-cross diagrams, for example, chlorine gas consists of chlorine molecules, Cl_2:

The attractions between halogen molecules are intermolecular forces called van der Waals forces. You read about these in Chapter 3. The electron distribution in a molecule or atom changes from one instant to the next, each time causing a temporary dipole, with one end of the molecule being more negative and the other end less negative. This induces a temporary dipole in an adjacent molecule and these intermolecular forces attract the atoms or molecules to each other. The more electrons an atom or molecule has, the larger the van der Waals forces. This is because the electron cloud is more easily distorted the bigger it is and the further it extends from the nucleus.

When halogens change from a liquid to a gas, enough energy has to be transferred from the surroundings to break these intermolecular forces. The boiling point of fluorine is – 188 °C. Fluorine atoms have nine electrons each and these produce very weak van der Waals forces. Little energy is needed to break them and fluorine's boiling point is low. With 17 electrons, chlorine has stronger van der Waal's forces and more energy is needed to break them. So the boiling point of chlorine is higher, at – 35 °C.

Similarly, the strength of the van der Waal's forces increases from bromine to iodine as more electrons are added. More energy is needed to break them and the boiling points increase down Group 7(17).

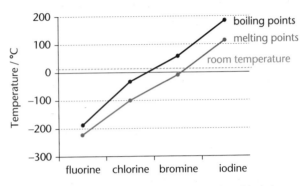

Figure 3 *Trends in the boiling points and melting points of the halogens*

QUESTIONS

1. Little is known about astatine's properties. Predict its:

   a. electronegativity

   b. boiling point.

KEY IDEAS

› The halogens, the Group 7(17) elements, are fluorine, chlorine, bromine, iodine and astatine.

› The boiling points of the halogens increase down Group 7(17) because of the increasing van der Waals forces.

› The electronegativity of the halogens decreases down Group 7(17) as the atomic radii and the number of electron shells increase.

12.2 TRENDS IN CHEMICAL PROPERTIES

Oxidising power of the halogens

When halogens react, they either form compounds that have covalent molecules, as in hydrogen chloride gas, or they form negative ions in an ionic compound. Compounds made from halogen atoms and atoms of another element are called **halides.** Many halides are salts, such as sodium chloride and potassium bromide. These salts are ionically bonded.

Halogen atoms accept electrons readily, making halogens very good **oxidising agents**. The order of oxidising power is:

$$F_2 > Cl_2 > Br_2 > I_2$$

This trend in oxidising power decreases down the group. This can be shown by the reactions of the halogens with aqueous halide ions.

Fluorine is the most reactive element known. It is extremely hazardous and should not be used in a laboratory unless suitable apparatus is available. Therefore, it is not used in the following test tube experiments.

If aqueous chlorine is added to potassium bromide solution, a yellow-orange solution of bromine is formed. The reaction is:

$$Cl_2(aq) + 2KBr(aq) \rightarrow 2KCl(aq) + Br_2(aq)$$

or

$$Cl_2(aq) + 2Br^-(aq) \rightarrow 2Cl^-(aq) + Br_2(aq)$$

The redox half-equations for oxidation and reduction, respectively, are:

$$2Br^- \rightarrow Br_2 + 2e^-$$
$$Cl_2 + 2e^- \rightarrow 2Cl^-$$

Bromine is formed because chlorine is a stronger oxidising agent than bromine and so withdraws an electron from each bromide ion to leave bromine atoms and negatively charged chloride ions. The bromine atoms then combine to form molecules of bromine, which give the solution its new colour. Chlorine displaces bromine from a solution of its salt.

The reaction is also called a **displacement reaction** because bromine is displaced from a solution of bromide ions by chlorine.

A similar reaction happens when aqueous bromine is added to potassium iodide solution. A brown solution of iodine is formed. The full equation is:

$$Br_2(aq) + 2KI(aq) \rightarrow 2KBr(aq) + I_2(aq)$$

or

$$Br_2(aq) + 2I^-(aq) \rightarrow 2Br^-(aq) + I_2(aq)$$

The oxidation state of bromine changes from -1 in Br^- to 0 in Br_2 and bromine is oxidised. The oxidation state of chlorine changes from 0 in Cl_2 to -1 in Cl^- and the chlorine is reduced.

Both chlorine and bromine displace iodine from solutions of iodide ions.

A reaction in which one halogen is displaced from a solution of its ions by another halogen is also called a displacement reaction.

Figure 4 *Aqueous chlorine is added to potassium iodide solution. The displaced iodine causes the brown colour.*

QUESTIONS

2. a. Write an overall redox equation for the reaction between aqueous chlorine and potassium iodide solution shown in Figure 4.

   b. Write half-equations for the reaction in a.

Stretch and challenge

3. Electron affinity is a measure of the attraction between the nucleus of an atom and the incoming electron. It is measured in $kJ \, mol^{-1}$ for gaseous atoms and gaseous ions.

   First electron affinity: $X(g) + e^- \rightarrow X^-(g)$

   Second electron affinity: $X^-(g) + e^- \rightarrow X^{2-}(g)$

   Electron affinity values for some halogens are: $Cl = 348 \, kJ \, mol^{-1}$, $Br = -340 kJ \, mol^{-1}$, $I = -297 \, kJ \, mol^{-1}$

   a. Explain why electron affinity values decrease from chlorine to iodine.

   b. Explain the difference between *electron affinity* and *electronegativity*.

ASSIGNMENT 1: EXTRACTING BROMINE

(PS 1.1, 1.2)

Figure A1 *The River Jordan is the only major water input to the Dead Sea. There are no major outlets. Millions of years of evaporation have left the Dead Sea with the highest concentration of minerals in all the Earth's seas and oceans. Estimates suggest there are more than 43 billion tonnes of salts in the Dead Sea. The concentration of bromine is 140 times higher than in normal sea water. This plant on the Dead Sea coast extracts bromine from sea water.*

Bromine can be extracted from water in the Dead Sea by displacement with chlorine. However, the bromide ion concentration in normal sea water is too low for this process to work. Bromine has to be extracted from the Dead Sea in three stages:

Step 1 A volume of sea water is acidified to give a pH of 3.5. This is necessary because at higher pH, chlorine and bromine react with water to form a mixture of acids. Chlorine gas is bubbled into the sea water to displace the bromine. Bromine forms and is removed as a vapour by blowing air through the water. The concentration is too low for bromine to form as a liquid.

Step 2 The concentration of bromide ions needs to be increased. This is achieved by reducing the bromine formed to hydrobromic acid by injecting sulfur dioxide into the reaction mixture. The reaction is:

$$Br_2(aq) + SO_2(g) + 2H_2O(l) \rightarrow$$
$$2HBr(aq) + H_2SO_4(aq)$$

Adding fresh water produces a fine mist of acids that can then be condensed and collected. However, the acid mixture still only contains about 13% bromine by mass.

Step 3 Chlorine gas is used to displace bromine from hydrogen bromide. Adding steam produces a mixture of hot gaseous products. An aqueous layer and a layer of bromine condenses out. The bromine can now be separated.

Questions

A1. Write a full redox equation to show the displacement of bromine from sea water.

A2. Write half-equations for:

 a. the oxidation reaction

 b. the reduction reaction.

A3. Why is bromine a liquid at room temperature and not a gas like chlorine and fluorine?

A4. Use oxidation states to explain what is oxidised and what is reduced when sulfur dioxide is added to bromine and water.

A5. a. Write a full balanced equation to show the displacement of bromine from hydrogen bromide.

 b. Use oxidation states to explain what is oxidised and what is reduced.

A6. Carry out an internet search to find out about:

 a. uses of bromine

 b. the safety precautions needed when transporting and handling bromine.

12.3 THE HALIDE ION

Metal halides form when halogens react with Group 1 and 2 metals. The bonding is ionic and the metal halides have giant lattice structures consisting of metal ions and **halide ions**. The chloride ion has the electron configuration:

All halide ions have a noble gas electronic configuration. The chloride ion, for example, has the same electronic configuration as an argon atom.

Trends in the reducing properties of the halide ion

Halogens are strong oxidising agents because they readily accept electrons to become halides. Halide ions, in turn, can act as reducing agents when they transfer electrons and are oxidised to halogens. The reducing ability of the halide ion shows a trend down Group 7(17), as the reactions between halide ions and concentrated sulfuric acid illustrate.

Solid ionic halides and concentrated H_2SO_4

Solid ionic halides can be identified by their reactions with concentrated sulfuric acid, H_2SO_4. The reactions of sodium halides with concentrated H_2SO_4 are summarised here:

NaCl(s) and concentrated H_2SO_4(aq)

H_2SO_4 acts as an acid, donating H^+ to Cl^-, to produce HCl. There are no changes of oxidation state and, therefore, it is not a redox reaction.

$$NaCl(s) + H_2SO_4(aq) \rightarrow HCl(g) + NaHSO_4(aq)$$

Observation: Evolution of a colour gas and formation of a colourless solution.

NaBr(s) and concentrated H_2SO_4(aq)

H_2SO_4 acts as an acid, donating H^+ to Br^-, to produce HBr. There are no changes of oxidation state and, therefore, it is not a redox reaction.

$$NaBr(s) + H_2SO_4(aq) \rightarrow HBr(g) + NaHSO_4(aq)$$

Some of the HBr formed is oxidised to Br_2

$$2HBr(g) + H_2SO_4(aq) \rightarrow 2H_2O(l) + SO_2(g) + Br_2(g)$$

Reduction: Sulfur changes from oxidation state $+6$ in H_2SO_4 to $+4$ in SO_2.

Oxidation: Bromine changes from oxidation state -1 in Br^- to 0 in Br_2.

Observation: Bromine is seen as a brown orange liquid. Sulfur dioxide, SO_2, is colourless, but has a pungent odour.

NaI(s) and concentrated H_2SO_4(aq)

H_2SO_4 acts as an acid, donating H^+ to I^-, to produce HI. There are no changes of oxidation state and, therefore, it is not a redox reaction.

$$NaI(s) + H_2SO_4(aq) \rightarrow HI(g) + NaHSO_4(aq)$$

Most of the HI formed is oxidised to I_2. There are three different reactions.

$$2HI(g) + H_2SO_4(aq) \rightarrow I_2(g) + SO_2(g) + 2H_2O(l)$$

Reduction: Sulfur change from oxidation state $+6$ in H_2SO_4 to $+4$ in SO_2.

Oxidation: Iodine changes from oxidation state -1 in I^- to 0 in I_2.

$$6HI(g) + H_2SO_4(aq) \rightarrow 3I_2(g) + S(s) + 4H_2O(l)$$

Reduction: Sulfur change from oxidation state $+6$ in H_2SO_4 to 0 in S.

Oxidation: Iodine changes from oxidation state -1 in I^- to 0 in I_2.

$$8HI(g) + H_2SO_4(aq) \rightarrow H_2S(g) + 4H_2O(l) + 4I_2(g)$$

Reduction: Sulfur change from oxidation state $+6$ in H_2SO_4 to -2 in H_2S.

Oxidation: Iodine changes for oxidation state -1 in I^- to 0 in I_2.

Observation: A mixture of products from all three reactions is obtained. Purple iodine vapour is seen. Sulfur dioxide, SO_2, is colourless with a punget odour. Sulfur is a pale yelow solid, but may be difficult to see in the reaction mixture. Hydrogen sulfide, H_2S, is colourless and smells of rotten eggs.

As you can see, the three sodium halides react with concentrated sulfuric acid to produce the corresponding hydrogen halide and sodium hydrogen sulfate. All hydrogen halides are gases at room temperature. The next reaction in each case depends on the reducing power of the hydrogen halide formed and the oxidising power of concentrated sulfuric acid.

Fluorides are not included in the explanation, but hydrogen fluoride is not a reducing agent and sodium fluoride gives similar results to sodium chloride.

Hydrogen fluoride and hydrogen chloride do not reduce sulfuric acid.

Hydrogen bromide reduces the sulfur in sulfuric acid to sulfur dioxide. The bromide ions are oxidised to bromine.

Hydrogen iodide reduces the sulfur in sulfuric acid to mainly hydrogen sulfide, but some sulfur dioxide and some sulfur are also formed. The iodide ions are oxidised to iodine (Figure 5). Hydrogen iodide is a stronger reducing agent than hydrogen bromide and hydrogen bromide is a stronger reducing agent than hydrogen chloride.

In summary, the trend in the reducing ability of the halide ion is: $F^- < Cl^- < Br^- < I^-$

Figure 5 *Sodium iodide reacts with concentrated sulfuric acid. The purple vapour produced is iodine.*

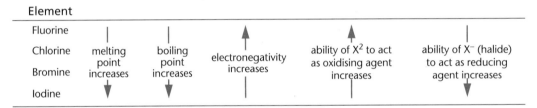

Element					
Fluorine					
Chlorine	melting point increases	boiling point increases	electronegativity increases	ability of X_2 to act as oxidising agent increases	ability of X^- (halide) to act as reducing agent increases
Bromine					
Iodine					

Figure 6 *Summary of physical and chemical trends in Group 7(17)*

QUESTIONS

4. Use oxidation states to explain why the first reaction in Table 3 between sodium halide and concentrated sulfuric acid is not a redox reaction.

5. For the reaction between HBr and concentrated H_2SO_4, use oxidation states to explain:

   a. what is oxidised

   b. what is reduced

   c. what is the reducing agent

   d. what is the oxidising agent.

6. Write half-equations for the reaction between HBr and concentrated H_2SO_4.

7. Write half-equations for the reaction between HI and concentrated H_2SO_4.

KEY IDEAS

› Halogens readily accept electrons and are strong oxidising agents.

› The oxidising power of the halogens decreases moving down Group 7(17): $F_2 > Cl_2 > Br_2 > I_2$

› Bromide and iodide ions can reduce concentrated sulfuric acid.

› The reducing power of the halide ion increases moving down Group 7(17): $F^- < Cl^- < Br^- < I^-$

Reactions of halide ions with silver nitrate solution

Unlike most ionic halides, silver halides are insoluble in water, with the exception of silver fluoride, which is soluble. When acidified silver nitrate solution is added to a solution containing halide ions, the silver halide is formed as a precipitate. A reaction in which mixing two solutions produces a solid is called a precipitation reaction. Silver halide precipitates are different colours, as shown in Figure 7.

Figure 7 *Test tubes showing AgI, AgBr and AgCl precipitates (yellow, cream and white).*

Silver nitrate solution is acidified with dilute nitric acid to prevent the formation of precipitates from other ions that might be in solution, for example, carbonate ions.

Reactions with acidified silver nitrate solution make it possible to distinguish between the halides in solution.

Silver fluoride is soluble and does not form a precipitate. This makes it easy to distinguish between fluoride ions in solution and other halide ions in solution, all of which form silver halide precipitates.

However, the colours of the silver halide precipitates (white, cream and yellow) are easily confused, so an additional test is often used. It is based on the differing solubilities of silver halides in ammonia solution.

› Silver chloride dissolves in dilute ammonia solution.

› Silver bromide dissolves in concentrated ammonia solution.

› Silver iodide does not dissolve in concentration ammonia solution.

The tests are summarised in Table 3.

Halide	Reaction with acidified silver nitrate solution	Reaction with aqueous ammonia
fluoride	$Ag^+(aq) + F^-(aq) \rightarrow AgF(aq)$ no precipitate: AgF is water soluble	no precipitate
chloride	$Ag^+(aq) + Cl^-(aq) \rightarrow AgCl(s)$ white precipitate of silver chloride	silver chloride dissolves in dilute ammonia solution: $AgCl(s) + 2NH_3(aq) \rightarrow [Ag(NH_3)_2]^+(aq) + Cl^-(aq)$ the diamine silver(I) ion, $[Ag(NH_3)_2]^+(aq)$, is colourless
bromide	$Ag^+(aq) + Br^-(aq) \rightarrow AgBr(s)$ cream precipitate of silver bromide	AgBr is sparingly soluble in dilute ammonia, but dissolves in concentrated $NH_3(aq)$: $AgBr(s) + 2NH_3(aq) \rightarrow [Ag(NH_3)_2]^+(aq) + Br^-(aq)$
iodide	$Ag^+(aq) + I^-(aq) \rightarrow AgI(s)$ yellow precipitate of silver iodide	silver iodide does not dissolve in concentrated $NH_3(aq)$

Table 3 *Testing for halides*

QUESTIONS

8. Write the electron configuration for a:

    a. fluoride ion

    b. chloride ion.

9. Which atoms have the same electron configuration as the fluoride and chloride ions?

10. Identify the sodium halide present:

    a. When acidified silver nitrate solution is added to a solution of the sodium halide, no precipitate forms.

    b. When concentrated sulfuric acid is added to the sodium halide, a brown gas, plus two colourless gases are given off.

    c. When acidified silver nitrate solution is added to a solution of the halide, a white precipitate forms which quickly turns violet and dissolves in dilute ammonia solution.

    d. When concentrated sulfuric acid is added to the sodium halide, violet fumes are given off, plus three other gases. There is a distinct smell of rotten eggs.

11. Silver nitrate solution used to identify halides is acidified with dilute nitric acid to remove ions other than halides that might also give a precipitate. Which ions does nitric acid remove?

KEY IDEAS

❯ Acidified silver nitrate solution can be used to identify aqueous halide ions.

❯ Aqueous fluoride ions are soluble and do not produce a precipitate.

❯ Aqueous chloride ions produce a white precipitate of silver chloride, which dissolves in dilute ammonia.

❯ Aqueous bromide ions produce a cream precipitate of silver bromide, which is sparingly soluble in dilute ammonia, but soluble in concentrated ammonia.

❯ Aqueous iodide ions produce a yellow precipitate of silver iodide, which is insoluble in concentrated ammonia.

REQUIRED PRACTICAL ACTIVITY 4: APPARATUS AND TECHNIQUES

(PS 2.1 and 4.1, AT b, d and k)

Carry out simple test tube reactions to identify:

> cations – Group 2, M^{2+}; ammonium, NH_4^+

> anions – Group 7(17), X^- (halide ions); hydroxide, OH^-; carbonate, CO_3^{2-}; sulfate, SO_4^{2-}

This practical gives you the opportunity to show that you can:

> use a water bath for heating

> use laboratory apparatus for a variety of experimental techniques, including qualitative tests for ions

> safely and carefully handle solids and liquids, including corrosive, irritant, flammable and toxic substances.

Apparatus

Qualitative analysis to identify cations and anions in a compound or mixture may require the preparation of a sample to be tested and the addition of suitable chemical reagents. Observed changes can confirm the presence of a particular cation or anion.

If the sample to be tested is a solid, you will need to make an aqueous solution. Dissolve the solid by adding it to deionised water in, for example, a clean glass beaker, stirring and warming if necessary. Store the sample in a covered container.

Reaction vessels

Although they are called 'test tube reactions' the identification tests can be carried out in a variety of apparatus, such as those shown in Figure P1:

Test tubes should be borosilicate glass or Pyrex™ glass – both can be heated safely. Plastic test tubes are available, but these tend to be translucent rather than transparent and can confuse the observations. Boiling tubes should be used if reactants are to be heated in a Bunsen flame. They are larger than normal laboratory test tubes and, therefore, there is less chance of the contents boiling over. A test tube rack, temperature resistant if needed, is used to store the test tubes.

You will also need a test tube holder – especially if the test tube and its contents are to be heated.

Well plates or microtiter plates are flat plates with multiple wells in which reactants can be mixed. They come in a variety of sizes and materials, the most common being poly(styrene). While mixing is easy, heating reaction mixtures is less easy. More commonly used in biology laboratories, they can be useful for microscale qualitative chemistry where small quantities are used. They make it easy to carry out and compare the results of a large number of tests.

Porcelain spot test plates or spotting tiles are useful for small scale analyses where just a few cubic millimetres of reactants are used. They have similar advantage to well plates, but are less versatile in that there is room only for a few drops of solution. Also, it is difficult to mix the reactants, sometimes confusing the result.

Reagents

Frequently used reagents are usually stored in labelled bottles with dropping pipettes. Dark glass bottles are used for light sensitive reagents, such as silver nitrate solution. They should be labelled with the name of the reagent, its concentration and relevant hazard warnings. If reagents have no droppers, then a separate dropping pipette is needed. This may be a simple glass pipette with a rubber teat, a one-piece plastic dropper pipette or a more sophisticated automated syringe pipette designed to release a specific volume (Figure P2).

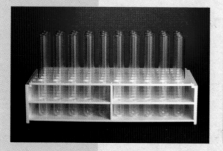

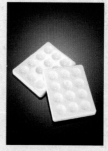

Figure P1 *You have a choice of apparatus to use when identifying cations and anions in aqueous solution, such as test tubes (left), well plates (middle) or spotting tiles (right).*

Figure P2 *Storing and transferring reagent solutions. Left to right: reagent bottles with dropper pipettes fitted in their stoppers, a plastic dropper pipette and an automated syringe pipette designed to deliver a more accurate volume of liquid.*

These are the concentrations and hazards of reagents commonly used in analysis at A-level.

Reagent	Concentration/ mol dm^{-3}	Hazard at this concentration
dilute ammonia solution	2.0	irritant
concentrated ammonia solution	14.4	corrosive, toxic
dilute hydrochloric acid	0.40	irritant
barium chloride solution	0.10	low hazard
dilute nitric acid	0.40	low hazard
silver nitrate solution	0.10	low hazard, can stain skin
sodium hydroxide solution	< 0.50	irritant
lime water	0.02	none

Table P1 *Reagents commonly used in analysis at A-level*

Technique

Preparing the sample
The volume of solution needed for testing depends on the scale of your analysis and the number of tests you are carrying out. Your instructions may tell you the quantity of solid to use. You need to check hazard labels.

You may be given aqueous solutions to analyse, but if you are given a solid you need to prepare a solution. Rinse a clean beaker or conical flask with deionised water, add the required volume of deionised water and then the solid. Do not use tap water as this contains many ions in solution. Stir the solution with a clean glass rod, or a mechanical or magnetic stirrer until the solid has dissolved. Label the beaker or flask using a permanent marker pen. Labels using temporary marker pens tend to rub off.

Figure P3 *Often, samples need to be collected in the field and analysed.*

The analysis

Label each test tube or well using a marker pen. You need to decide how to identify which reaction mixture is in which test tube or well.

A volume of the sample solution is placed in a clean test tube, spotting tile well or in a well plate. Use about 2 cm^3 in a test tube or well plate and a few drops on a spotting tile. Usually it is not necessary to measure the volume accurately, though sometimes the number of drops of reagent is specified.

Add the reagent according to the instructions. Make sure the dropper does not touch the sides of the container or the surface of the solution. This will contaminate the reagent and interfere with future tests. Replace the dropper in the reagent bottle and observe any changes carefully, noting any evolution of gas, colour change or precipitate formation. If using test tubes, the contents can be shaken gently. More reagent may be added as needed.

A water bath or Bunsen burner can be used for tests that require heating. If using a Bunsen burner, the open end of the boiling tube must face away from you and other students. Use a blue flame and gently move the boiling tube in the flame to distribute the heat evenly in the contents.

If a gas is given off, the instructions will describe how it can be identified.

QUESTIONS

P1. Why does silver nitrate solution need to be stored in a dark bottle?

P2. How does the concentration of ammonia solution affect its hazard warning?

P3. Why is there no need to measure the volume of sample solution accurately in a qualitative analysis?

Stretch and challenge

P4. You have learnt and read about other analysis techniques: flames tests and mass spectroscopy. You will find out about IR spectroscopy in Chapter 16. Carry out a search if needed and complete a table to show the type of information each method provides.

12.4 USES OF CHLORINE

Chlorine is a poisonous gas. It was used, along with other poisonous gases, in World War 1. Yet the title of the best life-saving element also has to go to chlorine. Millions of people are alive today because of its properties and uses.

Up to 160 years ago, thousands of people died every year in London from cholera alone. Cholera is a bacterial infection of the small intestine and is transmitted primarily by water. Adding chlorine to drinking water has cut the number of cholera deaths in London to zero.

Chlorine's powerful disinfecting properties are used to make swimming pools, hospitals, homes, hotels, restaurants and many other public places safe. 85% of all medicines are made from chlorine atoms and the atoms of other elements – chiefly carbon, hydrogen and oxygen. 30% of all chlorine produced is used to make chloroethene, the monomer used to make poly(chloroethene), commonly called (PVC).

Chlorine is a major component of many insecticides and herbicides, and is also used to make bleach. It has a major role in bleaching wood pulp to make paper.

All this success comes at a cost as chlorine residues have built up around the world in waste material. Dioxins and chlorofluorocarbons (CFCs) contain chlorine. Dioxins cause many problems, including genetic abnormalities in children and chloracne (a severe skin complaint). CFCs are held responsible for the thinning of the ozone layer in the stratosphere (see Chapter 13).

Figure 8 *Many anaesthetics are compounds of chlorine and other halogens.*

The use of chlorine in water treatment

At the treatment works, chlorine is pumped into water during the final stage of water treatment. The chlorinated water is stored for about two hours to allow disinfection to occur. The reaction of chlorine with water is:

$$Cl_2(aq) + H_2O(l) \rightleftharpoons HOCl(aq) + HCl(aq)$$

HOCl is chloric(I) acid (old name, hypochlorous acid). It is a weak acid, which means that a solution contains an equilibrium mixture:

$$HOCl(aq) \rightleftharpoons H^+(aq) + ClO^-(aq)$$

The position of the equilibrium lies far to the left, so a solution contains very few $H^+(aq)$ and $ClO^-(aq)$ ions (which are a bleach). We say that the acid partially dissociates. Chloric(I) acid and, to a lesser extent, the ClO^- ion, disrupt the cell membranes of bacteria. The chloric(I) acid can enter the cell and kill it. The action takes just a few seconds. Viruses are more resistant to chlorine. HOCl(aq) is about 80 times more effective at killing bacteria than $ClO^-(aq)$.

When chlorine is added to water, one chlorine in Cl_2 changes oxidation state from 0 to −1 in HCl and is reduced, the other changes oxidation state from 0 to +1 in HOCl and is oxidised. This is called **disproportionation.** A disproportionation reaction is one in which a substance is simultaneously oxidised and reduced.

The hydrochloric acid produced when chlorine is added to water dissociates completely to form chloride ions and hydrogen ions, $HCl(aq) \rightarrow H^+(aq) + Cl^-(aq)$. This time, the position of the equilibrium lies far to the right.

So chlorinating water produces a mixture of chloric(I) acid, hydrogen ions, chlorate(I) ions, and chlorine ions.

An aqueous solution of chloric(I) acid is not very stable. In the presence of sunlight, chloric(I) acid decomposes to produce hydrochloric acid and oxygen. The reaction is:

$$2HOCl(aq) \rightarrow 2HCl(aq) + O_2(g)$$

QUESTIONS

12. When damp litmus paper is added to a gas jar of chlorine, it turns red and is then bleached. Explain why.

ASSIGNMENT 2: THE BENEFITS AND RISKS OF USING CHLORINE FOR WATER TREATMENT

(MS 0.0, 4.1; PS 1.1, 1.2, 2.1, 3.2)

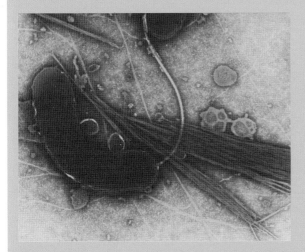

Figure A1 *These are* Vibrio cholerae *bacteria that cause cholera. These bacteria are common in untreated water. The water-borne diseases of cholera, dysentery, typhoid and gastroenteritis have killed more people between them than all the wars in history.*

A 2008 report stated that one billion people do not have a safe water supply within one kilometre of their homes. 4000 children under the age of five die every day from diarrhoeal diseases from drinking unsafe water (source: World Health Organisation). Most of these deaths could be prevented if the children had access to a clean, sterile water supply. Chlorine's toxic properties are put to use in water treatment by using concentrations that are safe for humans but fatal for bacteria. When it was first used in the UK in the 19th century, the cases of water-borne diseases immediately dropped. The importance of chlorine treatment was emphasised when a misunderstanding led to Peruvian authorities to suspend their water-treatment program in 1991. A cholera epidemic followed with 800 000 cases and 6000 deaths. The disease spread to almost all other Latin American countries.

Most governments around the world have concluded that the benefits of disinfecting water with chlorine far outweighs any negative effects. The concentration of chlorine needed to disinfect water is between two and 10 ppm.

But, chlorine is a toxic gas. At 1–3 ppm, it causes mild mucous irritation; at 30 ppm it causes chest pain and vomiting and at 430 ppm it is lethal after 30 minutes. Handling containers of chlorine gas is hazardous. Solid chlorine-containing compounds that produce chloric(I) acid when they dissolve in water are usually used in public places for safety reasons.

Various studies have highlighted the negative aspects of chlorine use. It has been suggested that showering or bathing in chlorine-treated water increases levels of chlorine exposure to between six and 100 times the dose of that received from drinking the water.

During the water-treatment process, chlorine can react with organic material, such as decaying plants and animals, and produce chloroalkanes. The most common of these formed is trichloromethane. Chloroalkanes are known carcinogens, but only at levels many thousands of times greater than those found in water supplies. Other studies suggest links with an increased risk of rectal, colon and bladder cancer.

More certain are the effects of chlorinated water on the skin and eyes. You may experience sore eyes in chlorinated swimming pools. Allergy sufferers may find chlorine irritates their eczema.

The alternatives to the use of chlorine in water treatment are:

❯ Ozone (O_3), which readily gives up one oxygen atom and is a powerful oxidising agent. It oxidises and destroys most water-borne pathogens.

❯ Ultraviolet (UV) radiation, which destroys most pathogens, as long as the water has a low level of colour and the UV radiation can pass through it.

The advantage of chlorine over these methods is that chlorine stays in the water supply from the treatment works to the tap and beyond. The protection is total. Water treated with ozone or radiated with UV can be re-contaminated between the treatment works and the user.

Fluoride ions are thought to strengthen tooth enamel. Calcium fluoride is naturally present in virtually all UK water supplies, but most levels fall short of that required for optimum dental health of 1 ppm. Five and a half million people have sodium fluoride or other fluoride compounds added to their water supplies to raise the fluoride concentration to 1 ppm. The downside of a higher fluoride concentration in drinking water is dental fluorosis, as shown in Figure A2.

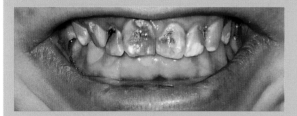

Figure A2 *A severe case of dental fluorosis*

Questions

A1. What are

    a. 1 ppm

    b. 30 ppm

    c. 430 ppm

as percentages?

A2. Trichloromethane is $CHCl_3$.

    a. What is the shape of the molecule?

    b. What are the bond angles in the molecule?

A3. a. Give one advantage and one disadvantage of using ozone to disinfect water.

    b. Give one advantage and one disadvantage of using chlorine to disinfect water.

A4. Describe a chemical test to detect the presence of chloride ions in water.

A5. Why is sodium fluoride, and not fluorine gas, used to add fluoride to water supplies not fluorine gas?

A6. A sample of water contains chloride and fluoride ions only. Use formulae and state symbols to show all the products when acidified silver nitrate solution is added.

A7. Carry out an internet search to make a list of reasons why fluoride should be added to water supplies and another list as to why it should not.

The reactions of chlorine with sodium hydroxide solution and the use of its products

Since chlorine reacts with water to produce a mixture of acids, it can be expected to react with bases. The products obtained depend on the concentration of sodium hydroxide solution and the temperature.

When chlorine is added to cold (15 °C) dilute sodium hydroxide solution, a mixture of chlorate(I) and chloride ions form:

$$Cl_2(g) + 2NaOH(aq) \rightarrow NaClO(aq) + NaCl(aq) + H_2O(l)$$

or, ionically:

$$Cl_2(g) + 2OH^-(aq) \rightarrow ClO^-(aq) + Cl^-(aq) + H_2O(l)$$

NaClO is sodium chlorate(I) (old name sodium hypochlorite).

If hot concentrated sodium hydroxide solution is used, the reaction is:

$$3Cl_2(g) + 6NaOH(aq) \rightarrow$$
$$NaClO_3(aq) + 5NaCl(aq) + 3H_2O(l)$$

$NaClO_3$ is sodium chlorate(V) and is used as weed killer.

Figure 9 *3-8% solutions of sodium chlorate(I) are sold as domestic bleach under a variety of trade names. More concentrated solutions are used in industry to bleach fabric and wood pulp for paper.*

QUESTIONS

13. Why are both reactions of chlorine with sodium hydroxide solution disproportionation reactions?

14. When bromine is added to cold dilute sodium hydroxide solution, a reaction similar to chlorine's takes place. Write an overall redox equation for the reaction.

ASSIGNMENT 3: A SWIMMING POOL PROBLEM

(MS 4.1; PS 1.1, 1.2)

Figure A1 *The amount of chloramine in swimming pool water is directly correlated to the cleanliness of the swimmers.*

Free available chlorine (FAC) is a measure of the amount of chloric(I) acid and chlorate(I) ions in swimming pool water. This is the chlorine available to kill bacteria.

The problem is that HOCl reacts with ammonia and ammonia-like compounds in swimming pool water, to produce chloramines and reduce the amount of free available chlorine to kill bacteria. Chloramines are poor disinfectants and some irritate mucous membranes, causing red stinging eyes and respiratory problems. They are also responsible for the strong swimming pool odour in badly managed pools. Pool users usually blame chlorine for the smell and their symptoms.

Swimming is energetic and swimmers perspire. Perspiration, and also urine, contain urea, $CO(NH_2)_2$, a compound that is related to ammonia.

Controlling the pH of pool water is important to control the free available chlorine in pool water. Chloramine formation is also pH dependent. Monochloramine forms at pH 7 to 9. Dichloramine and trichloramine, which are responsible for the bad odours and irritation problems, form between pH 1 and 6.

Figure A2 *Levels of HOCl vary with pH and the ideal pH is between 7 and 8. If the pH is too high, not enough chloric(I) acid will be present to disinfect the water.*

Questions

Stretch and challenge

A1. The shape of an ammonia molecule is based on a tetrahedron. Why are the bond angles 107° and not 109°?

A2. Chloric(I) acid reacts with ammonia in three stages to produce monochloramine, NH_2Cl, then dichloramine, $NHCl_2$, then trichloramine, NCl_3. Write full balanced equations to show the reactions.

A3. Use the equation HOCl(aq) $\rightleftharpoons$ H$^+$(aq) + ClO$^-$(aq) to explain why a high pH reduces the concentration of chloric(I) acid in a swimming pool?

A4. What type of compound needs to be added to restore the pH to 7?

A5. How does maintaining a pool pH of 7 limit the formation of chloramines?

A6. If the pH of pool water falls below 7, explain why pool managers will add more chlorine.

KEY IDEAS

> Chlorine reacts with water to form chloric(I) acid and hydrochloric acid.

> Chloric(I) acid kills bacteria in water supplies and swimming pools.

> Chlorine reacts with cold dilute sodium hydroxide solution to produce chloride ions and chlorate(I) ions.

> Household bleach is a solution of sodium chlorate(I).

PRACTICE QUESTIONS

1. A student investigated the chemistry of the halogens and the halide ions.

   a. In the first two tests, the student made the following observations.

Test		Observation
1	Add chlorine water to aqueous potassium iodide solution.	The colourless solution turned a brown colour.
2	Add silver nitrate solution to aqueous potassium chloride solution.	The colourless solution produced a white precipitate.

   Table Q1

   i. Identify the species responsible for the brown colour in Test 1.

   Write the simplest ionic equation for the reaction that has taken place in Test 1.

   State the type of reaction that has taken place in Test 1.

   ii. Name the species responsible for the white precipitate in Test 2.

   Write the simplest ionic equation for the reaction that has taken place in Test 2.

   State what would be observed when an excess of dilute ammonia solution is added to the white precipitate obtained in Test 2.

   b. In two further tests, the student made the following observations.

Test		Observation
3	Add concentrated sulfuric acid to solid potassium chloride.	The white solid produced misty white fumes which turned blue litmus paper to red.
4	Add concentrated sulfuric acid to solid potassium iodide.	The white solid turned black. A gas was released that smelled of rotten eggs. A yellow solid formed.

   Table Q2

   i. Write the simplest ionic equation for the reaction that has taken place in Test 3.

   Identify the species responsible for the misty white fumes produced in Test 3.

ii. The student had read in a textbook that the equation for one of the reactions in Test 4 is as follows.

$$8H^+ + 8I^- + H_2SO_4 \rightarrow 4I_2 + H_2S + 4H_2O$$

Write the two half-equations for this reaction.

State the role of the sulfuric acid and identify the yellow solid that is also observed in Test 4.

iii. The student knew that bromine can be used for killing microorganisms in swimming pool water.

The following equilibrium is established when bromine is added to cold water.

$$Br_2(l) + H_2O(l) \rightarrow HBrO(aq) + H^+(aq) + Br^-(aq)$$

Use Le Châtelier's principle to explain why this equilibrium moves to the right when sodium hydroxide solution is added to a solution containing dissolved bromine.

Deduce why bromine can be used for killing microorganisms in swimming pool water, even though bromine is toxic.

AQA Jun 2012 Unit 2 Question 9

2. Chlorine is a powerful oxidising agent.

a. Write the simplest ionic equation for the reaction between chlorine and aqueous potassium bromide.

State what is observed when this reaction occurs.

b. Write an equation for the reaction between chlorine and cold, dilute, aqueous sodium hydroxide.

Give a major use for the solution that is formed by this reaction.

Give the IUPAC name of the chlorine-containing compound formed in this reaction in which chlorine has an oxidation state of +1.

AQA Jan 2013 Unit 2 Question 10

3. **a.** Chlorine displaces iodine from aqueous potassium iodide.

i. Write the simplest ionic equation for this reaction.

ii. Give one observation that you would make when this reaction occurs.

b. In bright sunlight, chlorine reacts with water to form oxygen as one of the products. Write an equation for this reaction.

c. Explain why chlorine has a lower boiling point than bromine.

AQA Jul 2013 Unit 2 Question 6

4. **a. i.** Why are halogens good oxidising agents?

ii. State the trend in oxidising ability of the halogens from fluorine to iodine.

b. i. State what you observe when chlorine is added to potassium bromide solution.

ii. Write an equation to show the reaction.

iii. Explain, in terms of electrons, why chlorine is acting as an oxidising agent in this reaction.

c. Halide ions can act as reducing agents.

i. Explain in terms of electrons why halide ions can act as reducing agents.

ii. What is the trend in the reducing ability of halide ions?

Concentrated sulfuric acid is added to a solid sodium halide, X. Purple fumes, Y, are given off, plus three other gases. There is a strong smell of rotten eggs.

iii. Name X and Y. Identify the gas responsible for the rotten egg smell.

iv. Several reactions take place. In one reaction, sulfur in sulfuric acid is reduced to sulfur in sulfur dioxide. What is the change in oxidation states of sulfur?

AQA Jan 2007 Unit 2 Question 3

5. Table Q3 shows the boiling points and electronegativity values of the halogens.

Halogen	Boiling point/°C	Electronegativity
fluorine	−188	4.0
chlorine	−35	3.0
bromine	59	2.8
iodine	184	2.0

Table Q3

(Continued)

a. i. Explain the trend in boiling points.

   ii. Halogens exist as diatomic molecules. Name and describe the structure of iodine at room temperature.

   iii. Explain the trend in electronegativity.

b. Students used three different methods to identify bromide ions.

Method 1: They bubbled chlorine gas through aqueous bromide ions.

Method 2: They added acidified silver nitrate solution to aqueous bromide ions, followed by concentrated ammonia solution.

Method 3: They added concentrated sulfuric acid to solid potassium bromide.

   i. State what they observed for each method.

   ii. Write an equation for the reaction in method 1.

   iii. Write an equation for the reaction between bromide ions and acidified silver nitrate solution.

   iv. Explain how the reactions of halides with concentrated sulfuric acid (Method 3) can be used to distinguish between potassium bromide and potassium iodide.

AQA Jan 2005 Unit 2 Question 1

6. Table Q4 shows the atomic radii of some halogens.

Halogen	Atomic radius/nm
F	0.133
Cl	0.181
Br	0.196
I	0.219

Table Q4

a. Why does atomic radius increase down Group 7(17)?

b. Halogens exist as diatomic molecules.

   i. Explain why chlorine is a gas at room temperature while bromine is a liquid.

   ii. Describe the bonding in an iodine crystal.

c. Chloride and bromide ions can be identified by adding acidified silver nitrate solution, followed by dilute, then concentrated ammonia solution.

   i. Describe the observations when acidified silver nitrate solution is added to separate samples of chloride and bromide ions.

   ii. Describe the result when dilute ammonia solution is added to the products of (i), followed by concentrated ammonia solution.

AQA Jan 2005 Unit 2 Question 1

7. Chlorine is used to disinfect water supplies. The reaction of chlorine with water is:

$$Cl_2(g) + H_2O(l) \rightarrow HOCl(aq) + HCl(aq)$$

a. Copy and complete Table Q5 to show the oxidation states of chlorine.

Species	Oxidation state of chlorine
Cl_2	
HOCl	
HCl	

Table Q5

b. Describe what happens to the chlorine in terms of redox when chlorine dissolves in water.

c. HOCl is chloric(I) acid and is responsible for killing most of the bacterial pathogens in water. Ozone gas can be used as an alternative to chlorine to disinfect water supplies. Ozone does not react with water. Both are equally effective at killing pathogens.

   i. Give one advantage of using ozone to disinfect water, compared with chlorine.

   ii. Give one advantage of using chlorine to disinfect water supplies, compared with ozone.

   iii. Give one disadvantage of using chlorine to disinfect water supplies.

d. When chlorine dissolves in sodium hydroxide solution, the following reaction takes place.

$$Cl_2(g) + 2NaOH(aq) \rightarrow$$
$$NaClO(aq) + NaCl(aq) + H_2O(l)$$

(Continued)

i. Give the IUPAC name for NaClO.

ii. Give a use of NaClO solution.

8. Bromine is extracted from sea water, where it is present as bromide ions. Its concentration in ocean water is 0.07 g dm^{-3}. The first stage of the extraction involves passing chlorine gas through sea water.

a. Write an redox equation to show the reaction of bromide ions with chlorine gas.

b. Explain why this is called a redox reaction.

c. What type of reaction is this (apart from being a redox reaction)?

d. The second stage involves concentrating the bromine by blowing sulfur dioxide gas through the bromine solution. The reaction is:

$$Br_2(aq) + SO_2(g) + H_2O(l) \rightarrow 2HBr(aq) + H_2SO_4(aq)$$

The acids produced form a mist and can be removed. The bromine has been concentrated.

i. Identify which substances are oxidised and which are reduced.

ii. Write two half equations to show the oxidation and reduction reactions.

e. The final stage in the extraction of bromine involves displacing the bromide ions in HBr using chlorine, to produce bromine. Describe what you would see.

f. i. Describe a chemical test to show the presence of bromide ions.

ii. Give the result of this test.

Practical skills question

9. Students are identifying solutions of halides labelled as X, Y and Z. They use spotting tiles and the table below to add reactants. The reactants in each box are given in the heading above and to the left. They record their observations in Table Q6.

a. What is the advantage of using spotting tiles for these tests instead of test tubes?

b. What is a precipitate?

c. Use the results in the table to identify X, Y and Z.

d. Write an equation to describe the result in cell A2.

e. Write an equation to describe the reaction in cell D3.

f. Describe a test and the results that could be used to confirm the results from adding silver nitrate solution.

g. Explain why students do not need to use a measuring cylinder for these tests.

		A	B	C	D
	Solution	Chlorine solution	Bromine water	Iodine solution	Acidified silver nitrate solution
1	X	colour change: colourless to orange/brown	no change	no change	cream precipitate
2	Y	colour change: colourless to purple/ brown	colour change: colourless to orange/brown	no change	yellow precipitate
3	Z	no change	no change	no change	white precipitate

Table Q6

(Continued)

Stretch and challenge

10. Bleach is a solution of sodium chlorate(I), NaClO(aq). The 'available chlorine' in bleach can be determined by carrying out a titration.

    The available chlorine is the chlorine liberated when acid (H^+ ions) is added. The reaction is:

    $2H^+(aq) + 2ClO^-(aq) \rightarrow Cl_2(aq) + H_2O(l)$

    If potassium iodide solution is then added, iodine is produced.

    $Cl_2(aq) + 2I^-(aq) \rightarrow 2Cl^-(aq) + I_2(aq)$

    The iodine liberated can be titrated against sodium thiosulfate solution using a starch indicator. The reaction is:

    $2S_2O_3^{2-}(aq) + I_2(aq) \rightarrow S_4O_6^{2-}(aq) + 2I-(aq)$

    Students added excess sulfuric acid and excess potassium iodide solution to 10 cm³ bleach. They titrated this against 0.100 mol dm⁻³ sodium thiosulfate solution. They obtained a mean titre of 15.00 cm³.

    a. How many moles of sodium thiosulfate solution are needed to react with one mole of chlorine molecules?

    b. How many moles of sodium thiosulfate were used?

    c. How many moles of chlorine are in 10 cm³ bleach?

    d. How many moles of ClO^- ions are in 10 cm³ bleach?

    e. If the original sample of bleach had been diluted by a factor of ten, calculate the concentration of NaClO in bleach.

Multiple choice

11. Why are halogens good oxidising agents?
    A. They readily accept electrons
    B. They form covalent bonds
    C. They readily donate electrons
    D. They are reactive

12. What is a halide ion?
    A. A halogen that has gained one electron
    B. A halogen that has lost one electron
    C. A halogen atom
    D. A covalently bonded halogen

13. Which solution does not form a precipitate with acidified silver nitrate solution?
    A. Sodium fluoride solution
    B. Sodium chloride solution
    C. Sodium bromide solution
    D. Sodium iodide solution

14. What is the formula of bleach solution?
    A. NaClO (aq)
    B. $NaClO_2$ (aq)
    C. $NaClO_3$ (aq)
    D. NaCl (aq)

13 HALOGENOALKANES

PRIOR KNOWLEDGE

In Chapter 5, you read about types of formulae used to describe organic compounds and some reaction mechanisms. You read about the properties and reactions of halogens in Chapter 12.

LEARNING OBJECTIVES

In this chapter you will find out about some properties and chemical reactions of a homologous series of organic compounds that contain halogens, and that are called the halogenoalkanes. You will read about how some of these contribute to the depletion of the ozone layer.

(Specification 3.3.3.1, 3.3.3.2, 3.3.3.3)

In the 1970s, a British Antarctic Survey (BAS) team obtained data suggesting that the layer of ozone in the upper atmosphere over Antarctica was being destroyed. Its measurements were so far from what was expected that the scientists initially assumed that their instruments were faulty. Subsequent data confirmed a 35% drop in ozone concentration. The findings were reported in 1984. Research and government activity has been ongoing since then.

Ozone is an effective filter of UV, and a thinning ozone layer above Earth is dangerous because it means that more harmful UV radiation can reach Earth's surface. Today, satellite instruments monitor the amount of ozone over Antarctica daily. The data is used to create images, such as the one in Figure 1.

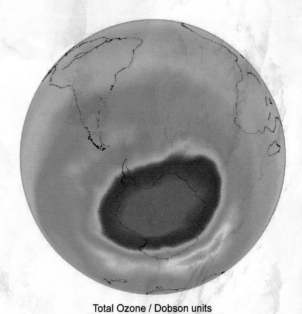

Total Ozone / Dobson units

150 200 250 300 350 400 450

Figure 1 *Dobson units (DU) are a unit of measure for total ozone. If there was no depletion, the ozone layer would be 0.3 cm thick. This is equivalent to 300 DU. The purple areas on the map are ozone-depleted areas.*

In the UK, amounts of ground-level UV radiation and stratospheric ozone are measured daily.

Each year, for the past few decades, chemical reactions of ozone with chlorine and bromine compounds have depleted the ozone layer, and rapidly during the Southern Hemisphere season of spring. The most severely depleted region is known as the ozone hole. Above Antarctica, this area has an ozone concentration of less than 220 DU.

13.1 WHAT ARE HALOGENOALKANES?

Halogenoalkanes are alkanes in which one or more hydrogen atoms has been replaced by a halogen atom. Methane, CH_4, is an alkane. If a chlorine atom is substituted for a hydrogen atom, then chloromethane, CH_3Cl, is produced. Chloromethane is a halogenoalkane.

The systematic naming of halogenoalkanes is covered in Chapter 5.

Halogenoalkanes that contain only one type of halogen atom are named collectively by the type of halogen atom; for example:

> a chloroalkane contains carbon, hydrogen and chlorine atoms

> a bromoalkane contains carbon, hydrogen and bromine atoms.

Halogenoalkanes that contain two types of halogen atom are similarly named, with the halogens listed in alphabetical order in the name. So 'bromo' always comes before 'chloro' in the name and a halogenoalkane that contains carbon, hydrogen, chlorine and bromine is called a bromochloroalkane.

Some examples of the names of halogenoalkanes are shown in Figure 2.

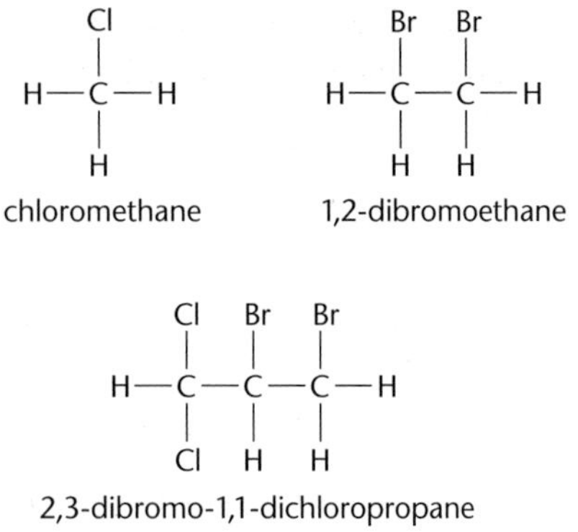

Cl
|
H—C—H
|
H

chloromethane

Br Br
| |
H—C—C—H
| |
H   H

1,2-dibromoethane

Cl Br Br
| | |
H—C—C—C—H
| | |
Cl H   H

2,3-dibromo-1,1-dichloropropane

Figure 2 *Examples of halogenoalkanes and their names*

QUESTIONS

1. Name these halogenoalkanes:
   a. CH_3CH_2Br
   b. CH_2BrCH_2Br
   c. $CH_3CH_2CCl_3$

d. $CHBr_2CH_2Br$
e. $CHBr_2CHClCH_3$
f. CF_3CHF_2

2. Give the elements that occur in:
   a. iodoalkane
   b. fluoroalkane
   c. chlorofluoroalkane

Bonding

Halogenoalkanes have covalent bonding, but the C–F, C–Cl and C–Br bonds are polar because the halogen atoms are more electronegative than carbon atoms (Table 1).

Atom	Electronegativity
carbon	2.5
hydrogen	2.1
fluorine	4.0
chlorine	3.0
bromine	2.8
iodine	2.5

Table 1 *Some electronegativities*

The halogen attracts the bonding pair of electrons more strongly than the carbon. As a result, the halogen has a small negative charge and the carbon has a small positive charge. The bond is polar and has a permanent dipole. This makes halogenoalkanes far more reactive than alkanes because they can attract other ions and molecules with polar covalent bonds.

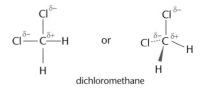

Figure 3 *Polar bonds in chloromethane and dichloromethane*

QUESTIONS

3. Why is the C–I bond non-polar?

4. Explain why CH_3X (where X=F, Cl or Br) molecules are polar, but CX_4 molecules are not.

5. What happens to the strength of the permanent dipole between a halogen atom and a carbon atom as you go down Group 7(17) and how does this affect the polarity of the molecule?

6. Halogenoalkanes also have van der Waals forces between their molecules.

   a. How many electrons are there in:

      i. one molecule of fluoromethane

      ii. one molecule of iodomethane?

Stretch and challenge

   b. Which out of fluoromethane and iodomethane will have the stronger van der Waals forces?

7. Look at the boiling points in Table 2. What do they suggest about the relative strength of intermolecular forces between molecules of halogenoalkanes derived from methane?

halogenoalkane	boiling point/K
CH_3F	194.7
CH_3Cl	248.9
CH_3Br	276.7
CH_3I	315.5

Table 2 Boiling points of halogenoalkanes derived from methane

KEY IDEAS

> Halogenoalkanes are alkanes in which one or more hydrogen atoms have been replaced by halogen atoms.

> Carbon–halogen covalent bonds (C–X) are polar.

> Symmetrical molecules such as CX_4 are non-polar.

> Unsymmetrical halogenoalkane molecules such as CH_3X are polar.

13.2 CHEMICAL REACTIONS OF HALOGENOALKANES

Halogenoalkanes undergo two types of reaction:

> **Substitution reactions:** a halogen atom is exchanged for another atom or group of atoms in the halogenoalkane.

> **Elimination reactions:** one hydrogen atom and one halogen atom are removed from neighbouring carbon atoms in the molecule and a double bond forms between the two carbons.

Substitution reactions

The carbon in a polar carbon–halogen bond carries a small positive charge. This attracts negative ions (anions) or the negative end of a polar covalent bond in another molecule. When these ions or molecules also have a lone pair of electrons they are called **nucleophiles**.

The nucleophile's lone pair forms a bond with the carbon atom as it replaces the atom or group of atoms that was bonded to the carbon. This is a substitution reaction, called **nucleophilic substitution**. Halogenoalkanes undergo nucleophilic substitution with hydroxide ions, ammonia and cyanide ions.

The reaction mechanism can be shown using curly arrows (see Chapter 5) to indicate the movement of electrons, as in the examples that follow. Overall, the carbon–halogen bond is broken and a carbon–nucleophile bond forms.

Nucleophilic substitution by hydroxide ions

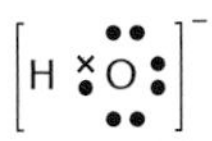

Figure 4 The hydroxide ion is an example of a nucleophile.

The oxygen in an hydroxide ion has three lone pairs of electrons. These are negatively charged areas of high electron density and are attracted to the small positive charge on the carbon of the polar carbon–halogen bond.

An example of this reaction is the reaction between 1-bromopropane and sodium hydroxide solution. The overall reaction is:

$$CH_3CH_2CH_2Br + OH^- \rightarrow CH_3CH_2CH_2OH + Br^-$$

1-bromopropane   propan-1-ol

The products of reaction are an alcohol and a bromide ion.

You show the mechanism as:

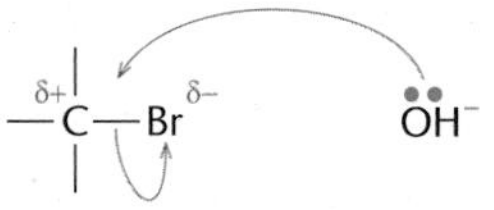

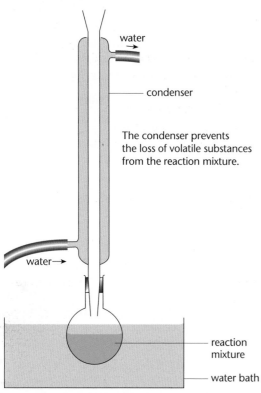

The curly arrow shows the movement of a pair of electrons. The OH$^-$ nucleophile donates one of its lone pairs of electrons to form a covalent bond with the carbon atom. The C–Br bond breaks heterolytically and a bromide ion forms.

Dilute sodium or potassium hydroxide can be used to provide the hydroxide ions. The reaction takes place when the reactants are refluxed in a 50/50 mixture of alcohol and water (Figure 5). Water can also be used to provide the hydroxide ions, as you will find out in the assignment *Detecting halogenoalkanes and confirming relative bond enthalpies*, but the reaction is very slow.

The condenser prevents the loss of volatile substances from the reaction mixture.

The reaction mixture is usually heated using either a water bath or an electric heater.

Figure 5 *The reflux apparatus keeps the reaction mixture at its boiling point without any losses due to evaporation for as long as is needed for the reaction to occur. You can read the Required practical activity in Chapter 15 to find out how distillation is used to separate the alcohol product from the reaction mixture.*

As the size of the halogen atom increases and the polarity of the carbon–halogen bond decreases, so the attraction for the nucleophile decreases. This would suggest that the reactivity of the carbon–halogen bond decreases from chlorine to iodine.

However, it is the strength of the carbon–halogen bond (Table 3) that determines reactivity between halogenoalkanes and nucleophiles. The carbon–fluorine bond is highly polar, but fluoroalkanes do not react with, for example, OH$^-$ ions because the C–F bond is very strong.

Bond	Bond enthalpy/ kJ mol^{-1}
C–F	467
C–Cl	346
C–Br	290
C–I	228

Table 3 *Bond enthalpies for carbon halogen bonds*

As the size of the halogen atom increases, the bond strength decreases. Weaker bonds are broken more easily and the rate of nucleophilic substitution of the halogenoalkanes increases from chloroalkanes to bromoalkanes to iodoalkanes.

Nucleophilic substitution by ammonia

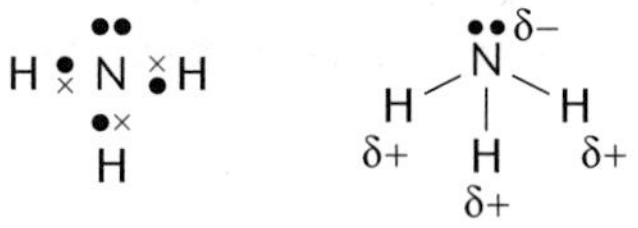

Figure 6 *The ammonia molecules is another example of a nucleophile. Each N–H bond is polar, giving a δ– charge on the nitrogen.*

Ammonia NH$_3$, has three bonding pairs of electrons and one lone pair and the nitrogen has a small negative charge because of the electronegativity difference between N and H. The reaction mechanism is the same as the mechanism for the reaction between a halogenoalkane and hydroxide ions.

For example, when ammonia reacts with 1-bromopropane, the mechanism is:

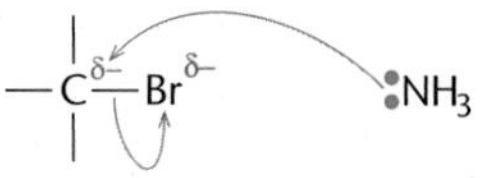

The C–Br bond breaks heterolytically and a covalent bond forms between the carbon atom and the nitrogen atom. The overall reaction is:

$$CH_3CH_2CH_2Br + NH_3 \rightarrow CH_3CH_2CH_2NH_2 + HBr$$

1-bromopropane   1-propylamine

The group –NH$_2$ in an organic compound is an **amine**. NH$_2$ is the functional group.

266

QUESTIONS

8. Which halogenoalkane reacts with ammonia to produce the following compounds? Write an equation for the reaction in each case.

   a. butylamine

   b. ethylamine

Nucleophilic substitution by cyanide ions

Figure 7 *The cyanide ion is a further example of a nucleophile.*

The ion is negatively charged and has lone pairs of electrons on the carbon and nitrogen. In this case, the lone pair on carbon is more important because it gained the extra electron to give the cyanide ion a negative charge. The negative charge of the lone pair of electrons on the carbon in the cyanide group is attracted to the slightly positive charge of the carbon in the carbon–halogen bond. When 1-bromopropane reacts with the cyanide nucleophile, the mechanism is:

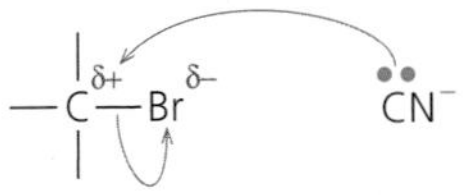

The carbon of the CN^- donates a pair of electrons to the C–Br bond and the C–Br bond breaks heterolytically. A **nitrile** called butanenitrile and a bromide ion form. The overall reaction is:

$$CH_3CH_2CH_2Br + CN^- \rightarrow CH_3CH_2CH_2CN + Br^-$$

1-bromopropane   1-butanenitrile

A solution of sodium cyanide or potassium cyanide in ethanol is often used to provide the cyanide ions. The reaction is particularly useful because it is a way of adding to a carbon chain. In this case, the reactant has three carbons in its chain and the product has four. When chemists manufacture an organic chemical, they prefer to start with a chemical made from the same number of carbon atoms as the product. But sometimes this is not possible – a carbon chain has to be lengthened or shortened. The reaction with the cyanide ion lengthens the chain by one carbon and is a valuable manufacturing route.

QUESTIONS

9. Bromomethane can be converted into ethanenitrile (commonly called acetonitrile).

   a. Write a full equation for the reaction.

   b. Explain the mechanism.

ASSIGNMENT 1: DETECTING HALOGENOALKANES AND CONFIRMING RELATIVE BOND ENTHALPIES

(PS 1.1, 1.2, 2.3)

A student carried out an experiment in which he placed 1 cm³ acidified silver nitrate solution in each of three test tubes, labelled A, B and C. He added 1 cm³ ethanol to each and placed the test tubes in a water bath at 50 °C for ten minutes to let them reach this temperature.

He then added five drops of 1-chlorobutane to test tube A, five drops of 1-bromobutane to test tube B and five drops of 1-iodobutane to test tube C. He placed the test tubes back in the water bath and

started a stop clock. He recorded the time when a precipitate appeared, and its colour. The results are given in Table A1.

Test tube	Time for precipitate to form/seconds	Colour of precipitate
A	920	white
B	324	cream
C	30	yellow

Table A1 *The student's observations and measurements*

Figure A1 *Silver nitrate solution can be used to detect the presence of (left to right) iodide, bromide and chloride ions.*

Questions

A1. Why does silver nitrate solution not form an immediate precipitate with the halogenoalkanes? (Hint: think about the type of bonding in halogenoalkanes. Silver nitrate solution is used to detect halide ions.)

A2. Why is ethanol needed?

A3. In this experiment, an OH⁻ ion is substituted for the halogen atom in the halogenoalkanes.

The OH⁻ comes from the partial ionisation of water molecules. The position of equilibrium is far to the left; in other words, there are very few ions in the equilibrium mixture. However, as these few ions get used up, more are formed (see Chapter 9).

$$H_2O(l) \rightleftharpoons H^+(aq) + OH^-(aq)$$

What is the mechanism of these reactions?

A4. Where did the water come from?

A5. Write an equation for the reactions in A3.

A6. How can the colour of the precipitate be used to distinguish between the halogenoalkanes?

A7. Write an ionic equation for the formation of a precipitate with each halide ion.

A8. Explain why the precipitate formed first in test tube C and last in test tube A.

A9. How does this experiment confirm the trend in bond enthalpies between carbon and halogen atoms?

Elimination reactions

In an elimination reaction, a halogenoalkane forms an alkene molecule and a hydrogen halide molecule is eliminated. Elimination reactions are usually carried out by reacting the halogenoalkane with a concentrated solution of OH⁻ ions, such as potassium hydroxide or sodium hydroxide in ethanol. The hydroxide ions with their lone pair of electrons act both as a nucleophile and as a base. A lone pair of electrons on the OH⁻ ion is attracted to the hydrogen on the carbon next to the carbon with the halogen bond. Strong bases are proton acceptors and the bond breaks heterolytically as a proton (H⁺) is removed. A double bond forms between the two carbon atoms and a halide ion is produced.

2-bromopropane reacts with hydroxide ions in an elimination reaction when alcoholic potassium hydroxide solution and a high temperature is used. Propene, water and bromide ions form.

The mechanism is:

H   H   H
| | |
H—C—C—C—H $\overset{\bullet\bullet}{O}H^-$
| | |
H   Br H

↓

H   H
| |
H—C—C=C—H   + H_2O + $\overset{\bullet\bullet}{B}r^-$
| |
H   H

Elimination or nucleophilic substitution by hydroxide ions?

You have just read that both nucleophilic substitution reactions and elimination reactions can occur between a halogenoalkane and hydroxide ions.

There is competition between these reaction types and the products depend mainly on the type of halogenoalkane and, to a lesser extent, on the reaction conditions.

Halogenoalkanes may be primary, secondary or tertiary (Figure 8).

› A primary halogenoalkane has a halogen atom attached to a carbon atom with two hydrogen atoms (or three hydrogen atoms in a halogenomethane).

› Secondary halogenoalkanes have the halogen atom attached to a carbon with only one hydrogen atom.

› Tertiary halogenoalkanes have no hydrogen atoms attached to the carbon with the halogen atom.

Halogenoalkane	Example
primary 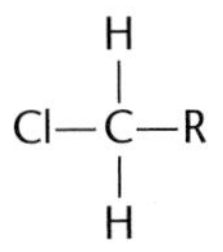 Cl—C—R with H top and H bottom	chloromethane Cl—C—H with H top and H bottom 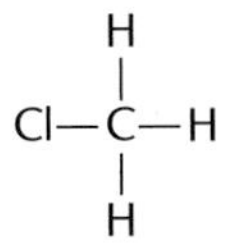
secondary Cl—C—H with R top and R bottom 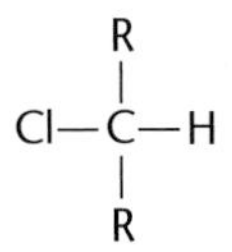	2-chloropropane Cl—C—H with CH_3 top and CH_3 bottom 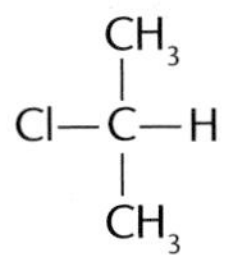
tertiary Cl—C—R with R top and R bottom 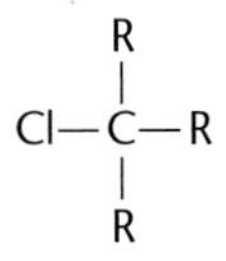	2-chloro-2-methylpropane 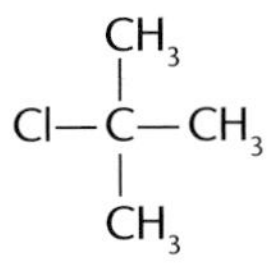 Cl—C—CH_3 with CH_3 top and CH_3 bottom

Figure 8 *Primary, secondary and tertiary halogenoalkanes*

When primary halogenoalkanes react with a solution of hydroxide ions the products are mainly primary alcohols, regardless of the reaction conditions. Nucleophilic substitution is strongly favoured over elimination.

Secondary halogenoalkanes can undergo both substitution and elimination reactions, depending on the conditions used. Frequently, a mixture of the appropriate alcohol (by substitution) and alkene (by elimination) is obtained. Table 4 shows the likely outcome from using different reaction conditions.

Tertiary halogenoalkanes produce mainly alkenes by elimination reactions.

Conditions	Elimination reaction	Substitution reaction
Base	high concentration of OH⁻	low concentration of OH⁻
Solvent	ethanol	water
Temperature	high	low

Table 4 *Predicting the products of reactions between secondary halogenoalkanes and hydroxide ions*

QUESTIONS

10. What is the minimum number of carbon atoms needed to make:
    a. a secondary halogenoalkane
    b. a tertiary halogenoalkane?

11. What type of reaction happens and what are the likely products when:
    a. 1-chloropropane is warmed with aqueous dilute sodium hydroxide solution
    b. 2-chloropropane in ethanol is heated with concentrated sodium hydroxide solution?

Figure 9 summarises the substitution and elimination reactions of halogenoalkanes.

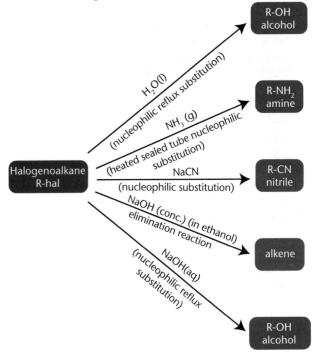

Figure 9 *The reactions of halogenoalkanes. 'R' denotes an alkyl group, C_nH_{2n+1}.*

KEY IDEAS

- Halogenoalkanes undergo nucleophilic substitution and elimination reactions.

- In nucleophilic substitution, another atom or group of atoms is substituted for the halogen atom.

- One hydrogen atom and one halogen atom are removed in an elimination reaction and an alkene is formed.

- A nucleophile has a lone pair of electrons and carries a negative charge (whole in an anion or partial in a polar molecule).

- Nucleophiles are attracted to the slightly positively charged carbon in the carbon–halogen bond.

- Halogenoalkanes undergo nucleophilic substitution reactions with OH^-, NH_3 and CN^-.

- Halogenoalkanes undergo elimination reactions with OH^- ions. Ethanol is often used as a solvent.

- The type of halogenoalkane and the conditions of the reaction determine whether an elimination reaction or a nucleophilic substitution reaction takes place.

ASSIGNMENT 2: A REALLY USEFUL MEDICINE

(MS 4.1; PS 1.1, 1.2)

In October 2013, the World Health Organization (WHO) published a list of minimum medical needs for a basic healthcare system. Phenobarbital (Figure A1) is one of its 204 medicines. It is the most commonly used anticonvulsant that is used to treat epilepsy, world wide.

Many medicines that we use today were first identified in plants. For example, a compound similar to aspirin was initially extracted from willow bark. Once scientists knew the structure of the aspirin molecule, it became possible to make aspirin in the laboratory. A series of manufacturing steps was needed to produce aspirin from readily available compounds. The process involves several steps, each of which produces an intermediate compound before the final step yields the required product. Halogenoalkanes are very useful in this series of steps because their nucleophilic and substitution reactions are frequently used to change one functional group for another.

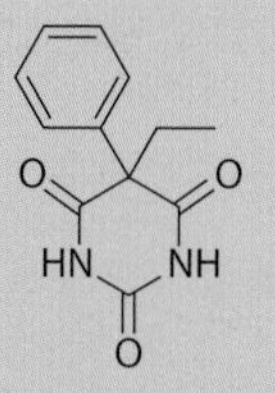

Figure A1 *The structure of phenobarbital*

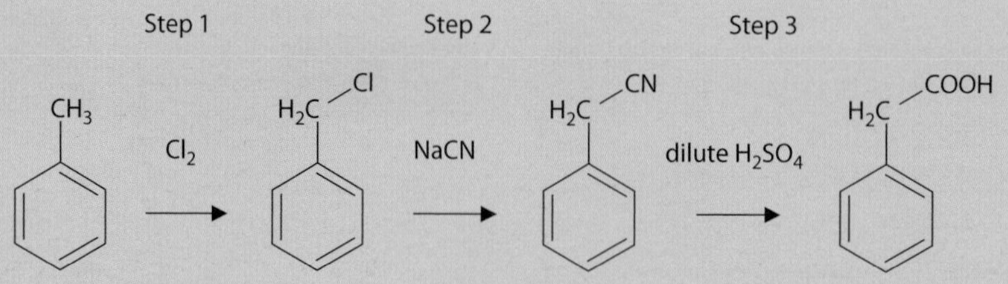

Figure A2 *The first three steps in the manufacture of phenobarbital. The starting compound is methylbenzene, which has a methyl group attached to a benzene ring.*

Questions

A1. a. What is the molecular formula of benzene?

Stretch and challenge

b. Write the molecular formula of methyl benzene and draw its displayed formula.

A2. The hydrogens in the methyl group on the benzene ring behave similarly to the hydrogens in methane.

a. What conditions are necessary to replace a hydrogen on the methyl group in methyl benzene with a chlorine atom?

b. What is this type of reaction called?

A3. a. Which step increases the length of the carbon chain?

b. Which reagents and what conditions are used in this step?

c. Name and describe the mechanism of this reaction.

A4. Which compound shown in Figure A2 is:

a. a hydrocarbon

b. an organic acid

c. a nitrile?

ASSIGNMENT 3: A FREE-RADICAL MECHANISM

(PS 1.1, 1.2, 2.2)

When methane and chlorine are mixed in the dark, there is no reaction. If the mixture is exposed to radiation from an ulltraviolet (UV) lamp, the mixture explodes. In sunlight, the UV is less intense and the reaction happens slowly to produce a mixture of halogenoalkanes.

UV radiation carries and transfers sufficient energy to split the chlorine molecule into two chlorine radicals. This is called photodissociation. Chlorine radicals can substitute for hydrogens in methane. The reaction is a free radical mechanism.

The products depend on the number of hydrogens substituted. If one hydrogen is substituted by a chlorine, chloromethane forms:

$$CH_4(g) + Cl_2(g) \rightarrow CH_3Cl(g) + HCl(g)$$

Questions

A1. a. Write an equation to show the photodissociation of chlorine. Use dots to show the mechanism.

b. Explain why this is an example of homolytic fission.

c. Why is this an initiation reaction?

A2. a. Write an equation to show:

i. the reaction between a chlorine radical and methane to produce a methyl radical and hydrogen chloride.

ii. the reaction between a methyl radical and a chlorine molecule to produce chloromethane and a chlorine radical. Note: Use dots to show the radicals.

b. Why is this a propagation reaction?

c. Curly arrows are used below to show the mechanism of the reactions in question A2a.

Use the diagram to explain the movement of electrons in the reaction.

A3. a. Subsequent propagation reactions occur until all hydrogens in methane have been replaced by chlorines. Write full equations for these propagation reactions.

b. Name all the halogenoalkanes formed.

A4. Termination reactions happen when two radicals collide and react. Give an example of a termination reaction.

13.3 OZONE IN THE STRATOSPHERE

Ozone is present in small amounts at ground level and in the troposphere. It is a pale blue gas with a sharp smell. You may have noticed the smell because ozone is formed around photocopiers. However, in the stratosphere, ozone is present in larger concentrations and is continually being formed and broken down by the intense UV radiation.

The formation of ozone happens in two steps. The initiation step is:

UV radiation
$$O_2 \quad \rightarrow \quad O\bullet + O\bullet \quad \Delta H^\ominus = +498 \text{ kJ mol}^{-1}$$

Ozone forms when an oxygen radical collides with an oxygen atom:

$$O\bullet + O_2 \rightarrow O_3 \qquad \Delta H^\ominus = -106 \text{ kJ mol}^{-1}$$

But UV radiation has enough energy to break the bonds in ozone:

UV radiation
$$O_3 \quad \rightarrow \quad O_2 + O\bullet \quad \Delta H^\ominus = +302 \text{ kJ mol}^{-1}$$

The frequencies absorbed to break the bonds in ozone are the frequencies that damage biological life on the surface of Earth. So the natural breaking down of the ozone in the stratosphere reduces the concentration of harmful UV radiation that reaches Earth's surface.

When there are no pollutants, the natural rate at which ozone forms in the stratosphere equals the rate at which it is broken down and the concentration of ozone is constant. But the rates are no longer equal. Ozone is currently being broken down faster than it is being formed.

What is destroying the ozone?

As part of the natural cycle, oxygen radicals break down the ozone molecules ($O\bullet + O_3 \rightarrow 2O_2$). But there are other radicals, such as chlorine and bromine radicals, that also break down ozone. Gaseous organic compounds containing chlorine and bromine atoms are made synthetically in large quantities because they are very useful. When they eventually diffuse into the stratosphere, the high-energy UV radiation breaks the C–Cl bond to produce a chlorine radical. The chlorine radicals are very reactive and can react with ozone:

$$Cl\bullet + O_3 \rightarrow ClO\bullet + O_2$$

Oxygen and a chlorine oxide radical are produced. The chlorine oxide radical is very reactive and reacts with another ozone molecule:

$$ClO\bullet + O_3 \rightarrow 2O_2 + Cl\bullet$$

The chlorine radical formed now reacts with other ozone molecules and the reaction starts over again. These are propagation steps.

You can write an overall equation by adding the two equations together:

$Cl\bullet + O_3$	$\rightarrow$	$ClO\bullet + O_2$
$ClO\bullet + O_3$	$\rightarrow$	$2O_2 + Cl\bullet$
$2O_3$	$\rightarrow$	$3O_2$

$Cl\bullet$ and $ClO\bullet$ on both sides cancel each other out.

Overall, ozone breaks down into oxygen. If it happens very slowly, it would not matter much, but if it happens rapidly it causes problems. The reaction of chlorine radicals with ozone has a lower activation energy than the reaction between oxygen radicals and ozone. Chlorine radicals break down ozone 1500 times faster than oxygen radicals do. It is estimated that one chlorine radical will react with 100 000 ozone molecules before colliding with another radical and the reaction terminating (see Figure 10).

The chlorine radical is regenerated at the end of each cycle so it is acting as a catalyst, speeding up the reaction while being unchanged itself at the end.

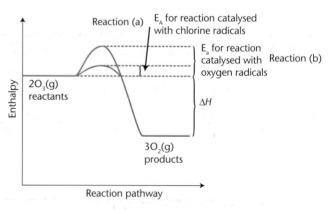

Figure 10 *Energy profiles for the reactions between (a) chlorine radicals and ozone, (b) oxygen radicals and ozone*

The chlorofluorocarbon (CFC) story

CFCs are a group of gaseous halogenoalkanes in which all the hydrogen atoms have been replaced by chlorine and fluorine atoms. Many different CFCs are possible.

The amounts produced and used rocketed during the 1960s and 1970s. You have to look at some of their properties to understand why.

Table 3 shows the bond enthalpies of the carbon–halogen bonds. The carbon–fluorine bond has the highest bond enthalpy. This means that substances that contain carbon and fluorine tend to be unreactive. The carbon–chlorine bond is not as strong as the carbon–fluorine one, but it is still strong and not very reactive. So CFCs are very stable and stay around for hundreds of years.

Together with the fact that they have no smell or taste and are usually gases at room temperature, their stability makes them very useful. For example, CFCs have been used in:

> the blow-foam plastics process (for example, expanding poly(styrene)); they do not react with the plastics nor do they decompose

> fridges, freezers and air-conditioning systems as refrigerants

> aerosol cans as ideal propellant gases.

And because they were so unreactive, CFCs were considered harmless to living organisms and the environment. Consequently, vast quantities of CFCs were released into the atmosphere.

Professor Sherry Rowland at the University of California first investigated the stability of CFCs in the atmosphere in the early 1970s. The link between ozone destruction and CFCs in the stratosphere soon became apparent. The terms CFC and ozone hole entered our vocabulary.

Most pollutants released into the atmosphere eventually break down in the troposphere by reacting with oxygen or water. However, the stability of CFCs gives them plenty of time to diffuse into the stratosphere. Here, conditions are different. The large amounts of UV radiation break the C–Cl bonds in the CFCs to produce chlorine radicals. It is an example of photodissociation. The C–F bonds do not break because their bond enthalpy is higher and UV radiation does carry sufficient energy.

When dichlorodifluoromethane interacts with UV radiation, the following reaction occurs:

$$F-CCl_2F \xrightarrow{\text{UV radiation}} F-\overset{\cdot}{C}ClF + Cl\cdot$$

Alternative chlorine-free compounds

In response to the scientific evidence, 68 nations met in Montreal and signed the Montreal Protocol in 1987. This was an international agreement that called for an immediate reduction in the production and use of CFCs and other compounds that destroyed ozone. As the size of the ozone hole continued to grow in the 1990s, a total ban on the manufacture of CFCs was implemented in 1996. The Montreal Protocol has been dubbed the most successful international environmental agreement ever and is now ratified by 196 countries. In the UK today, both EU and UK legislation is in place to control the manufacture, use and disposal of CFCs (Figure 11) and their replacements.

Figure 11 *Banning CFC manufacture still left an estimated 200 million fridges and freezers in the UK alone that contained about 100 000 tonnes of CFCs. When these were scrapped, the CFCs needed to be disposed of safely. In January 2002, it became illegal to dispose of fridges or freezers without safely disposing of the CFCs.*

The scientists' task was to find suitable replacements for CFCs so that the latter could be phased out. Ideally, these replacements would have similar properties as CFCs, but would decompose in the troposphere and not reach the stratosphere. Two major types of compounds have been used as replacements.

› Hydrochlorofluorocarbons (HCFCs), such as chlorodifluoromethane.

› Hydrofluorocarbons (HFCs), such as trifluoromethane.

The presence of hydrogen atoms in the HCFCs and HFCs increases their reactivity. The C–H bonds are attacked by OH radicals in the atmosphere. The products dissolve in atmospheric water and end up in rain. Hence, most chlorine atoms do not reach the stratosphere and HCFCs and HFCs have a shorter life (HFCs < 20 years).

However, for the few HCFCs that do reach the stratosphere, the stratospheric conditions extend their life span to 120 years, and it seems that HCFCs cause more ozone depletion than CFCs. HCFCs are also very potent greenhouse gases. Plans are to phase out HCFCs by 2030.

HFCs have no chlorine and do not break down ozone. They are viewed as a more long-term replacement for CFCs. However, the downside is that they are also very potent greenhouse gases.

Scientists estimate that ozone concentrations in the stratosphere will not return to pre-1980 levels until between 2060 and 2075.

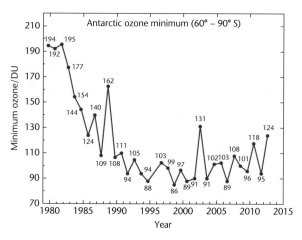

Figure 12 *The graph shows the changes in ozone concentration in the stratosphere above the Antarctic during the past 35 years. Ozone concentration is temperature dependent and there are seasonal variations.*

QUESTIONS

13. Draw the displayed formula and write the molecular formula for:
    a. chlorodifluoromethane
    b. trifluoromethane

14. 1,1,1,2-tetrafluoroethane is the most widely used HFC in fridges and air-conditioning.
    a. Draw its displayed formula.
    b. What is its molecular formula?

15. a. Draw a dot-and-cross diagram to show the bonding in the OH radical.
    b. Explain why it is a radical.

Stretch and challenge

16. Large amounts of chlorine from swimming pools is released into the atmosphere daily. Explain why this chlorine does not deplete the ozone layer but chlorine in CFCs does.

17. Why does most ozone depletion over Antarctica happen in September and October during the Southern Hemisphere spring?

KEY IDEAS

› Ozone in the stratosphere absorbs UV radiation and is continually being formed and destroyed in a natural cycle.

› UV radiation is harmful to biological life. Stratospheric ozone absorbs UV in its reactions and prevents a large percentage of UV radiation from reaching Earth.

› Chlorine radicals are formed when C–Cl bonds in CFCs are broken by UV radiation.

› Chlorine radicals catalyse the reaction of ozone, O_3, into oxygen, O_2.

ASSIGNMENT 4: HOW DO SUNSCREENS WORK?

(PS 1.1, 1.2, 2.3, 3.2)

UV radiation has wavelengths between 100 and 400 nm. These are divided into UVA (400–315 nm), UVB (315–280 nm) and UVC (280–100 nm).

As UV radiation enters Earth's atmosphere, all UVC radiation and about 90% of UVB radiation is absorbed by ozone in the stratosphere. The UV radiation reaching Earth is mostly UVA with small amounts of UVB.

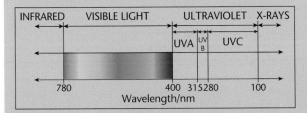

Figure A1 *Types of UV radiation*

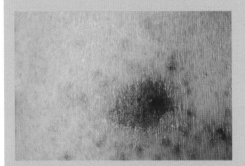

Figure A2 *This is a basal cell carcinoma, a form of skin cancer. Fortunately, it does not spread in the body and can be surgically removed. More serious and rarer are malignant melanomas, which can spread cancer to other parts of the body. Basal cell carcinoma and malignant melanoma are caused by excessive exposure to UV radiation.*

Good effective sunscreens protect the user from all wavelengths of UV radiation. Sunscreens contain two types of ingredient: sun blocks to physically reflect and scatter UV radiation, and chemicals that can absorb UV radiation.

Nanoparticles of titanium(IV) oxide and zinc oxide are commonly used for sun blocks; they reflect all UV wavelengths.

Figure A3 *The effectiveness of sunscreen is rated using a scale known as SPF. An SPF of 15 means that you can safely stay in the sun for 15 times longer than you could without sunscreen.*

Sunscreens may contain several chemicals to absorb the whole range of UV radiation, since one ingredient will only absorb a narrow range of UV. Chemical absorbers interact with the top layer of the skin and absorb UV when electrons become excited. As the electrons return to their ground state, the energy is emitted as less harmful IR radiation.

Questions

A1. How will decreases in the concentration of stratospheric ozone change the amount of UV radiation reaching Earth?

A2. Carry out an Internet search to find out the damaging effects on humans of:

   a. UVA radiation

   b. UVB radiation.

The WHO website has reliable information.

A3. The SPF scale used to rate sunscreens is an arbitrary scale and intended for guidance only. How the sunscreen is used is important in determining its effectiveness. What conditions may make the sunscreen less effective than its SPF rating?

Stretch and challenge

A4. Use Figure A4 to help work out which types of UV the following sunscreen ingredients protect against.

   a. Sunscreen containing octyl methoxycinnamate only.

   b. Sunscreen containing butyl methoxydibenzoylmethane only.

   c. Sunscreen containing zinc oxide and octyl methoxycinnamate.

A5. What is the difference between the way that zinc oxide and octyl methoxycinnamate work in a sunscreen?

A6. The emission spectrum of an element shows single absorbance lines for a particular frequency. Suggest why the absorbance spectra for molecules in sunscreen show wide bands of absorbance.

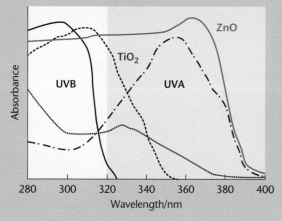

Figure A4 *A UV absorbance spectrum showing the wavelength of UV radiation absorbed (or reflected) by some different substances*

PRACTICE QUESTIONS

1. a. The refrigerant R112A (CCl_3CF_2Cl) has been banned because of concerns about ozone depletion. Give the IUPAC name for CCl_3CF_2Cl.

   b. Nitrogen monoxide (NO) catalyses the decomposition of ozone into oxygen.

   i. Write the overall equation for this decomposition.

   ii. Use the overall equation to deduce Step 3 in the following mechanism that shows how nitrogen monoxide catalyses this decomposition.

   Step 1 $O_3 \rightarrow O + O_2$

   Step 2 $NO + O_3 \rightarrow NO_2 + O_2$

   Step 3 _____

   AQA Jan 2013 Unit 2 Question 7 b and c

(Continued)

2. Chlorofluorocarbons (CFCs) were used as refrigerants. In the upper atmosphere, UV radiation breaks bonds in the CFCs to produce a reactive intermediate that catalyses the decomposition of ozone.

   a. An example of a CFC is 1,1,1-trichloro-2,2-difluoroethane. Draw the displayed formula of this CFC.

   b. Identify a bond in a CFC that is broken by UV radiation to produce a reactive intermediate.

      i. Give the name of this reactive intermediate that catalyses the decomposition of ozone.

      ii. Write an overall equation for this decomposition of ozone.

   AQA Jun 2011 Unit 2 Question 7b

3. Organic reaction mechanisms help chemists to understand how the reactions of organic compounds occur. The following conversions illustrate a number of different types of reaction mechanism.

   a. When 2-bromopentane reacts with ethanolic KOH, two structurally isomeric alkenes are formed.

      i. Name and outline a mechanism for the conversion of 2-bromopentane into pent-2-ene as shown here:

      ethanolic KOH
      $$CH_3CH_2CH_2CHBrCH_3 \rightarrow CH_3CH_2CH=CHCH_3$$

      ii. Draw the structure of the other structurally isomeric alkene produced when 2-bromopentane reacts with ethanolic KOH.

   b. Name and outline a mechanism for this conversion:

   $$NH_3$$
   $$CH_3CH_2CH_2Br \rightarrow CH_3CH_2CH_2NH_2$$

   AQA Jun 2011 Unit 2 Question 9 a and c

4. CCl_4 is an effective fire extinguisher but it is no longer used because of its toxicity and its role in the depletion of the ozone layer. In the upper atmosphere, a bond in CCl_4 breaks and reactive species are formed.

   a. i. Identify the condition that causes a bond in CCl_4 to break.

      ii. Deduce an equation for the formation of the reactive species.

   b. One of the reactive species formed from CCl_4 acts as a catalyst in the decomposition of ozone. Write two equations to show how this species acts as a catalyst.

   AQA Specimen Paper 2 2015 Question 5

5. One of the fractions obtained from petroleum can be thermally cracked to produce propene. Some of the reactions of propene are shown in Figure Q1.

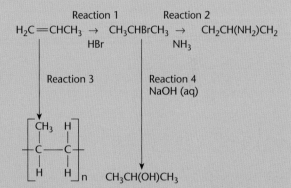

Figure Q1

   a. Explain why the carbon–bromine bond is polar.

   b. Reaction 2 is a nucleophilic substitution.

      i. Identify the nucleophile.

      ii. Which property do all nucleophiles have in common?

      iii. Outline the mechanism for reaction 2.

   c. i. Identify the reaction mechanism for reaction 4.

      ii. What feature of the $CH_3CHBrCH_3$ molecule makes it susceptible to attack by an ammonia molecule?

   d. $CH_3CHBrCH_3$ reacts with potassium hydroxide in ethanol to produce an alkene.

      i. Name the alkene produced.

      ii. Name the type of reaction to produce the alkene.

   AQA Jun 2004 Unit 3(a) Question 2

(Continued)

6. Consider the following reaction in which an alkene is formed from a halogenoalkane.

$$CH_3CHBrCH_2CH_3 + KOH \xrightarrow[\substack{ethanol \\ solvent}]{heat} CH_3CH = CHCH_3 + KBr + H_2O$$

a. Name the halogenoalkane used in this reaction.

b. Name and outline a mechanism for this reaction.

c. Another alkene, which is a structural isomer of but-2-ene, is also formed during this reaction.

  i. State what is meant by the term *structural isomers*.

  ii. Draw the structure of this other alkene.

d. i. Name the main organic product formed when the reaction is carried out using potassium hydroxide dissolved in water.

  ii. Identify the reaction mechanism for using aqueous potassium hydroxide.

AQA Jun 2005 Unit 3(a) Question 3

7. Two refrigerants, R12 and R13, are no longer used because of concerns about their possible effects on the ozone layer.

a. The refrigerant R12 is the compound dichlorodifluoromethane.

  i. Write the formula for dichlorodifluoromethane.

  ii. The compound R13 contains 11.5% carbon and 34.0% chlorine by mass. The remainder of the compound is fluorine. Calculate the empirical formula of R13.

b. i. Write an equation to show how a chlorine atom forms from dichlorofluoromethane in the upper atmosphere.

  ii. Name the conditions for this reaction.

  iii. Write two equations to show how chlorine atoms catalyse the break down of ozone.

c. Although ozone-damaging CFCs are banned in most countries, scientists expect ozone depletion to be a problem for many years. Explain why CFCs do not easily decompose in the atmosphere.

8. Scientists have developed alternative chlorine-free compounds to replace CFCs. One such compound is trifluoromethane.

a. i. Identify two properties of CFCs that led to their wide-scale use.

  ii. Which major discovery led to the banning of CFCs in most countries?

b. i. Write the molecular formula for trifluoromethane.

  ii. What properties must a replacement CFC have?

  iii. Explain why trifluoromethane does not produce fluorine atoms in the upper atmosphere.

AQA Jan 2007 Unit 3(a) Question 5

Stretch and challenge

9. Chloroethane and hexafluoroethane are two halogenoalkanes. Chloroethane gas breaks down slowly in the atmosphere by reacting with water. This is a nucleophilic substitution reaction. Hexafluoroethane does not break down in the lower atmosphere.

a. Describe the features of a water molecule that enable it to act as a nucleophile.

b. Explain the mechanism for the nucleophilic substitution of chloroethane by water and name the organic product.

c. Explain why hexafluoroethane does not react easily in the lower atmosphere or in the stratosphere (upper atmosphere).

Multiple choice

10. What does 1,2-dibromo-1,1,2,2-tetrafluorethane contain?

A. 4 fluorine atoms and 2 bromine atoms only

B. 2 fluorine atoms and 4 bromine atoms and 2 carbon atoms

C. 4 fluorine atoms and 2 bromine atoms and 2 carbon atoms

D. 4 fluorine atoms and 2 bromine atoms and 4 carbon atoms

11. Which definition is correct?

A. In an elimination reaction, two halogen atoms are removed from a halogenoalkane.

(Continued)

B. In an elimination reaction, an electrophile is attracted to the halogen atom in a halogenoalkane.

C. In an elimination reaction, a nucleophile is attracted to the halogen atom of a halogenoalkane.

D. In an elimination reaction, a hydrogen atom and a halogen atom are removed from a halogenoalkane.

12. What is the organic product formed when ammonia reacts with a halogenoalkane in a nucleophilic substitution reaction?

   A. An amine

   B. An amide

   C. An amino acid

   D. An aldehyde

13. When is the amount of ozone in the stratosphere constant?

   A. When the rate at which ozone breaks down naturally is less than the rate at which it is being formed.

   B. When the rate at which ozone breaks down naturally is greater than the rate at which it is being formed.

   C. When the concentration of chlorine atoms in the stratosphere increases.

   D. When the rate at which ozone breaks down naturally equals the rate at which it is being formed.

14 ALKENES

PRIOR KNOWLEDGE

You may have learnt about alkenes at GCSE and how they are used in polymerisation reactions to make polymers. You read in Chapter 5 that alkenes are a homologous series and you found out how different types of formulae can be used to describe them.

LEARNING OBJECTIVES

In this chapter you will build on these ideas and find out about the addition reactions of alkenes.

(Specification 3.3.4.1, 3.3.4.2, 3.3.4.3)

Worldwide production of ethene exceeds that of any other organic compound. We use billions of tonnes of ethene annually just to make plastic bags, and it is used extensively to make many other polymers and as an intermediate in the chemical industry. Ethene also has a major role in plant chemistry and is produced in every cell of higher plants.

You may think that hormones are complicated molecules – and many are. However, the simple compound ethene C_2H_4 is a very important plant hormone. Hormones are chemical messengers in living organisms. Since ethene is a gas, relatively large amounts are produced by plants to compensate for loss by diffusion. Ethene stimulates the growth of roots and shoots, flower opening and fruit ripening. Controlling ethene is important for vegetable, fruit and flower producers who need to get their produce to the consumer in the best condition with as long a shelf life as possible.

Bananas can be picked green and transported and stored in an environment where the concentration of ethene is controlled so that they do not ripen too quickly and they reach the supermarket just ready to eat. Unfortunately for the producer, ethene's action does not stop there and continued production of ethene causes fruits and vegetables to over-ripen and die, so limiting their shelf life. Some fruits and vegetables produce more ethene than others, but all are sensitive to its presence. Careful positioning of produce in a supermarket can help prolong shelf life; for example, tomatoes produce relatively large amounts of ethene and produce displayed near tomatoes will be affected by the higher ethene concentration.

14.1 STRUCTURE AND BONDING IN ALKENES

Alkenes have the general formula C_nH_{2n}. Figure 1 shows the structure of the first six alkenes in the series – ethene, propene, but-1-ene, pent-1-ene, hex-1-ene and hept-1-ene. Alkenes can also have a cyclic structure, such as cyclohexene.

The first three members of the alkene series are gases at room temperature and pressure. Longer chain alkenes are liquids and solids (Table 1). An increase in the boiling point as the number of carbon atoms increases in the molecule is typical for all hydrocarbons and is explained by the increasing strength of the van der Waals forces as the length of the carbon chain increases.

Saturated or unsaturated?

The difference between an alkene and an alkane made from the same number of carbon atoms is the number of hydrogen atoms needed to make the molecule. An ethane molecule is made from two carbon atoms and six hydrogen atoms. It is **saturated** because each carbon has been bonded to four other atoms. Remember, a carbon atom has four electrons in its outer shell and so it can form single covalent bonds with four other atoms.

Ethene, the corresponding alkene, is made from two carbon atoms and only four hydrogen atoms. It is **unsaturated** because two of the carbons only form bonds with three other atoms. The bond between these two carbon atoms is a double covalent bond.

QUESTIONS

1. Look at the alkenes in Figure 1. The general formula for a non-cyclic alkene is C_nH_{2n}. What is the general formula for a cyclic alkene?

Alkene	Structural formula	Melting point/°C	Boiling point/°C
ethene	$CH_2 = CH_2$	– 169	– 105
propene	$CH_3CH = CH_2$	– 185	– 48
but-1-ene	$CH_3CH_2CH = CH_2$	– 185	– 6
pent-1-ene	$CH_3CH_2CH_2CH = CH_2$	– 165	30
hex-1-ene	$CH_3CH_2CH_2CH_2CH = CH_2$	– 140	63

Table 1 *Melting and boiling points of some alkenes*

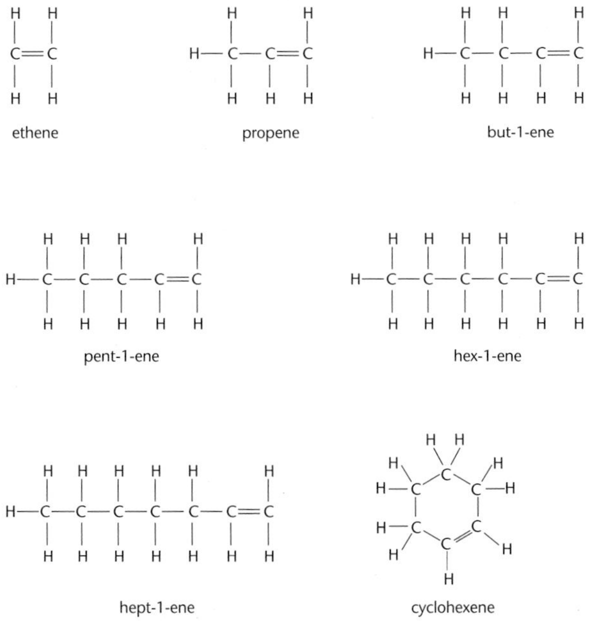

Figure 1 *The displayed formulae of some alkenes*

Bonding in alkenes

An alkene contains one double bond between one pair of its carbon atoms. This is the functional group in the alkene series. It explains why alkenes are such reactive and useful compounds.

Figure 2 shows the arrangement of electrons in ethene. The double bond, C=C, consists of two bonding pairs of electrons.

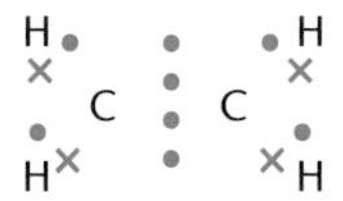

Figure 2 *Dot-and-cross diagram to show the bonding in an ethene molecule*

A double bond consists of two different types of bonds. One electron from each carbon atom is used to form a sigma bond (σ bond) between the two carbon atoms, and two electrons are used to form σ bonds with two hydrogen atoms. All single covalent bonds are σ bonds, formed from the end-on overlap of orbitals. The fourth electron in each carbon atom is shared in a pi (π) bond. This is formed from the sideways overlapping of orbitals.

Double bonds are areas of high electron density. A C=C double is much stronger than a C–C single bond (Table 2). However, the electrons that make up the π bond are held less tightly in the molecule than the electrons within the σ bond. This means that the π bond attracts positively charged ions or the positive end of polar molecules that come near the alkene molecule. This makes alkenes far more reactive than alkanes.

Bond	Bond energy/kJ mol^{-1}	Bond length/nm
C–C	346	0.154
C=C	598	0.134

Table 2 Comparison of single and double carbon bonds

QUESTIONS

2. Why is the bond energy for C=C not double the bond energy for C–C?

The pi and sigma bonds in the C=C bond hold the carbon atoms in a fixed position and they are not free to rotate. The C=C group has a planar shape, and can have *E-Z* isomers (see Chapter 5).

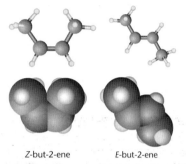

Z-but-2-ene   E-but-2-ene

Figure 4 *E-Z isomers of but-2-ene: ball-and-stick and space-filling models*

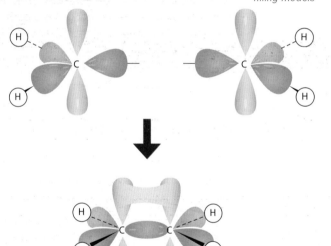

These shapes represent orbitals – regions in which there is a high probability of finding an electron.

The white spheres are hydrogen orbitals and the shaded areas are carbon orbitals.

The orbitals overlap end-to-end, forming   bonds or side-to-side, forming π bonds.

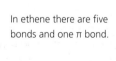

In ethene there are five bonds and one π bond.

Figure 3 *Pi and sigma bonds in ethene*

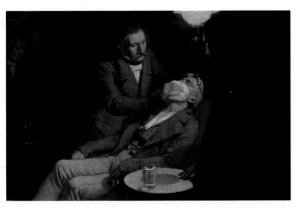

Figure 5 *Alkenes were used as anaesthetics during the 19th and 20th centuries. However, Ether (not an alkene) was more common; this image depicts its first use, in 1846, by William Morton.*

KEY IDEAS

> Unbranched alkenes have the general formula C_nH_{2n}.

> Alkenes are unsaturated.

> The double bonds in alkenes are areas of high electron density.

14.2 ADDITION REACTIONS OF ALKENES

Alkenes undergo addition reactions in which another molecule adds across the double bond. The σ bond between the two carbons remains intact, but the two electrons in the π bond rearrange to form bonds with atoms from the incoming molecule. For example:

$$CH_2 = CH_2 + Cl_2 \rightarrow CH_2ClCH_2Cl$$

The double bond is an area of high electron density and this area of negative charge attracts ions or polar molecules that carry a positive charge. Molecules and groups attracted by regions of negative charge are called **electrophiles** ('electron-loving'). Addition reactions involving an electrophile are called **electrophilic addition** reactions.

The reaction mechanism can be shown using curly arrows to indicate the rearrangement of electrons. The following three examples of addition reactions of ethene illustrate this.

Ethene and hydrogen bromide

Alkenes react with the hydrogen halides to form halogenoalkanes. Ethene reacts readily with hydrogen iodide and hydrogen bromide, but more slowly with hydrogen chloride. The reaction of ethene with

hydrogen bromide takes place in the gas phase or in concentrated aqueous solution, to form bromoethane:

$$CH_2 = CH_2 + HBr \rightarrow CH_3CH_2Br$$

Hydrogen bromide has a permanent dipole:

$$H^{\delta+}-Br^{\delta-}$$

The hydrogen in the hydrogen bromide molecule has a small positive charge because of the electronegativity difference between bromine and hydrogen. The $H^{\delta+}$ end of the hydrogen bromide molecule is the electrophile.

In the gas phase the positively charged hydrogen in the polar hydrogen bromide molecule is attracted to the electron density of the C=C double bond in the ethene molecule. Two electrons in the C=C double bond (that formed the π bond) transfer to the $H^{\delta+}$ in HBr. Simultaneously, the pair of electrons in the covalent H–Br bond transfer to the $Br^{\delta-}$ of HBr to form Br^-. This rearrangement of electrons can be represented using curly arrows:

A positive ion called a **carbocation** forms, together with a bromide ion. A carbocation is a positively charged molecular ion, where one carbon carries a single positive charge.

The bromide ion reacts immediately with the carbocation to make 1-bromoethane:

In aqueous solution, hydrogen bromide ionises:

$$HBr(g) + aq \rightarrow H^+(aq) + Br^-(aq)$$

The $H^+(aq)$ is the electrophile and reacts with ethene to make the carbocation, which immediately reacts with $Br^-(aq)$ to produce bromoethane.

QUESTIONS

5. Which electrons in the C=C bond are attracted to the hydrogen ion?

6. Why is the reaction between ethene and hydrogen chloride slower than the reaction between ethene and hydrogen bromide?

7. **a.** Write an overall equation to show the reaction between hydrogen iodide and ethene. Name the product.

   b. Name and describe the mechanism of the reaction.

8. What is the difference between a nucleophile and an electrophile?

Ethene and sulfuric acid

Alkenes react with cold, concentrated sulfuric acid to form alkyl hydrogen sulfates.

The displayed formula of sulfuric acid is:

Because of the difference in the electronegativities of oxygen and hydrogen, the O–H bonds are polar and hydrogens are positively charged. They act as an electrophile. When ethene gas reacts with concentrated sulfuric acid, ethyl hydrogen sulfate is produced. Figure 6 shows the mechanism.

Sulfuric acid is an acid because it dissociates to release hydrogen ions (H^+). A hydrogen ion acts as an electrophile and is attracted to the high electron density of the double bond.

The carbonium ion, $CH_3\text{-}CH_2^+$, forms. The negative charge of $-SO_3H$ is attracted to the positive charge of the carbonium ion, forming ethyl hydrogen sulfate.

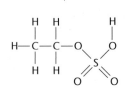

Figure 6 *Hydration of ethene by sulfuric acid*

QUESTIONS

Stretch and challenge

9. If ethyl hydrogen sulfate from the reaction in Figure 6 is warmed in dilute sulfuric acid, hydrolysis occurs and ethanol and sulfuric acid are produced.

   a. Write a full balanced equation for the reaction to produce ethanol.

   b. Since sulfuric acid is regenerated at the end of the reactions, what role does it play in the reaction?

   c. How can propene be used to produce propanol?

Ethene and bromine

When ethene is bubbled into an aqueous solution of bromine, commonly called bromine water, the orange-yellow colour of bromine disappears as the colourless compound 1,2-dibromoethane (CH_2BrCH_2Br) is formed (Figure 7):

$$CH_2 = CH_2 + Br_2 \rightarrow CH_2BrCH_2Br$$

The same reaction happens in non-aqueous solvents such as hexane or cyclohexane.

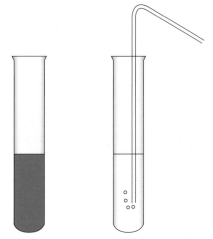

Figure 7 *The bromine water on the left is decolourised when ethene gas is bubbled through. This is a commonly used test for alkenes.*

The reaction involves the electrophilic addition of bromine to ethene and happens in three stages.

Stage 1: Dipole induced in bromine

Normally, the electrons in the Br–Br bond are distributed evenly between the two bromine atoms. In the presence of ethene, the high electron density of the double bond induces a dipole in the bromine molecule.

$Br^{\delta+}–Br^{\delta-}$

One bromine atom then has a small positive charge while the other has a small negative charge.

Stage 2: Electrophilic attack

The positively charged bromine atom is an electrophile and attracts electrons from the double bond in ethene. A carbon–bromine bond, C–Br, and a Br^- ion form. The redistribution of electrons (the mechanism) is similar to that for the reaction of an alkene with hydrogen bromide.

Stage 3: Nucleophilic attack

The newly formed C–Br bond is made from one of the electron pairs of the C=C bond. The other carbon now has a single positive charge. This ion is a carbocation. Br^- ion acts as a nucleophile and a pair of electrons forms a new covalent bond with the carbocation:

ASSIGNMENT 1: COMPARING DEGREES OF UNSATURATION IN OILS

(MS 0.0, 2.3; PS 1.1, 1.2, 2.1, 2.3, 4.1)

Unsaturated fats have long carbon chains with some double bonds between the carbon atoms. Saturated fats have single carbon–carbon bonds. Bromine water can be used to compare the degree of unsaturation of different vegetable oils.

The degree of unsaturation of an oil is measured in terms of its **bromine number**. This is the mass of bromine (in grams) that reacts with 100 g of the oil. Most industrial laboratories measure the iodine number, however. This is equivalent to the bromine number, but the method gives more accurate results, although it is more complicated.

Here is a description of a titration method for measuring the bromine number of oils and fats.

Fill a burette with 0.02 mol dm^{-3} bromine water. Bromine water is toxic and, therefore, the titration should be carried out in a fume cupboard. Measure 5 cm^3 of Volasil^{TM} (a non-aqueous solvent of low toxicity) into a conical flask and add five drops of vegetable oil. Swirl the contents to dissolve the oil. Add bromine water from the burette until there is a permanent colour change.

An experiment was carried out by students using a number of different vegetable oils. Their titration results are shown in Table A1.

Vegetable oil	Initial burette reading/cm^3	Final burette reading/cm^3	Titre/cm^3
coconut oil	0.50	2.50	2.00
palm oil	2.50	7.00	4.50
olive oil	7.00	13.50	5.50
sunflower oil	13.50	20.50	7.00

Table A1 The titration of five drops of vegetable oil against 0.02 mol dm^{-3} bromine water

Questions

A1. What do the students' results show?

A2. What is the colour change when all the C=C bonds have reacted with bromine?

A3. In the reaction between bromine water and the vegetable oils:

a. what is the electrophile?

b. why is it attracted to the long chains of carbon atoms in oil molecules?

c. explain the mechanism.

A4. This experiment compares the degree of unsaturation of different vegetable oils, but if we know the volume of a drop, we can work out the number of moles of unsaturated bonds in a litre bottle of oil (1 litre = 1 dm³).

a. If a drop has a volume of 0.05 cm³, what volume of oil was used in each titration?

b. How many moles of bromine, Br_2, did this react with?

c. Since one bromine molecule is added to each double bond, how many moles of double bonds were present in five drops of oil?

d. How many moles of double bonds are present in a one litre (1 dm³) bottle of coconut oil?

A5. The density of vegetable oils is about 0.92 g cm⁻³. The relative molecular mass of Br_2 is 160. From the students' results, calculate the bromine number of sunflower oil. Show your working.

A6. How could the students' experiment be modified to increase the accuracy?

Predicting the products of electrophilic addition reactions to unsymmetrical alkenes

Alkenes can be symmetrical or unsymmetrical. Ethene is a symmetrical alkene. Both carbon atoms have two hydrogens attached to them:

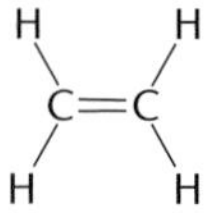

Propene has a hydrogen and a methyl group at one end of its double bond and two hydrogens at the other end. It is unsymmetrical. Similarly, but-1-ene has two hydrogens at one end of its double bond and an ethyl group and a hydrogen at the other end. But-1-ene is also unsymmetrical.

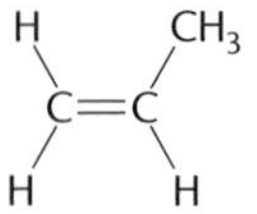

propene 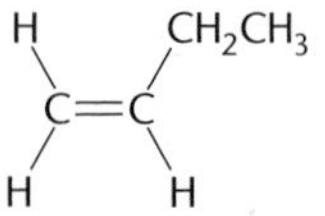   but-1-ene

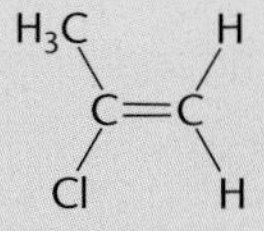

2-chloropropene 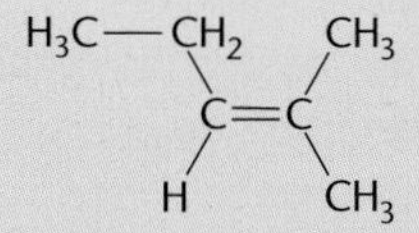   2-methyl-Z-pent-2-ene

The electrophilic addition of a hydrogen halide to an unsymmetrical alkene can give more than one addition product and this is dependent on which end of the double bond reacts with the electrophile. For example, when propene reacts with hydrogen bromide, there are two possible products. Either:

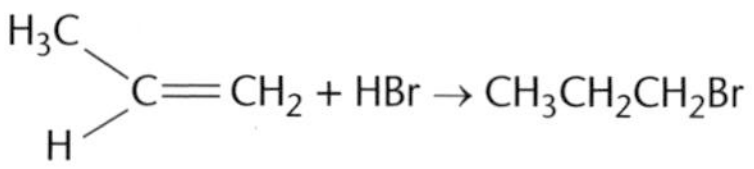

$$H_3C\!\!\diagdown$$
$$C\!\!=\!\!CH_2 + HBr \rightarrow CH_3CH_2CH_2Br$$
$$H\!\!\diagup$$

or

$$H_3C\!\!\diagdown$$
$$C\!\!=\!\!CH_2 + HBr \rightarrow CH_3CHBrCH_3Br$$
$$H\!\!\diagup$$

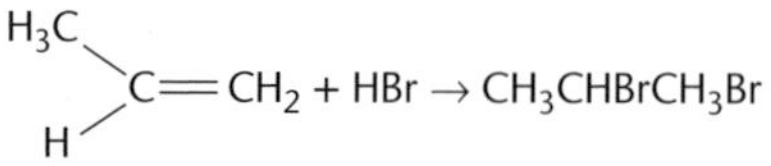

The bromine atom can be bonded to either the first or second carbon atom in the chain. In practice, it is mainly 2-bromopropane that forms.

The reason is that the addition product of an unsymmetrical alkene depends on the stability of the intermediate species formed, the carbocation (Table 3). A primary carbocation, a species in which a single positive charge on a carbon is bonded to just one other carbon atom, is much less stable than a secondary carbocation, where two other carbon atoms are bonded. As you may expect, this trend would

QUESTIONS

10. Which of these four alkenes are symmetrical and which are unsymmetrical?

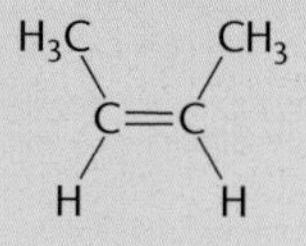

Z-but-2-ene   2,3-dichloro-Z-but-2-ene

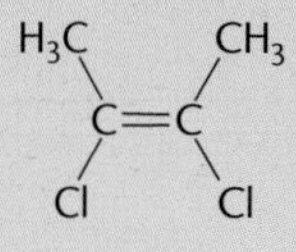

suggest that a tertiary carbocation is the most stable of all, which it is. The reason for the extra stability is that the alkyl groups donate electrons towards the positive carbon atom, so minimising the positive charge on the carbocation.

Type of carbocation	Structure	Description
primary carbocation	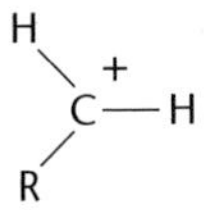	The positive carbon atom is attached to one other carbon atom.
secondary carbocation	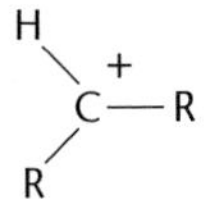	The positive carbon atom is attached to two other carbon atoms.
tertiary carbocation	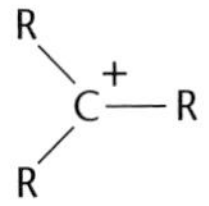	The positive carbon atom is attached to three other carbon atoms.

Table 3 *Primary, secondary and tertiary carbocations (R represents an alkyl group)*

In an addition reaction that involves an unsymmetrical alkene, the product most likely to be formed is the one that is produced from the most stable carbocation. This is the major product. The minor product is the alternative product.

To summarise, the tendency is for an electrophile to add to an unsymmetrical C=C bond so that the most stable carbocation is formed as an intermediate.

You can see how this applies to a real reaction by looking at the reaction between propene and hydrogen bromide, shown in Figure 8. If bromine bonds with C^1, a primary cation forms. If it bonds with C^2, a secondary cation forms. Since the secondary cation is more stable, the major product is 2-bromopropane.

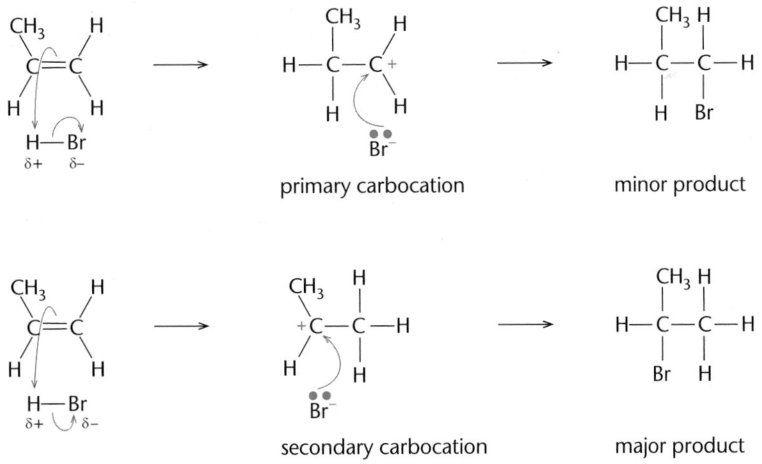

Figure 8 *The reaction of propene with hydrogen bromide*

QUESTIONS

11. Predict the major product from the reactions between:
    a. propene and hydrogen chloride
    b. pent-1-ene and hydrogen bromide
    c. but-1-ene and hydrogen bromide.

ASSIGNMENT 2: MANUFACTURING HALOGENOALKANES

(MS 4.1, 4.2; PS 1.1, 1.2, 2.2)

The chemicals industry makes many compounds in bulk from raw materials such as crude oil. These compounds are the starting materials from which chemical compounds used in commercial products such as medicines, dyes and cosmetics are made. 2-bromopentane is the starting material in the pharmaceutical industry. It can be manufactured from an alkene (obtained by cracking crude oil fractions) and hydrogen bromide.

Figure A1 *The chemicals industry is a major industry worldwide. Many of its products are used to make other chemicals that eventually end up in our shops, homes, farms and places of work.*

Questions

A1. 1-bromopentane, 2-bromopentane and 3-bromopentane are all isomers.

a. Draw their displayed formulae.

b. What type of isomerism do they have?

A2. Hydrogen bromide and pent-1-ene or hydrogen bromide and pent-2-ene are possible reactants for manufacturing 2-bromopentane.

a. What is the mechanism of the reaction?

b. What is the electrophile?

c. Explain why your answer in b. is an electrophile.

A3. The yield of 2-bromopentane when using either pent-1-ene or pent-2-ene depends on the stability of the intermediate carbocation formed.

a. What is a carbocation?

b. Write the structural formulae of the possible products when pent-1-ene is used.

c. Which product would you expect to be the major product? Why?

d. Write the structural formulae for the possible products when pent-2-ene is used as a reactant.

e. Which product would you expect to be the major product? Why?

ASSIGNMENT 3: OMEGA-3 OR OMEGA-6?

(MS 4.1, 4.2, 4.3)

Figure A1 *Fish oil is an important source of omega-3 essential fatty acids. Some people choose to supplement their diet with omega-3 supplements.*

Note: You may wish to revisit assignment 3 in Chapter 5 before starting this assignment.

Omega-3 and omega-6 are types of unsaturated essential fats. They are described as 'essential' because they have to be obtained from food. Your body cannot make them. Polyunsaturated means that they have more than one double bond in their long carbon chains. The main source of omega-3 fatty acids is fish oil (Figure A1). Omega-6 oils come from plant oils such as corn oil, soybean oil and sunflower oil. There is increasing evidence that both omega-3 and omega-6 oils help lower the risk of heart disease, but scientists are still debating the ideal proportion of each in a healthy diet.

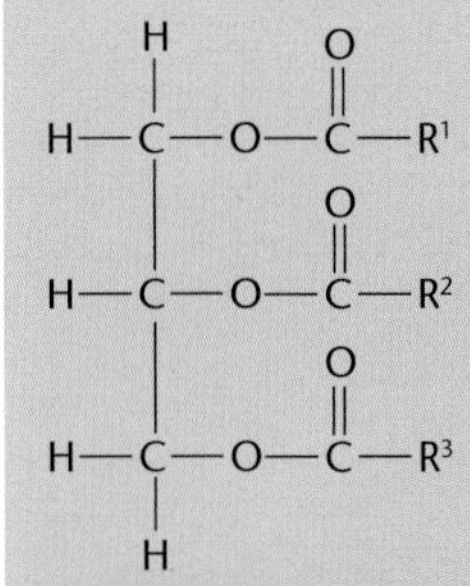

Figure A2 *The chemical structure of a fat or oil*

Fats and oils have the basic structure shown in Figure 2. R^1, R^2 and R^3 are long hydrocarbon chains, typically up to 18 carbon atoms, with a carboxylic acid group at one end, hence the name fatty acid. Fatty acids are either saturated or unsaturated.

The names omega-3 and omega-6 describe the position of one of the double bonds. Omega, being the last letter of the Greek alphabet, describes the carbon atom in the fatty acid chain furthest away from the carboxylic acid group. Omega-3 fatty acids have a double bond that is three carbon atoms from the omega carbon and between carbon atoms 3 and 4 (the lowest number is given, as in naming alkenes) and omega-6 fatty acids have a double bond on carbon number 6.

Questions

A1. Draw the skeletal structure of an omega-6 fatty acid with 18 carbon atoms in its chain, including the carbon in the COOH group. Include one double bond only.

A2. Why is this called an omega-6 fatty acid?

A3. Coconut oil contains the saturated fatty acid octanoic acid. Flax seed oil contains omega-3 and omega-6 fatty acids.
   a. Draw the structural formula for octanoic acid.
   b. Describe what you would see when bromine water is added to samples of each fatty acid.
   c. What type of reaction is this?
   d. Explain the mechanism.

A4. Some food manufacturers of spreads prefer to set unsaturated plant oils in a gel-type structure and add vitamins, such as those found in dairy products. Why is this a healthier option than spreads made with saturated oils?

Stretch and challenge

A5. Unsaturated fatty acids can be converted into margarine by the addition of hydrogen. A nickel catalyst is used at temperatures of 0–70 °C. Double bonds that survive this process are usually converted from *Z*- into *E-steroisomers*.
   a. What effect does the addition of hydrogen have on the molecular structure of the fatty acid?
   b. What effect does the change from *Z*- to *E*-stereoisomer have on the shape of the fatty acid?
   c. Is this a healthy option?

KEY IDEAS

› The double bond in an alkene is an area of high electron density.

› Alkenes react with hydrogen bromide, sulfuric acid and bromine by electrophilic addition.

› An electrophile is an electron-deficient species and has a full or small positive charge.

› Unsymmetrical alkenes have different groups of atoms attached to each of the carbons in the C=C bond.

› The product formed by addition reactions of unsymmetrical alkenes is determined by the stability of the intermediate carbocation.

› The order of stability of carbocations is tertiary (most stable) to primary (least stable).

14.3 ADDITION POLYMERS

A polymer is a long molecule. It is made of many smaller molecules called **monomers**, which are joined together. The process of forming a polymer is called **polymerisation**.

Addition polymers are formed from alkenes and substituted alkenes. They are unreactive and have numerous uses.

Molecules of the monomer join together to form very long chains. You can see that the percentage atom economy of addition polymerisation reactions is 100% since there are no other products.

The reactions can be summarised like this, with the monomer on the left of the reaction arrow and the polymer on the right:

$$n\underset{\underset{H}{|}}{\overset{\overset{X}{|}}{C}}=\underset{\underset{H}{|}}{\overset{\overset{H}{|}}{C} \rightarrow \left(\underset{\underset{H}{|}}{\overset{\overset{X}{|}}{C}} - \underset{\underset{H}{|}}{\overset{\overset{H}{|}}{C}} \right)_n$$

where, for example, X = H (ethene), CH_3 (propene), Cl (chloroethene) or C_6H_5 (phenylethene).

Other hydrogens may also be replaced. For example, tetrafluoroethene, C_2F_4, polymerises to make poly(tetrafluoroethene), $-(C_2F_4)-$, which is also known as PTFE and is the non-stick coating on frying pans and other products.

The **repeat unit** is that section of the polymer chain that is added when one molecule of the monomer is added.

Figure 9 *A space-filling model of a section of a poly(ethene) chain*

Polymers are named by writing the name of the monomer in brackets preceded by 'poly' outside the brackets.

 ▶ poly(ethene)

 ▶ poly(propene)

 ▶ poly(chloroethene)

 ▶ poly(tetrafluoroethene)

 ▶ poly(phenylethene)

Poly(ethene)
Monomer: ethene

Poly(ethene) is commonly known as polythene (Figure 9).

Poly(ethene) was discovered by accident by two British chemists, Eric Fawcett and Reginald Gibson, who worked for ICI (a large chemical company) in 1933. They were trying to produce a ketone from the reaction of benzaldehyde with ethene gas under pressure. Their reaction vessel sprang a leak, so they added more ethene gas and left it over the weekend. By Monday morning, no ketone was left, but there was a waxy white solid. Tests showed that it had the empirical formula CH_2. They tried to repeat their experiment several times. Sometimes the result was the same, and sometimes the reaction vessel exploded because of the high pressure used. Later they discovered that benzaldehyde was not needed to produce the solid, but oxygen was. By 1939 poly(ethene) was being produced commercially. The monomer for poly(ethene) is ethene, C_2H_4. The first poly(ethene) produced was made by polymerising ethene at 200 °C and 120 000 kPa, with a small amount of oxygen present.

Under the conditions above, the poly(ethene) produced has branched chains, so the chains cannot pack closely together. This is known as low-density poly(ethene), LDPE, which softens at temperatures of below 100 °C. This is still the primary type of poly(ethene) produced today. It is used to make polythene bags, carrier bags, the insulation around electrical wiring, and so on.

In 1963, Karl Ziegler, a German chemist, developed a way to produce poly(ethene) using a catalyst of titanium chloride and triethylaluminium at a temperature of 60 °C and atmospheric pressure. The poly(ethene) produced in this way has more straight-chain alkane molecules and the chains can pack closer together. This is known as high-density poly(ethene), HDPE, which softens at a higher temperature. This makes it more useful at higher temperatures, so it is used to make buckets and containers for liquids such as shampoo and kitchen products.

Poly(propene)
Monomer: propene

There are two possible arrangements of the H and CH_3 groups in poly(propene). If the CH_3 groups are ordered and facing in the same direction, the polypropylene is called isotactic poly(propene). Isotactic poly(propene) is a hard plastic and is used to make car bumpers, battery cases and crates (Figure 10), and it can be drawn into fibres to make carpets, clothing and ropes. If the H and CH_3 groups are randomly arranged, the poly(propene) is called atactic poly(propene).

Atactic poly(propene) is used to weatherproof materials and in sealants. Poly(propene) has properties mid-way between HDPE and LDPE and is the most versatile of plastics.

Figure 10 *Isotactic poly(propene) is a hard plastic and can be used to make crates.*

Poly(chloroethene)
Monomer: chloroethene

Commonly known as PVC, from its old name polyvinylchloride, poly(chloroethene) has a range of uses (see Figure 11).

People have used clay to make pottery for about 7000 years. Water was used as a plasticiser to soften the clay and make it easier to work with. Many plastics are hard and brittle and in the same way that water softens clay, other types of plasticisers are used

Figure 11 *These window frames are made of poly(chloroethene) (or polyvinylchloride – PVC). PVC is unreactive and these window frames can withstand extreme weather. PVC is also used to make guttering, drainpipes, plumbing, credit cards, footwear and leather-look fabric.*

to soften plastics such as PVC and make them more flexible. The plasticisers used are a group of chemical compounds called phthalates. Adding them to PVC makes it more suitable for use in industries such as construction, automotive, and wire and cables.

Poly(phenylethene)
Monomer: phenylethene

Commonly known as polystyrene, poly(phenylethene) is a rigid, tough plastic that is used to make disposable plastic plates, cups and cutlery, and many other items.

Expanded poly(phenylethene) (expanded polystyrene) is made from polystyrene beads. These are heated with steam (the most common method) or hot air. They expand and are then left for 24 hours or more to allow air to diffuse into the beads. This cools them and makes them harder. Expanded poly(phenylethene) is a very good insulator that has a range of uses (see Figure 12).

Figure 12 *Expanded polystyrene is used to make, among other things, food and drink containers, and for insulation in buildings.*

291

QUESTIONS

12. Draw the repeat unit in poly (tetrafluoroethene).

Stretch and challenge

13. Van der Waals forces operate along the length of the alkane chains in poly(ethene). These hold the poly(ethene) molecules together.

   a. Explain why LDPE softens at a lower temperature than HDPE does.

   b. Poly(ethene) is a mixture of long-chained alkane molecules. When it is heated, it softens over a range of temperatures rather than melting at a specific temperature. Why?

KEY IDEAS

> Addition polymers are produced by the polymerisation of alkenes.

> Alkene molecules become the repeating unit in an addition polymer.

> Ethene is polymerised to produce poly(ethene).

> Propene is polymerised to produce poly(propene).

> Addition polymers are unreactive, giving rise to a range of useful applications.

> The properties of polymers can be controlled by the reaction conditions used or by additives.

PRACTICE QUESTIONS

1. Consider the following reactions.

$$\underset{\text{Reaction 1}}{H_2C=CHCH_3 \xrightarrow[HBr]{}} \underset{\text{Reaction 2}}{CH_3CHBrCH_3 \xrightarrow[NH_3]{}} CH_2CH(NH_2)CH_2$$

X | Reaction 3

Figure Q1

   a. Name and outline a mechanism for Reaction 1.

   b. Name and outline a mechanism for Reaction 2.

   c. i. State the type of reaction in Reaction 3.

   ii. Give the name of substance X.

   d. The halogenoalkane produced in Reaction 1 can be converted back into propene in an elimination reaction using ethanolic potassium hydroxide.

$$\underset{}{CH_3CHBrCH_3 \xrightarrow{KOH} H_2C=CHCH_3}$$

Outline a mechanism for this conversion.

   AQA Jan 2013 Unit 2 Question 8

2. a. Bromine reacts with alkenes, even though bromine is a non-polar molecule.

   i. Explain why bromine molecules react with the double bonds in alkenes.

   ii. Name the type of mechanism involved in this reaction.

   iii. Draw the structure of the compound with $M_r = 387.6$ formed when penta-1,4-diene ($H_2C=CHCH_2CH=CH_2$) reacts with an excess of bromine.

   b. Two products are formed when propene reacts with hydrogen bromide.

   i. Draw the structure of the intermediate that leads to the formation of the major product in the reaction of propene with hydrogen bromide.

   ii. Give the name of this type of intermediate.

   AQA Jan 2011 Unit 2 Question 7 b, c

3. The alkene (Z)-3-methylpent-2-ene reacts with hydrogen bromide as shown in Figure Q2.

(Continued)

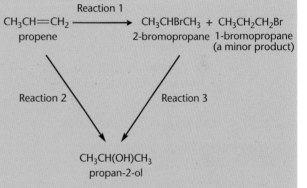

Figure Q2

a. i. Name the major product P.

ii. Name the mechanism for these reactions.

iii. Draw the displayed formula for the minor product Q and state the type of structural isomerism shown by P and Q.

iv. Draw the structure of the (E)-stereoisomer of 3-methylpent-2-ene.

AQA Jun 2010 Unit 2 Question 6a

4. A student carried out an experiment to determine the number of C=C double bonds in a molecule of a cooking oil by measuring the volume of bromine water decolourised. The student followed these instructions:

▶ Use a dropping pipette to add 5 drops of oil to 5.0 cm^3 of inert organic solvent in a conical flask.

▶ Use a funnel to fill a burette with bromine water.

▶ Add bromine water from the burette to the solution in the conical flask and swirl the flask after each addition to measure the volume of bromine water that is decolourised.

The student's results are shown in Table Q1.

Experiment	Volume of bromine water/cm^3
1	39.40
2	43.50
3	41.20

Table Q1

a. State two practical steps that the student should follow to ensure that the burette measures an accurate volume of bromine water.

b. Other than incorrect use of the burette, suggest a reason for the inconsistency in the student's results.

c. Suggest an alternative method the student could use in order to give more consistent results.

d. The oil has a density of 0.92 g cm^{-3} and each drop of oil has a volume of 0.050 cm^3. The approximate M_r of the oil is 885. The concentration of bromine water used was 0.020 mol dm^{-3}.

Use these data and the results from Table Q1 to deduce the number of C=C double bonds in a molecule of the oil. Show your working.

AQA Specimen paper 2 2015 Question 7

5. Consider the following reaction scheme.

Reaction 1

$CH_3CH\!\!=\!\!CH_2$ $\longrightarrow$ $CH_3CHBrCH_3$ + $CH_3CH_2CH_2Br$
propene   2-bromopropane 1-bromopropane
                                     (a minor product)

Reaction 2   Reaction 3

$CH_3CH(OH)CH_3$
propan-2-ol

a. Name the mechanism for Reaction 1.

b. Explain why 1-bromopropane is only a minor product in Reaction 1.

c. Name the reaction mechanism in Reaction 3.

d. State a difference between an electrophile and a nucleophile.

AQA Jun 2004 Unit 3(a) Question 2

6. The alkene C_4H_8 has four isomers. One structure is:

H_3C H
    $C\!\!=\!\!C$
H_3C H

a. Give the displayed formula of the product when it reacts with concentrated sulfuric acid.

b. Name and describe the mechanism of the reaction.

(Continued)

c. Circle the carbon atoms that lie in a flat plane in the displayed formula above.

d. Give the displayed formulae of three other isomers.

AQA Jan 2007 Unit 3(a)

7. Propene reacts with bromine to produce 1,2-dibromopropane.

a. i. Give the displayed formula for 1,2-dibromopropane.

   ii. Draw two other isomers of 1,2-dibromopropane.

   iii. State the type of isomerism shown in ii.

b. The reaction between propene and bromine is an example of electrophilic addition.

   i. Explain why bromine, a non-polar molecule, is able to react with propene.

   ii. Describe the mechanism for the electrophilic addition of bromine to propene.

AQA Jan 2002 Unit 3(a) Question 3

8. But-1-ene and methylpropene are isomers with the molecular formula C_4H_8.

a. i. Give the empirical formula of C_4H_8.

   ii. State whether these isomers of C_4H_8 are symmetrical or unsymmetrical.

b. Methylpropene reacts with hydrogen bromide to produce 2-bromo-2-methylpropane as the major product.

   i. Name and outline the mechanism for this reaction.

   ii. Draw the structure of another product of this reaction and explain why it is formed in smaller amounts.

c. i. Name the major product formed when but-1-ene reacts with hydrogen bromide.

   ii. Name the minor product formed and explain why it is the minor product.

AQA Jan 2004 Unit 3(a) Question 3

Stretch and challenge

9. Iodine reacts with ethene through a similar mechanism to that of the reaction between bromine and ethene.

a. i. Name the type of mechanism.

   ii. Explain the mechanism of the reaction between iodine and ethene and name the product formed.

   iii. Describe how you would carry out the reaction in the laboratory.

b. The reaction between iodine and ethene is slower than the reaction between bromine and ethene. Use bond enthalpies of carbon-halogen bonds to explain why.

Multiple choice

10. Which are alkenes?

    A. Cyclohexane

    B. Cyclohexene

    C. Benzene

    D. Hexane

11. Which of the following have both pi and sigma bonds?

    A. Ethane

    B. Ethene

    C. Chloromethane

    D. Ethanol

12. Which of these statements is true?

    A. An electrophile adds to an unsymmetrical C=C bond so that the most stable carbocation is formed.

    B. A nucleophile adds to an unsymmetrical C=C bond so that the most stable carbocation is formed.

    C. An electrophile adds to a symmetrical C=C bond so that the most stable carbocation is formed.

    D. An electrophile adds to an unsymmetrical C=C bond so that the least stable carbocation is formed.

13. Which property do electrophiles have?

    A. They have an area of negative charge.

    B. They have no charge.

    C. They have an area of positive charge.

    D. They are attracted to areas of positive charge.

15 ALCOHOLS

PRIOR KNOWLEDGE

You may have learnt about fermentation reactions to produce ethanol at GCSE. In Chapter 5, you read about homologous series and the use of different types of formulae to describe organic compounds.

LEARNING OBJECTIVES

In this chapter, you will build on these ideas and find out about alternative methods of producing ethanol. You will also learn about some chemical reactions of alcohols.

(Specification 3.3.5.1, 3.3.5.2, 3.3.5.3)

Figure 1 *Brazil is number one in the world when it comes to bioethanol. These Flex Fuel motorcycles can use blends of up to 100% ethanol. All car fuel sold in Brazil contains at least 25% ethanol. A hot sunny climate and large sugar cane plantations produce an economical supply of plant sugars for fermentation. In the UK, petrol sold at the pumps contains 5% ethanol.*

Using biomass to produce fuels as an alternative to fossil fuels is the focus of numerous research projects. Presently, the two biofuels commonly used as a petrol replacement are biodiesel and bioethanol.

Biodiesel can be produced by growing crops that have a high oil content, such as oilseed rape. In the manufacturing process the oil is reacted with short-chained alcohols. Biodiesel can be used in any diesel engine, either by itself or blended with fossil fuel diesel.

Fermentation of sugars in sugar cane and other crops produces ethanol. Ethanol can be used as a petrol replacement, but this has not been as successful as biodiesel. 100% bioethanol fuels can only be used in specially modified engines. It releases less energy per litre than petrol and since ethanol absorbs water, it causes corrosion in the engine. However, mixtures of ethanol and petrol have been more successful and existing car engines can use blends containing up to 15% ethanol.

Biofuel technology is still young and researchers are working to develop a second generation of biofuels. The aim is to develop a hydrocarbon fuel with a similar chemical structure to petrol that can be used in existing engines. To limit the use of food crops to produce fuel, researchers are investigating the use of agricultural wastes such as wheat straw, and woody biomass.

15.1 ALCOHOLS

'Alcohol' is commonly used to describe ethanol, CH_3CH_2OH, but alcohols are a homologous series containing the hydroxyl, OH, functional group. Ethanol is a member of this series. Figure 2 shows some examples of alcohols.

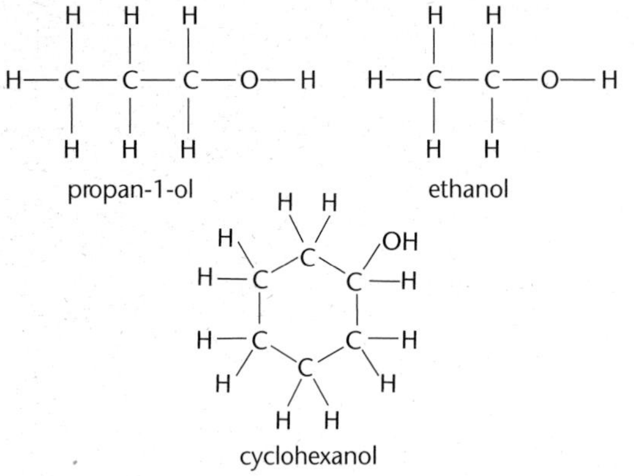

propan-1-ol   ethanol

cyclohexanol

Figure 2 *Examples of alcohols. Their common functional group is –OH (the hydroxyl group).*

Naming alcohols

The names of alcohols have the suffix *-ol*. Numbers are used, if necessary, immediately before the suffix *-ol* to show the position of the alcohol group in the carbon chain. The rest of the name comes from the number of carbon atoms in the carbon chain.

CH_3CH_2OH or 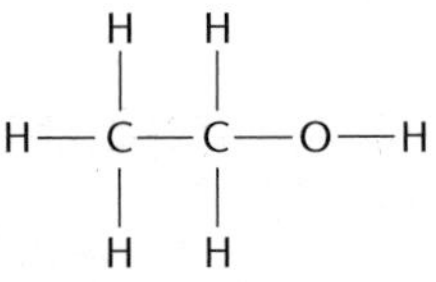
is ethanol

$CH_3CH_2CH_2OH$ or 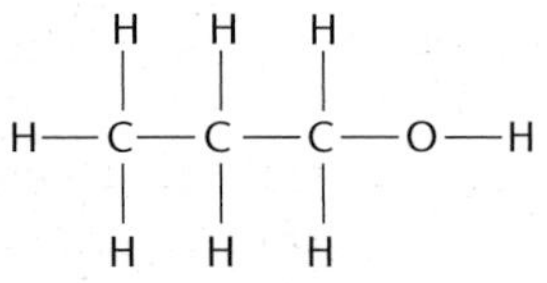
is propan-1-ol

$CH_3CH(OH)CH_3$ or 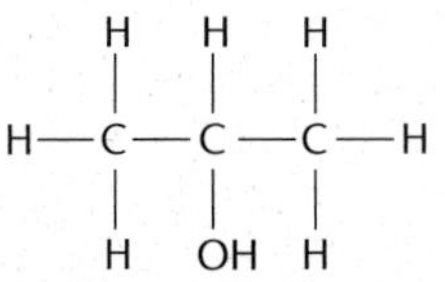
is propan-2-ol.

Alcohols can be classified as primary, secondary or tertiary, depending on the number of carbons attached to the carbon with the hydroxyl group.

Primary alcohols, such as ethanol, have one carbon attached to the carbon with the hydroxyl group. The one exception is methanol, which only has one carbon anyway.

Secondary alcohols, such as propan-2-ol, have two carbons attached to the carbon atom with the hydroxyl group.

Tertiary alcohols, such as 2-methylpropan-2-ol, have three carbons attached to the carbon with the hydroxyl group.

Figure 3 summarises the general structures of primary, secondary and tertiary alcohols.

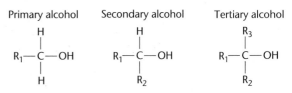

Primary alcohol   Secondary alcohol   Tertiary alcohol

Figure 3 *Primary, secondary and tertiary alcohols. R_1, R_2, R_3 are alkyl groups.*

KEY IDEAS

› Alcohols are a homologous series with hydroxyl, −OH, functional groups.

› The chemical names of all alcohols end in -ol.

› Alcohols can be classified as primary, secondary or tertiary depending on the number of carbon atoms bonded to the carbon that has the hydroxyl group attached.

ASSIGNMENT 1: EXPLAINING TRENDS

(PS1.1, 1.2, 2.3)

As with alkanes, alkenes and halogenoalkanes, trends in boiling points and melting points of alcohols depend on the intermolecular forces between the molecules (see Table A1).

There are two types of intermolecular forces between alcohol molecules: van der Waals forces and hydrogen bonds.

The carbon chains of the alcohols are attracted to each other by weak van der Waals forces that operate along the lengths of the chains. Secondary and tertiary alcohols have structures that do not allow alcohol molecules to lie so closely together. They have weaker van der Waals forces than the primary isomer.

Hydrogen bonds also exist between the lone pairs on the oxygen in the hydroxyl group of one molecule and the hydrogen in the hydroxyl group of another molecule. The hydroxyl group is polar with oxygen having a small negative charge (δ^-) and hydrogen a small positive charge (δ^+). This results in an attractive force between them.

Remember: hydrogen bonds are stronger than van der Waals forces. However, the strength of van der

Figure A1 *Wine and beer consist of a mixture of water and ethanol with flavourings and sugars. Hydrogen bonds between the water molecules and the ethanol molecules allow the liquids to mix.*

Waal forces increases when the areas of molecules laying alongside one another increases.

Before an alcohol can boil, enough energy has to be transferred from the surroundings to overcome these intermolecular forces (van der Waals and hydrogen bonds).

Alcohol	Structural formula	Boiling point/°C	Melting point/°C
methanol	CH_3OH	65	−98
ethanol	CH_3CH_2OH	78	−115
propan-1-ol	$CH_3CH_2CH_2OH$	98	−127
propan-2-ol	$CH_3CH(OH)CH_3$	83	−89
butan-1-ol	$CH_3CH_2CH_2CH_2(OH)$	117	−90
butan-2-ol	$CH_3CH_2CH(OH)CH_3$	100	−115
2-methylpropan-2-ol	$CH_3C(OH)(CH_3)CH_3$	82	24

Table A1 *Physical data for some alcohols*

15.2 ETHANOL PRODUCTION

Traditionally, ethanol was made by the fermentation of sugars in plant material and this method is still used to produce alcoholic drinks and biofuel. But fermentation is a relatively slow process and does not keep up with the demands of manufacturing and chemical industries. An industrial alternative is to produce ethanol from ethene and steam.

Other alcohols can also be produced from steam and the appropriate alkene:

$$C_nH_{2n} + H_2O \rightarrow C_nH_{2n+1}OH$$

Alkenes are a product of thermal cracking of crude oil fractions, together with shorter-chain alkanes that are used, for example, in petrol.

The hydration of ethene to make ethanol

Alcohols can be manufactured from alkenes. Since the reaction involves the addition of water, it is called **hydration**.

The overall reaction to convert ethene into ethanol is:

$$C_2H_4(g) + H_2O(g) \rightarrow C_2H_5OH(g) \quad \Delta H^\ominus = -45 \text{ kJ mol}^{-1}$$

The conditions used are a temperature of 600 K, a pressure of 6000–7000 kPa and a phosphoric(V) acid catalyst held in silica pellets.

You can read about the mechanism of this addition reaction in Chapter 14.

Theoretically, this reaction has a percentage atom economy of 100%. In practice, some side reactions occur, producing methanol, ethanal, poly(ethene) and ethoxyethane. The reaction is reversible.

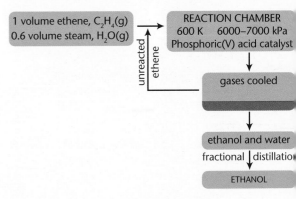

Figure 4 *Producing ethanol from ethene and steam*

The conversion rate of ethene into ethanol is about 5%, but by removing the ethanol and recycling the unreacted ethene, as shown in Figure 4, it is possible to achieve an overall 95% conversion rate.

Similar methods can be used to make other alcohols, provided that the correct alkene is used. Propan-2-ol can be made from propene and steam.

Figure 5 *The ethanol produced from ethene and steam is used as a solvent and as a feedstock to make other products like car tyres. A chemical feedstock is a raw material used in an industry.*

QUESTIONS

5. a. From the equation, what is the ratio of the reacting volumes of ethene and steam?

b. Phosphoric(V) acid is soluble in water. Suggest why equal volumes of ethene and steam are not used in the reaction mixture?

Stretch and challenge

6. a. Since the forward reaction to produce ethanol is exothermic, explain why the temperature of 600 K is a compromise temperature.

b. According to Le Châtelier's Principle, how does an increase in pressure affect the reaction between ethene and steam?

c. Ethene polymerises to poly(ethene) at high pressures. How does this affect the pressure used?

d. How does phosphoric(V) acid affect the rate of reaction and the position of the equilibrium?

Fermentation of glucose to make ethanol

Fermentation is the biological process in which glucose is converted into ethanol and carbon dioxide. The reaction is catalysed by the enzymes, called zymase, in yeast. It is an example of anaerobic respiration. Like all enzymes, zymase functions best at the optimum temperature of 37–40 °C.

The reaction is complex and involves a number of stages, but it can be summarised as:

$$C_6H_{12}O_6(aq) \rightarrow 2CH_3CH_2OH(aq) + 2CO_2(g)$$

glucose   ethanol   carbon dioxide

The conditions for the process are:

• the absence of oxygen

• the presence of yeast and sugar (glucose) solution

• a temperature of 37–40 °C.

Figure 6 *Making cider and wine from fruit juices with high sugar concentrations is a process that is thousands of years old. The best wines are made from grapes with sufficient sugar to produce high levels of ethanol.*

The biofuel industry

A biofuel is a fuel made from biomass. The plants that make up the biomass used carbon dioxide from the atmosphere to make sugars in photosynthesis. When biofuels are burnt carbon dioxide is produced and returned to the atmosphere.

Ethanol reacts exothermically with oxygen to produce carbon dioxide and water. The reaction is:

$$C_2H_5OH(g) + 3O_2(g) \rightarrow 2CO_2(g) + 3H_2O(l)$$

$$\Delta H^\ominus = -1367 \text{ kJ mol}^{-1}$$

The energy released is used to power vehicle engines.

It may be argued that ethanol made by fermentation is a carbon neutral fuel. The following equations demonstrate this. Making ethanol by fermenting a solution of glucose and then combusting the ethanol produces the same amounts of carbon dioxide and water as are used in photosynthesis:

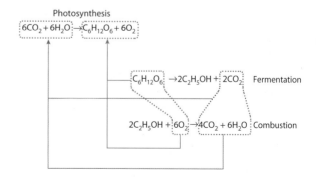

However, it requires energy to obtain ethanol from the solution produced by fermentation. Power is also needed for various other operations in the manufacturing process. This energy may come from burning fossil fuels such as natural gas or electricity produced in fossil-fuel power stations. This also produces carbon dioxide. Therefore, the statement that ethanol produced by fermentation is a carbon neutral fuel is not valid.

Crops such as sugar cane, wheat or corn can be used to provide sugars and starches for use in fermentation. The process to produce ethanol fuel involves three stages.

Stage 1: Starches are converted to glucose using enzymes. These digest the starches in a similar way to the enzymes in your digestive system.

Stage 2: Fermentation reactions convert glucose into ethanol and carbon dioxide. A mixture of ethanol, water, unreacted glucose and dead yeast cells is obtained.

Stage 3: Fractional distillation is used to separate the ethanol. Fractional distillation is used to separate liquids with different boiling points. The boiling point of ethanol is 78 °C and water 100 °C. By heating the reaction mixture, ethanol can be boiled off at 78 °C and the vapour condensed and collected.

Figure 7 *The Vivergo biofuels plant in Hull opened in July 2013. It will be the UK's biggest buyer of wheat and aims to produce 420 million litres of ethanol a year – that's a third of the UK's demand for bioethanol.*

Comparing methods of manufacturing ethanol

Selecting the appropriate method to produce ethanol needs both economic and environmental aspects to be considered, as well as the intended use of the ethanol. Table 1 compares ethanol production using fermentation with ethanol production from ethene and steam.

	Fermentation	Production from ethene and steam
Economic aspects	fermentation is a batch process; the reaction vessels have to be emptied, cleaned and restocked after each batch	this is a continuous process; ethene and steam are continuously fed over the catalyst and unreacted reactants recycled; continuous processes are more cost-effective than batch processes
	the reaction is very slow	the reaction is rapid
	the reaction occurs at a relatively low temperature and atmospheric pressure	a high temperature and pressure are used; these are expensive to maintain
	the process is labour-intensive from the growing of the crops to the purification of the final product	few workers are needed to operate the system, so labour requirements are low
	fractional distillation is needed to purify the product; this uses energy	the ethanol produced is between 98% and 99% pure; for most purposes, this does not need purification
Environmental aspects	fermentation uses renewable resources	ethene is produced from crude oil and the process uses a finite resource
	land used for growing biomass for fermentation is also needed for food production	the process does not use farming land

Table 1 *A comparison of two processes for manufacturing ethanol*

7. Why is fermentation used rather than ethene and steam to produce:

   a. ethanol for the drinks industry

   b. biofuel?

8. Researchers are using genetic engineering to design a microbe that will tolerate high levels of ethanol. What change will this make to the percentage yield of ethanol obtained from fermentation?

9. If ethanol can be produced from non-nutritional biomass such as wheat straw, how does this affect the considerations listed in Table 1?

10. Why do producers of bioethanol claim that using bioethanol fuel produces less greenhouse gas emissions overall?

11. Why does the production of ethanol from sugar cane require less energy than the production of ethanol from wheat?

KEY IDEAS

➤ Alcohols can be produced by hydrating alkenes.

➤ Ethanol can be produced by reacting ethene and steam using a phosphoric(V) acid catalyst.

➤ Ethanol can also be produced by the fermentation of biomass.

➤ Enzymes in yeast, glucose and water are needed to carry out fermentation.

➤ Since ethanol reacts exothermically with oxygen to produce carbon dioxide and water, it can be used as a biofuel.

➤ There are advantages and disadvantages to both methods used to produce ethanol.

5.3 CHEMICAL REACTIONS OF ALCOHOLS

When alcohols react, the reaction can involve the hydroxyl group, the carbon skeleton or both. This means that they can undergo several different types of reactions, making alcohols very valuable to the chemical industry. Two examples of alcohol reactions are oxidation and elimination.

Oxidation reactions

Acidified potassium dichromate(VI) solution can be used as an oxidising agent to oxidise alcohols. It is a dilute solution of potassium dichromate(VI) with dilute sulfuric acid added.

The products of the oxidation reaction depend on whether the alcohol is a primary, secondary or tertiary alcohol (Figure 3).

Oxidation of primary alcohols

Ethanol is a primary alcohol and can be oxidised to give an aldehyde. The aldehyde can then be further oxidised to give a carboxylic acid. The products obtained depend on the reaction conditions used.

−CHO is the functional group of an aldehyde

−COOH is the functional group of a carboxylic acid

Figure 8 *Structures of aldehydes and carboxylic acids*

An aldehyde is produced if excess alcohol is used and the aldehyde is distilled off as it forms, so as to prevent further oxidation. Two hydrogens are removed from the alcohol – one from the carbon skeleton and one from the hydroxyl group.

For example, ethanol is oxidised to ethanal. Water is the other product of the reaction:

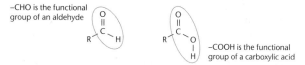

ethanol ethanal (an aldehyde)

The names of all aldehydes end in 'al'. The first part of the name is taken from the alkyl group present. So the aldehyde formed by the oxidation of ethanol is ethanal. The complete equation is very complicated, so we use [O] to represent the oxidising agent. The square brackets indicate that the oxygen is made available for the chemical reaction.

The equation can summarised as:

$$CH_3CH_2OH(l) + [O] \rightarrow CH_3CHO(l) + H_2O(l)$$

Distillation is used to remove the aldehyde as soon as it forms. The distillation apparatus is shown in Figure 9.

301

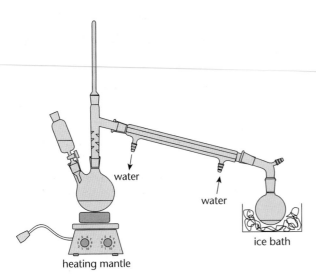

Figure 9 *Oxidising a primary alcohol to an aldehyde*

If an excess of acidified potassium dichromate(VI) solution is used and the reaction mixture is heated under reflux a carboxylic acid is produced. Two atoms of hydrogen are removed and one atom of oxygen gained.

The carboxylic acid formed is ethanoic acid. The equation can be written as:

$$CH_3CH_2OH(l) + 2[O] \rightarrow CH_3COOH(l) + H_2O(l)$$

Oxidation of secondary alcohols
Propan-2-ol is a secondary alcohol. When it is refluxed with acidified potassium dichromate(VI), two hydrogens are removed and a ketone is produced.

For example, propan-2-ol is oxidised to propanone. Water is the other product of the reaction:

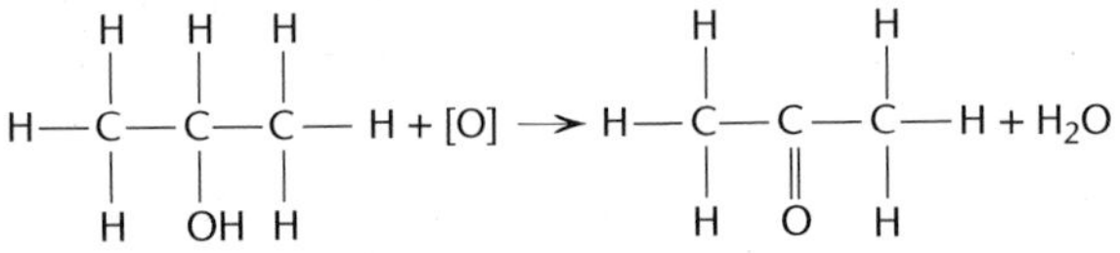

H—C—C—C—H + [O] → H—C—C—C—H + H₂O

propan-2-ol   propanone (a ketone)

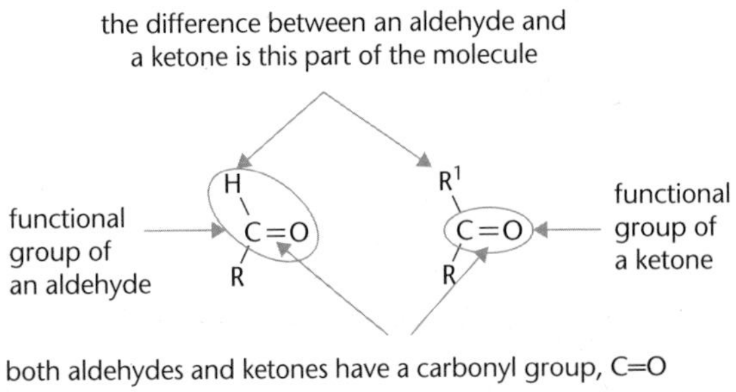

the difference between an aldehyde and
a ketone is this part of the molecule

functional group of an aldehyde   →   C=O   C=O   ←   functional group of a ketone

both aldehydes and ketones have a carbonyl group, C=O

Figure 10 *Structure of a ketone. The names of all ketones end in -one. Like an aldehyde it has a carbonyl group, >C=O, but unlike an aldehyde is has two alky groups bonded to the >C=O carbon.*

Tertiary alcohols
Primary and secondary alcohols both have hydrogens bonded to the carbon that carries the hydroxyl group. These hydrogens are removed in the oxidation process. But tertiary alcohols have no hydrogens on the hydroxyl-carrying carbon. Consequently, tertiary alcohols cannot be easily oxidised.

2-Methylpropan-2-ol is a tertiary alcohol. It has this structure:

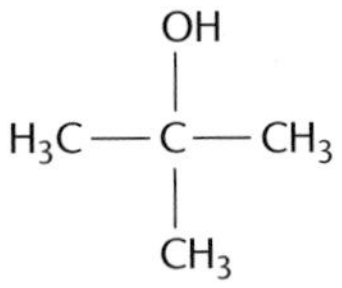

OH
|
H_3C — C — CH_3
|
CH_3

The carbon bonded to the hydroxyl group is bonded to three other carbon atoms. There are no hydrogen atoms available for oxidation.

The overall pattern is:

primary alcohols $\xrightarrow{\text{oxidation}}$ aldehyde $\xrightarrow{\text{oxidation}}$ carboxylic acid

secondary alcohols $\xrightarrow{\text{oxidation}}$ ketone

tertiary alcohols do not oxidise easily.

QUESTIONS

12. a. Classify butan-1-ol and butan-2-ol as primary, secondary or tertiary alcohols.

b. Write equations for all reactions when butan-1-ol is oxidised.

c. Write an equation for the reaction when butan-2-ol is oxidised.

Identifying alcohols by oxidation
Since primary, secondary and tertiary alcohols all give different products with oxidising agents, these reactions, or the lack of them, can be used to distinguish tertiary alcohols from primary and secondary alcohols. When primary and secondary alcohols are oxidised, the oxidising agent is reduced. When acidified potassium dichromate(VI) solution is the oxidising agent, the colour changes from orange to green because chromate(VI) ions (orange) are reduced to chromium(III) ions (green). The colour change is shown in Figure 11.

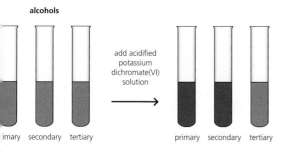

alcohols

add acidified potassium dichromate(VI) solution

imary secondary tertiary

primary secondary tertiary

Figure 11 During the oxidation of primary and secondary alcohols, using acidified potassium dichromate(VI) solution, the colour changes from orange to green. There is no reaction with a tertiary alcohol.

Figure 12 Some vinegars are made from cheap wine. Ethanol oxidised to ethanoic acid by acetobacter bacteria. The overall reaction is the same as the oxidation of ethanol using potassium dichromate(VI) solution.

Distinguishing between aldehydes and ketones

Test-tube reactions can be used to distinguish between the aldehydes and ketones formed when alcohols oxidise.

The carbon in the carbonyl bond, C=O, of an aldehyde is bonded to a hydrogen and another carbon. In a ketone, the carbon in the C = O bond is bonded to two other carbons. This makes the ketone more difficult to oxidise than the aldehyde.

Tollen's reagent and Fehling's solution may be used to distinguish between an aldehyde and a ketone. They both give positive results for aldehydes.

Tollen's reagent is prepared by mixing aqueous ammonia, NH_3(aq), and silver nitrate solution, $AgNO_3$(aq). The resulting solution contains diamine silver(I) ions, $[Ag(NH_3)_2]^+$(aq).

When Tollen's reagent is warmed with an aldehyde, silver atoms coat the inside of the test tube, producing a silver mirror. The silver mirror forms because silver(I) ions are reduced to silver atoms by the aldehyde, which is oxidised to a carboxylic acid. The equation reaction of ethanal with $[Ag(NH_3)_2]^+$(aq) is:

$$CH_3CHO(aq) + 2[Ag(NH_3)_2]^+(aq) + H_2O(l) \rightarrow$$
$$CH_3COOH(aq) + 2Ag(s) + 4NH_3(aq) + 2H^+(aq)$$

often written simply as:

$$CH_3CHO(aq) + 2Ag^+(aq) + H_2O(l) \rightarrow$$
$$CH_3COOH(aq) + 2Ag(s) + 2H^+(aq)$$

Fehling's solution contains deep blue copper(II) ions dissolved in dilute ammonia. The complex ions are a mixture, but the predominate one is $[Cu(NH_3)_5H_2O]^{2+}$. The complex copper(II) ions act as a weak oxidising agent and oxidises an aldehyde to give a carboxylic acid.

Figure 13 The tube on the left shows Tollen's reagent before the aldehyde is added. The tube on the right shows that the aldehyde reduces the $[Ag(NH_3)_2]^+$ to Ag(s) to give the silver mirror effect.

For simplicity, we usually write Cu^{2+}(aq) rather than the formula of the ammine complex ion. When ethanal is heated with Fehling's solution, the following reaction occurs:

$$CH_3CHO(aq) + 2Cu^{2+}(aq) + 2H_2O(l) \rightarrow$$
$$CH_3COOH(aq) + Cu_2O(s) + 4H^+(aq)$$

Copper(I) oxide is brick red and insoluble in water. So a brick-red precipitate forms when copper(II) ions are reduced.

Ketones do not affect Fehling's solution or Tollen's reagent. This allows these reagents to be used to distinguish between an aldehyde and a ketone.

QUESTIONS

Stretch and challenge

13. A primary and a secondary alcohol are both oxidised using acidified potassium dichromate(VI) solution and the products distilled off immediately. Both give a colour change of orange to green. How can Fehling's solution be used to identify which alcohol is the primary alcohol?

REQUIRED PRACTICAL ACTIVITY 5: APPARATUS AND TECHNIQUES

(PS 4.1; AT b, d, k)

Distillation of a product from a reaction

Practicals that involve distillation give you the opportunity to show that you can:

> use an electric heater for heating

> use laboratory apparatus for a variety of experimental techniques, including distillation and setting up glassware using retort stands and clamps

> safely and carefully handle liquids, including corrosive, irritant, flammable and toxic substances.

Apparatus

To distil a product from a reaction mixture you need to heat the mixture so that it boils. A thermostatically controlled electric heating mantle is ideal. It enables a constant temperature to be maintained and, because there is no naked flame, it is safer than using a Bunsen burner when using flammable substances. Alternatively, a hot water bath could be used.

The glassware most commonly used is Quickfit™ apparatus. Quickfit™ apparatus has ground glass joints that enable individual pieces of apparatus to be connected together in different arrangements. Different sized joints are available and adapters with gas tight joints can be used to change from one joint size to another. Using Quickfit™ apparatus eliminates the need for rubber bungs, glass tubing and corks and produces less contamination of the product from rubber bungs and corks.

Figure P1 *An example of apparatus assembled for distillation*

A suitable sized flask is needed to contain the mixture to be distilled, usually so that it is no more than one-third full. It is connected to a condenser via a still head.

The still head connector is usually designed to take a thermometer as well. The thermometer bulb sits just opposite where vapours pass from the flask to the condenser. Quickfit™ thermometers have a ground glass joint so that they fit into the still head connector.

The still head connector connects to a condenser. This is where the vapour condenses to a liquid. There are two common types: Leibeg condensers and coiled condensers. Both work on the same principle – a central tube that is kept cold by running water. Importantly, cold water is always run into the lower end of the condenser.

Distillate that forms in the condenser is directed into a receiving vessel, for example, a beaker or conical flask, by a delivery tube. Sometimes a Quickfit™ flask is used

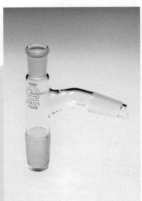

Figure P2 *Flasks, either round bottomed or pear shaped, are fitted to a condenser with a still head connector.*

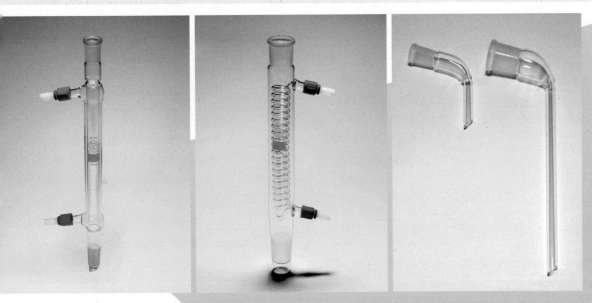

Figure P3 *Two types of condenser. Both are connected to a delivery tube that directs the distillate into the receiver container.*

Retort stands and clamps are needed to hold the assembled apparatus in place. The clamps have rubber or cork insulation to prevent damage to the glassware. The screw should be above the joint being clamped and should not be tightened too hard as this may break the glassware.

Techniques

You will need access to water, drainage and electricity. The distillation mixture is placed in the flask and a few anti-bumping granules added. These are small, rough-edged ceramic pieces that help the contents to boil smoothly. 'Bumping' occurs when larger gas bubbles collect in the flask and then rise to the surface, causing the distillation mixture to rise up the apparatus. The flask is placed in the heating mantle or water bath.

Three retort stands and clamps are needed to support the apparatus, as shown in Figure P5. Remember, do not over tighten the clamps as this will break the glass.

Water is turned on to allow a steady stream through the condenser. The electric heating mantle is turned on maximum to allow the contents of the distillation flask to boil. As vapour is given off, heating is lowered to give gentle boiling and to monitor the temperature. You will see the products rising through the apparatus and condensing. You need to collect the distillate given off at ± 2 °C of the boiling point of the desired product. Any distillate collected at temperatures below this temperature is discarded and a clean collecting flask used. Stop collecting the product when the temperature starts to rise above this range.

Figure P4 *Ground glass joints connect the apparatus. The photo on the left shows a joint between an Erlenmeyer flask and a condenser; the photo on the right shows the same joint closed. A very light smear of silicone grease can be used to seal the joints and the apparatus is assembled. Two pairs of hands are better than one here.*

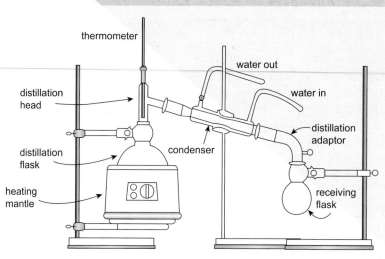

Figure P5 *Distilling the product*

thermometer

water out

distillation
head

water in

distillation
adaptor

distillation
flask

condenser

heating
mantle

receiving
flask

QUESTIONS

P1. What are the advantages of Quickfit™ distillation apparatus over conventional laboratory glassware?

P2. Why must the lower tube on the condenser be connected to the water supply and not the upper tube?

P3. Why must the collecting flask be open to the atmosphere?

P4. When you have collected the product, why should the heater be turned off, but the water left running for several minutes?

15.4 ELIMINATION REACTIONS OF ALCOHOLS

Dehydration

Elimination reactions involve the loss of a molecule of water from an alcohol to form an alkene. The reaction is often called **dehydration**, because water is lost from the molecule. During the reaction, the C–O bond in the alcohol breaks and one hydrogen atom is lost from the carbon chain. The reaction for ethanol is:

$$H-\underset{\underset{H}{|}}{\overset{\overset{H}{|}}{C}}-\underset{\underset{H}{|}}{\overset{\overset{H}{|}}{C}}-O-H \xrightarrow{\text{conc. } H_2SO_4} \underset{\underset{H}{|}}{\overset{\overset{H}{|}}{C}}=\underset{\underset{H}{|}}{\overset{\overset{H}{|}}{C}} + H_2O$$

The conditions for the reaction are heating with a suitable catalyst, such as concentrated sulfuric acid or concentrated phosphoric(V) acid. The mechanism may be shown as:

$$CH_3-CH_2-\overset{\bullet\bullet}{O}-H \longrightarrow CH_3-CH_2-\overset{+}{O}-H$$
$$\overset{|}{H^+} \qquad \qquad \overset{|}{H}$$

$$CH_3-CH_2-\overset{+}{O}-H \longrightarrow CH_3-\overset{+}{C}H_2 \quad H_2O$$
$$\overset{|}{H}$$

$$CH_2-\overset{+}{C}H_2 \longrightarrow CH_2=CH_2$$
$$\overset{|}{H} \qquad \qquad H^+$$

However, a more detailed description can help to understand the role of the concentrated acid more fully. It explains the source of H^+ and how H^+ is removed in the final stage.

You learned about the displayed formula of sulfuric acid in Chapter 14:

H^+ is transferred from sulfuric acid to the alcohol:

$$CH_3CH_2-\overset{\bullet\bullet}{O}H \longrightarrow CH_3CH_2-\overset{+}{O}H$$

This is followed by a simultaneous series of electron rearrangements:

$$CH_2-CH_2-\overset{+}{O}H \longrightarrow CH_2=CH_2 \quad H_2O$$

The hydrogen sulfate ion, HSO_4^-, accepts a proton (H^+), reforming sulfuric acid. So although sulfuric acid takes part in the reaction it is recovered chemically unchanged at the end. It catalyses the reaction.

In the laboratory, the dehydration of ethanol to produce ethene can be done using a different reaction. Ethanol vapour is passed over hot pumice stone or aluminium oxide and the ethene produced collected over water, as in Figure 14.

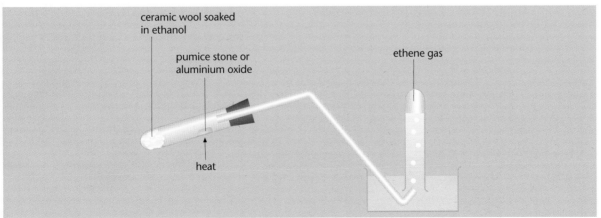

Figure 14 *Dehydration of ethanol over hot pumice stone or aluminium oxide*

QUESTIONS

14. Which processes are used to convert:

    a. plant starch to plant sugar

    b. plant sugar to ethanol

    c. ethanol to ethene

    d. ethene to poly(ethene)?

15. a. Write an equation to show the elimination reaction of propan-1-ol using a phosphoric(V) acid catalyst.

    b. Name the alkene formed.

    c. Write an equation to show the polymerisation of this alkene.

    d. Name the polymer formed.

ASSIGNMENT 2: PROPENE AND PROPAN-2-OL

(PS 1.1, 1.2, 2.3)

Propan-2-ol is a secondary alcohol with the molecular formula: $CH_3CH(OH)CH_3$. Its displayed formula is:

```
      H    H    H
      |    |    |
  H — C — C — C — H
      |    |    |
      H    O    H
           |
           H
```

Propan-2-ol is used as a solvent in industrial processes and medical disinfectants. It is manufactured from propene by a hydration reaction. Propene is a by-product of cracking the longer alkane chains in crude oil. In terms of quantities produced, propene is second only to ethene as a starting point in the chemical industry. Worldwide, 80 million tonnes are produced annually, that is 8×10^{10} kg. About 65% is used to make poly(propene) and the rest make compounds such as propan-2-ol, propanal, butanal, epoxypropane and cumene.

Propan-2-ol is also known as 'rubbing alcohol'. It has a number of pharmaceutical and medical uses. These include being an ingredient in some antiseptics, foot fungicides, and house and garden insecticides.

Questions

A1. Explain why propan-2-ol is a secondary alcohol.

Figure A2 *This antibacterial hand gel contains 62% ethanol and 10% propan-2-ol.*

A2. The manufacture of propan-2-ol involves two steps. First, propene gas is reacted with concentrated sulfuric acid. Second, water is added to the reaction mixture.

   a. What type of reaction happens between propene and concentrated sulfuric acid?

   b. Use structural formulae to describe the reaction between propene and concentrated sulfuric acid.

   c. Explain the position of the ethyl hydrogen sulfate group in the molecule.

A3. When water is added, propan-2-ol is produced. Explain why the overall reaction is a hydration reaction.

A4. Propan-2-ol can be converted to propene by an elimination reaction.

a. Write an equation for the reaction using structural formulae.

b. Why is this called an elimination reaction?

A5. Propan-2-ol was refluxed with acidified potassium dichromate(VI) solution. The colour changed from orange to green.

a. What type of reaction has taken place?

b. Name the product formed.

c. How would the result differ if propan-1-ol was used instead of propan-2-ol?

ASSIGNMENT 3: PREPARATION OF CYCLOHEXENE FROM CYCLOHEXANOL

(PS 1.1, 1.2, 2.1)

Cyclohexanol is an alcohol. Like other alcohols, it has elimination reactions. When cyclohexanol is heated with phosphoric acid, cyclohexene is produced. The reaction is:

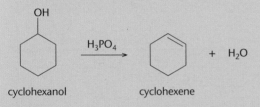

A group of students followed these worksheet instructions:

> Pour 10 cm^3 cyclohexanol into a 50 cm^3 round-bottomed flask and carefully add 3 cm^3 phosphoric(V) acid a few anti-bumping granules.

> Place the flask in an electric heating mantle and set the distillation apparatus.

> Heat the flask. White fumes will be given off. Stop heating when 10 cm^3 of distillate have been collected.

> Pour the distillate into a separating funnel. Allow the contents to settle and run off the lower aqueous layer into a conical flask.

> Add deionised water to the conical flask, shake to mix, allow to settle and run of the lower layer.

> Add 10% sodium carbonate solution to the conical flask. When no more gas is given off, run off the lower layer.

> Wash again with deionised water

> Transfer the cyclohexene to a stoppered conical flask and add a few granules of calcium chloride. Leave for 15 minutes and decant the cyclohexene into a clean, stoppered conical flask.

Figure A3 *A separating funnel is used to separate two immiscible liquids. The denser liquid settles below the less dense liquid and can be run off. You can see the dividing line between the two immiscible liquids.*

Questions

A1. Which types of formulae are used to show cyclohexanol and cyclohexene in the equation?

A2. What are the molecular formulae for cyclohexanol and cyclohexene?

A3. Why is this called an elimination reaction?

A4. What is the role of the phosphoric acid?

A5. Which impurities will be present in the distillate?

A6. Which is denser, cyclohexene or water?

A7. Why are the following used in the purification stages:

a. water

b. sodium carbonate solution

c. calcium chloride?

KEY IDEAS

> The products formed when a primary alcohol oxidises depend on the conditions used.

> Oxidation of a primary alcohol produces an aldehyde. Further oxidation produces a carboxylic acid.

> Oxidation of a secondary alcohol produces a ketone.

> Tertiary alcohols do not oxidise easily.

> Aldehydes give a positive silver mirror test with Tollen's reagent, but ketones do not.

> Aldehydes reduce the blue Cu^{2+} ions in Fehling's solution to Cu^+ ions in Cu_2O. This is a brick-red precipitate. Ketones have no effect.

> Alcohols undergo elimination reactions to produce an alkene and water.

PRACTICE QUESTIONS

1. Table Q1 shows the structures and names of three compounds with $M_r = 72.0$.

Compound	Formula	Name
1	$CH_3CH_2CH_2CHO$	butanal
2	$CH_3CH_2CH_2CH_2CH_3$	pentane
3	$CH_3CH_2COCH_3$	butanone

Table Q1

a. Explain why M_r values, measured to five decimal places, cannot distinguish between compounds 1 and 3 but can distinguish between compounds 1 and 2.

b. A simple chemical test, using either Fehling's solution or Tollen's reagent, can be used to distinguish between compound 1 and compound 3. Choose one of these two reagents and state what you would observe with each of compound 1 and compound 3.

AQA Jan 2012 Unit 2 Questions 6

2. A student devised an experiment to investigate the enthalpies of combustion of some alcohols. The student chose the following series of primary alcohols.

Name	Formula
Methanol	CH_3OH
Ethanol	CH_3CH_2OH
propan-1-ol	$CH_3CH_2CH_2OH$
Butan-1-ol	$CH_3CH_2CH_2CH_2OH$
Pentan-1-ol	$CH_3CH_2CH_2CH_2CH_2OH$
Alcohol X	$CH_3CH_2CH_2CH_2CH_2CH_2OH$
Heptan-1-ol	$CH_3CH_2CH_2CH_2CH_2CH_2CH_2OH$

Table Q2

a. i. Name alcohol X.

ii. State the general name of the type of series shown by these primary alcohols.

iii. Draw the displayed formula of the position isomer of butan-1-ol.

iv. Using [O] to represent the oxidising agent, write an equation for the oxidation of butan-1-ol to form an aldehyde.

v. Draw the displayed formula of a functional group isomer of this aldehyde.

(Continued)

b. The student carried out a laboratory experiment to determine the enthalpy change when a sample of butan-1-ol was burned. The student found that the temperature of 175 g of water increased by 8.0 °C when 5.00×10^{-3} mol of pure butan-1-ol was burned in air and the energy released produced was used to warm the water.

Use the student's results to calculate a value, in kJ mol^{-1}, for the enthalpy change when one mole of butan-1-ol is burned.

(The specific heat capacity of water is 4.18 J K^{-1} g^{-1}.)

c. i. Give the meaning of the term standard enthalpy of combustion.

ii. Use the standard enthalpy of formation data from the table and the equation for the combustion of butan-1-ol to calculate a value for the standard enthalpy of combustion of butan-1-ol.
$CH_3CH_2CH_2CH_2OH(l) + 6O_2(g) \rightarrow$
$$4CO_2(g) + 5H_2O(l)$$

AQA Jan 2011 Unit 2 Questions 9

3. Growing crops with a high sugar content provides an industrial route to produce ethene gas. Glucose, $C_6H_{12}O_6$, can be converted into ethanol. Ethanol can be converted into ethene in an acid-catalysed elimination reaction.

a. i. State three essential conditions for the conversion of glucose into ethanol.

ii. Name the process.

iii. Give an equation for the reaction that takes place.

b. Some ethanol produced from plant sugars is used as biofuel.

i. Write an equation for the complete combustion of ethanol.

ii. State one advantage of using ethanol as a biofuel in car engines instead of petrol.

c. i. Explain what is meant by the term elimination reaction.

ii. Identify a catalyst that could be used in the acid-catalysed elimination of ethanol.

iii. Write an equation for the reaction that takes place.

AQA Jan 2006 Unit 3(a) Question 5

4. A and B are two alcohols which are position isomers of each other. Their reactions are shown in Figure Q1.

a. State what is meant by the term *position isomers*.

b. Name compound A and compound B.

c. Each of the reactions above is of the same type and uses the same reagent.

i. State the type of reaction.

ii. Name a suitable reagent.

iii. State how you would ensure that compound A is converted into pentanoic acid rather than into pentanal.

iv. Draw the structure of an isomer of compound A that does not react with this combination of reagents.

v. Draw the structure of the carboxylic acid formed by the reaction of methanol with this combination of reagents.

d. i. State a reagent that could be used to distinguish between pentanal and compound C.

ii. Draw the structure of another aldehyde that is an isomer of butanal.

AQA Jan 2005 Unit 3(a) Question

$CH_3CH_2CH_2CH_2CH_2OH \rightarrow CH_3CH_2CH_2CH_2CHO \rightarrow CH_3CH_2CH_2CH_2COOH$

A   pentanal   pentanoic acid

$CH_3CH_2CH_2CH(OH)CH_3 \rightarrow CH_3CH_2CH_2COCH_3$

B   C

Figure Q1

(Continued)

5. A student has read in a text book that:
"Some alcohols can be oxidised to form aldehydes, which can then be oxidised further to form carboxylic acids. Some alcohols can be oxidised to form ketones, which resist further oxidation. Other alcohols are not easily oxidised."

a. i. Draw the structures of the two straight-chain isomeric alcohols with molecular formula, C_3H_8O.

   ii. State the type of isomerism shown by the two straight chain alcohols.

b. i. Draw the structures of the three oxidation products obtained when the two alcohols from part a. are oxidised separately using acidified potassium dichromate(VI).

   ii. Write equations for any reactions which occur, using [O] to represent the oxidising agent.

c. Draw the displayed formula of an alcohol containing four carbon atoms that does not oxidise easily.

   AQA Jun 2005 Unit 3(a) Question 6

6. a. Ethanol can be manufactured by the hydration of ethene and by the fermentation of sugars. The reaction to produce ethanol from ethene is:

$$C_2H_4(g) + H_2O(g) \rightleftharpoons C_2H_5OH(g)$$

$$\Delta H = -45 \text{ kJ mol}^{-1}$$

The conditions for the reaction are a temperature of 573 K and pressure of 6000 to 7000 kPa.

   i. Name a suitable catalyst to catalyse the reaction to produce ethanol from ethene.

   ii. The forward reaction is exothermic. Explain why the temperature used is a compromise temperature.

   iii. Use Le Châtelier's Principle to explain why a high pressure increases the yield of ethanol.

   iv. What effect does the catalyst have on the position of the equilibrium?

   v. What effect does the catalyst have on the rate of reaction?

   vi. Under the above conditions, the conversion rate to ethanol is 5%. How do manufacturers increase the conversion rate to 95% without changing the conditions?

b. Give one advantage and one disadvantage of manufacturing ethanol by fermentation rather than by hydration of ethene.

   AQA Jan 2003 Unit 3(a) Question

7. An alcohol containing carbon, hydrogen and oxygen only has 64.9% carbon and 13.5% hydrogen by mass.

a. Use this data to calculate the empirical formula of the alcohol.

b. i. Draw the displayed formulae of the four possible isomers of this alcohol.

   ii. Label your displayed formulae in b i as either primary, secondary or tertiary alcohols.

c. A primary alcohol was oxidised by adding it dropwise to acidified potassium dichromate(VI) solution and immediately distilling off the product. When this product was warmed with Fehling's solution, a brick-red precipitate was formed.

   i. State the type of product distilled off during the oxidation by acidified potassium dichromate(VI) solution.

   ii. Name and draw a structure for the organic product formed by its reaction with Fehling's solution.

   AQA Jun 2002 Unit 3(a) Question 4

Practical skills question

8. Ethanol consumed in alcoholic drinks is broken down in the liver. The reaction is catalysed by enzymes. Ethanol is oxidised to ethanal, which is then oxidised to ethanoic acid.

The reaction can be carried out in the laboratory using potassium dichromate(VI) instead of enzymes. To obtain ethanal and prevent further oxidation, the product must be removed as it forms. The apparatus in Figure Q2 can be used.

(Continued)

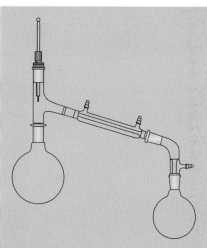

Figure Q2

a. What is this procedure called?

b. Explain why

  i. a thermometer is needed

  ii. the thermometer bulb must be opposite the outlet on the still head

  iii. a condenser is used.

  iv. antibumping granules are added to the reaction mixture

  v. the collection flask must be vented to the atmosphere.

c. The boiling point of ethanal is 21 °C. Describe how you would carry out this procedure.

Stretch and challenge

9. Students are determining the percentage composition by volume of each alcohol in a mixture of propan-1ol and propan-2-ol. The boiling points of the alcohols are:

propan-1-ol: 97. 0 °C

propan-2-ol: 82.6 °C.

Students aim to use distillation to separate the mixture.

a. Describe the procedure to separate a mixture of propan-1-ol and propan-2-ol using distillation.

b. Explain two safety precautions they need to take.

c. What measurements do they need to make in order to determine the percentage composition by volume of the mixture?

d. Why is the boiling point of propan-2-ol lower than the boiling point of propan-1-ol?

Multiple choice

10. Which of these is a tertiary alcohol?

A. butan-2-ol

B. 2-methylbutan-3-ol

C. 2-methylbutan-2-ol

D. pentan-2-ol

11. Which is the correct formula of 2-methylbutan-2-ol?

A. $CH_3C(CH_3)(OH)CH_2CH_3$

B. $CH_3CH(OH)CH(CH_3)CH_3$

C. $CH_3CH_2CH(CH_3)CH_2OH$

D. $CH_3CH(OH)CH_2CH_2CH_3$

12. What is reflux apparatus used for?

A. Distilling a product

B. Fermentation

C. Separating two immiscible liquids

D. Maintaining a reaction mixture at its boiling point

13. Ketones are not easily oxidised because:

A. The carbon with the OH group attached is attached to two other carbons and C–C bonds are strong bonds.

B. The carbon with the OH group attached is attached to two other carbons and C–C bonds are weak bonds.

C. The C=O bond is a strong bond.

D. The C=O bond is polar.

16 ANALYTICAL TECHNIQUES

PRIOR KNOWLEDGE

You may have carried out tests to identify metal ions using flame tests and reactions with sodium hydroxide solution in your GCSE course. You probably know how to test for some common gases such as carbon dioxide and oxygen. You may also have carried out tests to identify carbonates and halides. In earlier chapters you read about tests for functional groups in organic compounds and the uses of mass spectrometry as an identification tool.

LEARNING OBJECTIVES

In this chapter, you will build on these ideas and develop a strategy for testing for functional groups. You will find out how mass spectrometry is used to find the relative molecular mass of a compound and learn about a different type of spectroscopy, infrared (IR) spectroscopy.

(Specification 3.3.6.1, 3.3.6.2, 3.3.6.3)

Anabolic steroids such as testosterone, the naturally produced male hormone, build muscle. Analgesics or painkillers allow athletes to continue training and competing when injured. Diuretics increase urine production so that drugs in the body can be excreted more quickly. Growth hormones increase body mass. Stimulants, including caffeine in coffee, increase our energy levels and allow you to exercise longer.

The use of many substances is banned for Olympic athletes and drug testing is a regular procedure.

In each event at the London 2012 Olympics, the first five finishers were drug tested plus another two at random. An athlete fails a drugs test if the results show the presence of a banned substance or an excessive amount of a substance naturally produced in the body, such as testosterone. The International Olympic Committee (IOC) has set levels for permitted amounts of testosterone and athletes who exceed these levels are assumed to have taken testosterone supplements and are banned.

High-performance liquid chromatography (HPLC) linked to high-resolution mass spectrometers can detect anabolic steroid concentrations of one part per billion in urine. They also detect other substances taken to improve performance, such as stimulants, analgesics and diuretics.

Figure 1 *Athletes' urine samples are taken (occasionally blood samples also). The substances present are separated by chromatography and fed directly into a mass spectrometer to identify them.*

16.1 IDENTIFYING FUNCTIONAL GROUPS

Techniques such as mass spectroscopy and other types of spectroscopy are very useful tools for identifying unknown substances. They are used in a wide range of analytical laboratories, including forensic, public health and industrial.

Mass spectrometry provides information about relative atomic mass and relative molecular mass. High-resolution mass spectrometry enables empirical formulae to be determined. Fragmentation patterns also provide clues about the bonds and functional groups present, which can be used to work out the structure of a substance. Other instrumental techniques also allow functional groups to be identified.

However, the instrumentation for these techniques is expensive. It is used most often for the routine analysis of large numbers of samples, called high-throughput analysis. More often than not it is used

also for quantitative analysis – so 'how much?' as well as 'what?'.

For simple, quick and relatively inexpensive identification of functional groups, qualitative analysis can be done using test-tube reactions. You read about a number of these in earlier chapters. The functional groups include those that characterise the following types of organic compounds: alkenes, halogenoalkanes, alcohols, aldehydes, ketones and carboxylic acids.

This section provides a summary and reminder of those tests.

Testing for alkenes

The electrophilic addition of bromine to an alkene provides a test for the presence of C=C bonds in a molecule. Dilute bromine solution is orange and rapidly decolourises when reacted with an alkene, as shown in Figure 2.

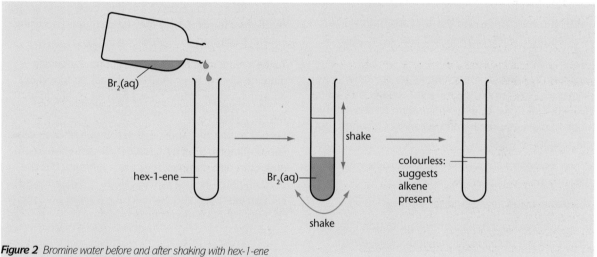

Figure 2 Bromine water before and after shaking with hex-1-ene

QUESTIONS

1. What is the electrophile present in dilute bromine solution?
2. Name the product when
   a. hex-1-ene reacts with bromine
   b. but-1-ene decolourises aqueous bromine solution.

Testing for halogenoalkanes

As you have read earlier, the carbon–halogen, C–X, bond in a halogenoalkane molecule is polar and susceptible to nucleophilic substitution. The hydroxide ion is a nucleophile. Reaction with a halogenoalkane breaks the C–X bond and a halide ion is one of the products formed.

The halide ion can be identified using acidified silver nitrate solution followed by ammonia solution. Table 1 shows the results that you can expect. Fluoroalkanes are not detected because they do not hydrolyse because the C–F bond is very strong and cannot be broken easily.

Procedure	Chloroalkane	Bromoalkane	Iodoalkane
warm with dilute sodium hydroxide solution	reactants mix on warming	reactants mix on warming	reactants mix on warming
add acidified silver nitrate solution	white precipitate of silver chloride	cream precipitate of silver bromide	yellow precipitate of silver iodide
add dilute ammonia solution, followed by concentrated ammonia solution	white precipitate dissolves to form a colourless solution when dilute ammonia is added	cream precipitate dissolves to form a colourless solution only when concentrated ammonia is added	precipitate is insoluble in both dilute and concentrated ammonia solution

Table 1 *Tests for halogenoalkanes*

QUESTIONS

3. Write an equation for the hydrolysis of a general halogenoalkane, RX (where X represents a halogen atom and R represents the alkyl group), by the hydroxide ion.

Testing for compounds containing oxygen

Alcohols, aldehydes, ketones and carboxylic acids all contain oxygen.

Alcohols may have primary, secondary or tertiary structures. You can distinguish between them by using two different tests. Acidified potassium dichromate(VI) solution oxidises primary and secondary alcohols, but not tertiary alcohols.

Primary alcohols are oxidised to aldehydes, which in turn are oxidised to carboxylic acids.

Secondary alcohols are oxidised ketones.

In all of the oxidation reactions dichromate(VI) ions (orange) are reduced to chromium(III) ions (green).

Tertiary alcohols are not oxidised by acidified potassium dichromate(VI) solution and, therefore, are distinguished by their lack of reaction.

As you see, acidified potassium dichromate(VI) solution does not help to distinguish a primary alcohol, a secondary alcohol or an aldehyde. Both are oxidised.

Aldehydes and ketones both contain a carboxyl group, $=C=O$. Because of this they have some similar reactions. However, they may be distinguished using Fehling's solution or Tollen's reagent.

Tollen's reagent is an alkaline solution containing diamine silver(I) ions, $[Ag(NH_3)_2]^+$(aq) dissolved in dilute ammonia. When mixed with an aldehyde and warmed, a silver mirror is produced.

Fehling's solution is an alkaline solution of deep blue copper(II) ions, $[Cu(NH_3)_5H_2O]^{2+}$, dissolved in dilute ammonia. The complex ions are a mixture, but the predominate one is $[Cu(NH_3)_5H_2O]^{2+}$. When mixed with an aldehyde and warmed, a brick red precipitate of copper(I) oxide is produced.

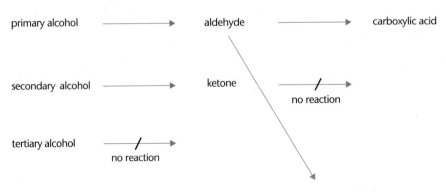

oxidation with acidified potassium dichromate(VI)

primary alcohol ⟶ aldehyde ⟶ carboxylic acid

secondary alcohol ⟶ ketone ⟶ no reaction

tertiary alcohol ⟶ no reaction

reaction with Fehling's or Tollen's

Figure 3 *Tests for oxygen-containing compounds*

Carboxylic acids are typical acids with pH values less than 7 and react with carbonates to produce carbon dioxide gas.

These various tests may be used to distinguish between alcohols, aldehydes, ketones and carboxylic acids (see Table 2).

Test	Add water and test the pH of the solution	Add sodium hydrogen carbonate solution	Warm with acidified potassium dichromate(VI) solution	Warm with Fehling's solution	Warm with Tollen's reagent
primary alcohol	neutral	no reaction	orange to green colour change	remains blue	remains colourless
secondary alcohol	neutral	no reaction	orange to green colour change	remains blue	remains colourless
tertiary alcohol	neutral	no reaction	remains orange	remains blue	remains colourless
aldehyde	neutral	no reaction	orange to green colour change	brick red precipitate forms	silver mirror forms on test tube
ketone	neutral	no reaction	remains orange	remains blue	remains colourless
carboxylic acid	pH about 3	carbon dioxide given off	remains orange	remains blue	remains colourless

Table 2 *Tests to identify alcohols, aldehydes, ketones and carboxylic acids*

QUESTIONS

4. Draw the functional group in
   a. an alcohol
   b. an aldehyde
   c. a ketone
   d. a carboxylic acid.

5. Classify the following as a primary alcohol, secondary alcohol, tertiary alcohol, aldehyde, ketone or carboxylic acid:
   a. propan-2-ol
   b. propanal
   c. propanoic acid
   d. 3-methylpentan-3-ol
   e. butanone

6. Which chemical tests would you use to identify the functional group(s) in these molecules?
   a. $CH_3CH_2CH_2COOH$
   b. $CH_3CH_2CH_2OH$
   c. $C_6H_5CH_2Br$
   d. $CH_3CH_2CH_2CH_2CHO$
   e. $CH_3CH = CHCH_2CH_2CH_3$
   f. $CH_3C(CH_3)(OH)CH_2CH_3$

7. What chemical tests would you apply to distinguish between the following pairs of molecules?
   a. $CH_3CH_2CH_2CH_2CH_2OH$ and $CH_3C(CH_3)(OH)CH_2CH_3$
   b. $CH_3CH_2CH_2CH_2Cl$ and $CH_3CH_2CH_2CH_2Br$
   c. $CH_3CH_2CH_2CH_3$ and $CH_3CH = CHCH_3$
   d. $CH_3CH_2COCH_3$ and $CH_3CH_2CH_2CHO$

REQUIRED PRACTICAL ACTIVITY 6: APPARATUS AND TECHNIQUES

(PS 2.1 and 4.1; AT b, c, d, k)

Tests for alcohol, aldehyde, alkene and carboxylic acid

Using tests to identify organic compounds gives you the opportunity to show that you can:

- use a water bath or electric heater for heating
- measure pH using pH charts, or pH meters, or a pH probe on a data logger
- use laboratory apparatus for a variety of techniques, including qualitative tests for organic functional groups
- safely and carefully handle solids and liquids, including corrosive, irritant, flammable and toxic substances.

Apparatus

Several alternatives are available to measure pH. pH paper or indicator solution, together with calibrated charts, can be used. A variety is available, designed to

Figure A1 *Two options for measuring pH: a pH probe and digital meter (left) and pH papers that detect pH values from 1 to 14 (right).*

cover a wide range of pH value or a narrow range. pH probes could be used, but not usually for spot tests.

Spot tests are chemical reactions carried out on a small scale in test tubes or, sometimes in well plates or spotting plates. The choice of apparatus is covered in Chapter 12. Similarly, the use of water baths and other heating methods is covered in Chapter 8.

You will use a number of reagents. Table P1 shows their concentrations.

Hazcards and Student Safety Sheets produced by CLEAPSS provide details about the hazards and advice for handling silver compounds.

Technique

To obtain accurate results it is essential to avoid contamination of the samples being tested, the reagents and the apparatus being used. Use clean test tubes for each test and separate droppers for each reagent. It is not necessary to measure volumes accurately because the tests are qualitative, but you need to familiarise yourself with what 1 cm^3 and 2 cm^3 of liquid looks like in a test tube.

Testing for an alkene involves adding bromine water. If the alkene is a liquid, the two reagents can be added in equal volumes (about 1 cm^3 each) in a clean test tube. The mixture has to be shaken. The way to do this is by holding the test tube just below its lip and shaking it with a jiggling motion. It is surprising how vigorously it can be shaken without liquid coming out of the tube, but practice with some water until you have mastered the technique. If the alkene is a gas, it can either be bubbled through bromine water or, if it is collected in test tubes, bromine water can be added to the test tube and the test tube shaken.

Reagent	Concentration commonly used	Hazard
bromine water	0,06 mol dm^{-3}	harmful
potassium dichromate(VI) solution	0.02 mol dm^{-3}	harmful
dilute sulfuric acid	1.0 or 2.0 mol dm^{-3}	irritant
Fehling's solution	contains copper sulfate at about 0.3 mol dm^{-3} and a high concentration of sodium hydroxide (about 2.5 mol dm^{-3})	copper sulfate is harmful, sodium hydroxide is corrosive at this concentration
sodium hydrogen carbonate solution	0.10 mol dm^{-3}	no hazard

Table P1 *Approximate concentrations of reagents used to identify alcohols, aldehydes, alkenes and carboxylic acids*

Carboxylic acids in water have a pH of about 3. The pH can be measured by dipping the end of a glass rod in the liquid, withdrawing it with a drop of liquid clinging to it and touching this on to a piece of pH paper. A pH probe could be used, but to simply identify the liquid as an acid, pH paper is much simpler. Since carbon dioxide is produced when acids react with carbonates or hydrogen carbonates, about 1–2 cm^3 sodium hydrogen carbonate solution can be added and limewater used to test any gas given off.

Alcohols may be primary, secondary or tertiary. Add about 1 cm^3 acidified potassium dichromate(VI) solution to the substance being tested, shake the test tube to mix the contents and then place it in a hot water bath (about 60 °C) for a few minutes.

Fehling's solution is unstable. It is stored as Fehling's A and Fehling's B solutions, which have to be mixed before use. Place about 1 cm^3 each of Fehling's A and Fehling's B in a test tube. Add a few drops of the sample being tested. Place the tube in a hot water bath (60 °C) for a few minutes.

When using Tollen's reagent, place 2 cm^3 of the sample being tested in a test tube and add about 5 cm^3 Tollen's reagent. Leave mixture at room temperature for a few minutes.

QUESTIONS

P1. Explain why you do not need a measuring cylinder for these tests.

P2. Which tests would you carry out to distinguish between a primary and a secondary alcohol?

P3. What is the advantage of using a water bath instead of a Bunsen burner when testing for an alcohol?

> Test-tube reactions may be used to identify the functional group in an organic compound and, therefore, the type of organic compound – including alkenes, halogenoalkanes, alcohols, aldehydes, ketones and carboxylic acids.

> The test for C=C in alkenes is the decolorisation of bromine water.

> The test for C–X in halogenoalkanes is hydrolysis followed by the silver nitrate test to identify halides ions.

> The test for –COOH in carboxylic acids is pH < 7 of aqueous solutions (alcohols, aldehydes and ketones and neutral, pH = 7).

> The tests for alcohols, aldehydes and ketones involve comparing their ease of oxidation.

16.2 MASS SPECTROMETRY

In Chapter 1 you found out how the mass numbers of an element's isotopes and their relative abundance are shown in a mass spectrum. You used this information to calculate the relative atomic mass of the element. Mass spectroscopy can also be used to determine the relative molecular mass and the molecular formula of a compound.

Interpreting a mass spectrum

A sample of ethanol injected into a time of flight mass spectrometer vaporises. In the ionisation area, the gas is bombarded with a stream of electrons. There are two possibilities:

1: electrons are knocked off the molecules and positive ions form. As with atoms of elements, most molecules lose one electron. Ethanol molecules, C_2H_5OH, lose an electron to become $[C_2H_5OH]^+$. The equation for the reaction in the ionisation area is:

$$C_2H_5OH(g) \rightarrow [C_2H_5OH]^+(g) + e^-$$

2: some of the molecular ions break into smaller fragments. The charged fragments will also move through the spectrometer and produce lines in the spectrum.

Because of its overall positive charge, ethanol ions and the fragments are accelerated by the electric field so that they travel at the same speed. The ion detector measures the flight time of the different ions and shows the information on a mass spectrum.

$[C_2H_5OH]^+$ ions are the heaviest ions that can be produced from a sample of ethanol and the peak they produce on a mass spectrum is called the **molecular ion peak**. It is also called the 'parent ion' because other molecular ions are made from it when it fragments in the mass spectrometer. Its it is given the symbol M^+. The m/z value of the molecular ion peak is the molecular mass of the compound. So the relative mass of the ethanol molecular ion is 46 and the molecular ion peak will be at $m/z = 46$.

Figure 4 shows the molecular ion peak on a mass spectrum of ethanol. You can also see peaks due to the fragmentation of the molecular ion.

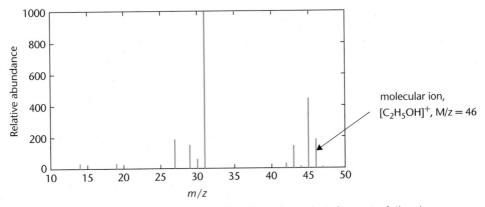

Figure 4 *The molecular ion peak on the mass spectrum of ethanol. The other peaks are due to fragments of ethanol.*

QUESTIONS

8. Write the formula for the molecular ion and gives its relative mass for

   a. ethanal, CH_3CHO

   b. ethanoic acid, CH_3COOH

9. Look at the mass spectrum of pentane, C_5H_{12}, in Figure 5.

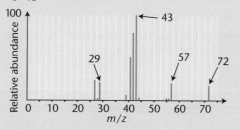

Figure 5 *The mass spectrum of pentane*

   a. Identify the molecular ion peak.

   b. What is the relative molecular mass of pentane?

10. Figure 6 shows the mass spectrum of 2-methylbutane.

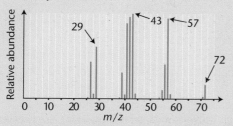

Figure 6 *The mass spectrum of 2-methylbutane*

   a. Draw the full structural formula for 2-methylbutane.

   b. Identify the molecular ion peak.

   i. What is the formula for the ion that produces this peak.

   ii. What does this peak tell you?

Stretch and challenge

   c. If carbon and hydrogen isotopes are present in the sample of 2-methylbutane, how will this change the mass spectrum?

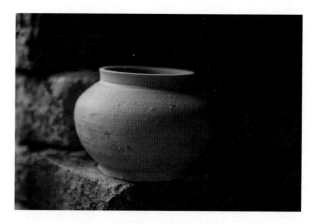

Figure 7 *Mass spectroscopy has detected traces of cocoa in 2000-year-old Mayan pots found in burial sites. This is thought to be the earliest recorded use of chocolate. It was also mixed with chilli, maize and honey.*

The molecular ion peak

The molecular ion peak gives the relative molecular mass of the compound. This should enable you to suggest the molecular formula, but frequently there are several possible molecular formulae for a particular mass. For example, propane, C_3H_8, has a relative molecular mass of 44. Ethanal, CH_3CHO, also has a relative molecular mass of 44. Using a **low-resolution mass spectrometer** gives these values and a mass spectrum of an unknown compound with a molecular ion peak at 44 could be either compound.

You need to know the mass of the molecular ion more accurately. A **high-resolution mass spectrometer** provides this information. You are familiar with relative isotopic masses of the following isotopes to one decimal place:

$$_1^1H = 1.0$$

$$_7^{14}N = 14.0$$

$$_8^{16}O = 16.0$$

If you write these relative isotopic masses to four decimal places, they are

$$_1^1H = 1.0078$$

$$_7^{14}N = 14.0031$$

$$_8^{16}O = 15.9949$$

and, of course, $_6^{12}C$ is 12.0000 because this is the mass that all others are compared to.

If you now calculate the relative molecular mass of propane, C_3H_8, to six significant figures, then it is:

$(3 \times 12.0000) + (8 \times 1.0078) = 44.0624$

The relative molecular mass of ethanal, CH_3CHO is:

$(2 \times 12.0000) + (4 \times 1.0078) +$
$\qquad\qquad\qquad 15.9949 = 44.0261$

A high-resolution mass spectrometer produces a molecular ion peak for propane at $m/z = 44.0624$, and a molecular ion peak for ethanal at $m/z = 44.0261$. This allows an accurate identification of the molecular formula from the molecular ion peak.

QUESTIONS

11. What is the relative mass of the molecular ion peak for methane using:

    a. a low-resolution mass spectrometer

    b. a high-resolution mass spectrometer?

12. Nitrogen monoxide (NO) and methanal (HCHO) both give peaks at $m/z = 30$ in a low-resolution mass spectrometer. Calculate the m/z values for the molecular ion peaks you would expect in a high-resolution mass spectrometer.

KEY IDEAS

> The m/z ratio of a molecular ion (or parent ion) is equal to the value of the compound's relative molecular mass.

> High-resolution mass spectrometry enables the empirical formula of a compound to be worked out.

16.3 INFRARED SPECTROSCOPY

Infrared radiation is the region of the electromagnetic spectrum that has wavelengths greater than the red end of the visible spectrum and less than the wavelengths of microwaves. If infrared radiation is passed through a compound, some of the radiation is absorbed and some is transmitted. Infrared spectroscopy measures the wavelengths of infrared radiation that have been absorbed. These are shown in the form of a chart, with wavelength plotted along the x-axis and the amount of radiation that passes through (is transmitted) on the y-axis. This is called an infrared spectrum.

Infrared spectra can be used to identify bonds in a compound. At GCSE you probably used frequency or wavelength to describe radiation. In infrared spectroscopy it is more common to use wavenumber. Wavenumber is the number of wave cycles per cm (its units are cm^{-1}).

$$\text{wavenumber} = \frac{1}{\text{wavelength} / cm}\ cm^{-1}$$

Figure 9 shows the relation between wavenumber, wavelength and frequency. You may notice that this spectrum is the other way round to the spectrum you are probably familiar with at Key Stage 4 GCSE – the infrared region is to the right of the visible region.

Figure 8 *Alleged drug traffickers are arrested by Colombian naval forces in this still frame from an infrared video camera from a U.S. Navy helicopter. Infrared cameras are sometimes called 'heat-seeking cameras'; they detect the infrared radiation emitted by different objects, including humans.*

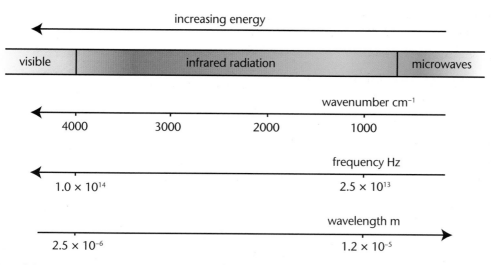

Figure 9 *Wavenumber, wavelength and frequency ranges of infrared radiation*

ASSIGNMENT 1: WHAT HAPPENS WHEN MOLECULES ABSORB INFRARED RADIATION?

(MS 3.1, 4.1; PS 1.1, 1.2, 2.3)

Different types of radiation interact in different ways with atoms and molecules. Microwave radiation affects how fast bonds in molecules rotate. Energy carried by microwave radiation is absorbed when this happens. The absorbed energy is transferred to the food, which is how food is cooked in a microwave oven.

Absorbing infrared radiation increases the speed at which bonds vibrate in molecules. Bonds vibrate all the time, but if they absorb the correct wavenumber of radiation, the bonds will be pushed into a higher state of vibration. If the molecule has more than two atoms, it can vibrate in different ways.

The energy absorbed when bonds vibrate depends on the mass of the atoms that are bonded together, so different bonds absorb different wave numbers of infrared radiation. Stronger bonds need more energy to increase their vibration than weaker bonds. So, strong bonds absorb a higher wavenumber of infrared radiation than weaker bonds. You read in Chapter 12 that strength of the hydrogen–halide bond decreases down Group 7(17). The Table shows the wavenumbers of infrared radiation needed to make these bonds vibrate. The higher the bond energy, the higher the wavenumber of the infrared radiation needed to make it vibrate more.

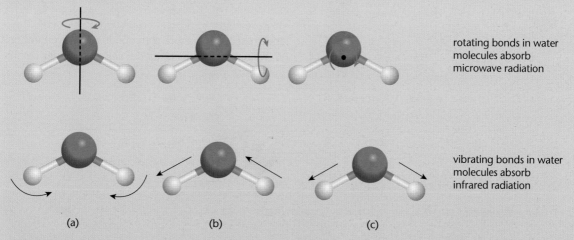

rotating bonds in water molecules absorb microwave radiation

vibrating bonds in water molecules absorb infrared radiation

(a)   (b)   (c)

Figure A1 *Bonds in molecules and the molecules themselves can rotate. Bonds in molecules can stretch, while molecules themselves can stretch, twist and bend (vibrate). These movements allow electromagnetic radiation to be absorbed.*

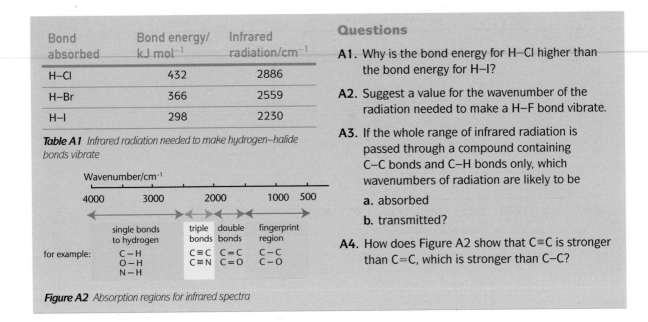

Bond absorbed	Bond energy/ kJ mol⁻¹	Infrared radiation/cm⁻¹
H–Cl	432	2886
H–Br	366	2559
H–I	298	2230

Table A1 *Infrared radiation needed to make hydrogen–halide bonds vibrate*

Wavenumber/cm⁻¹

4000   3000   2000   1000 500

single bonds to hydrogen — triple bonds — double bonds — fingerprint region

for example:
C – H   C ≡ C   C = C   C – C
O – H   C ≡ N   C = O   C – O
N – H

Figure A2 *Absorption regions for infrared spectra*

Questions

A1. Why is the bond energy for H–Cl higher than the bond energy for H–I?

A2. Suggest a value for the wavenumber of the radiation needed to make a H–F bond vibrate.

A3. If the whole range of infrared radiation is passed through a compound containing C–C bonds and C–H bonds only, which wavenumbers of radiation are likely to be

  a. absorbed

  b. transmitted?

A4. How does Figure A2 show that C≡C is stronger than C=C, which is stronger than C–C?

The infrared radiation transmitted by different wavenumbers is shown on an infrared spectrum. The absorbances can be used to identify different bonds, and hence the functional groups in a molecule.

Table 3 shows some characteristic infrared absorptions of different bonds.

Group	Location	Wavenumber range/cm⁻¹	Intensity
O–H	alcohols	3230–3550	strong, broad
	carboxylic acids	2500–3000	medium/strong
N–H	primary amines	3300–3500	medium/strong
C–H	alkanes, alkenes, arenes	2850–3300	medium/strong
C≡N	nitriles	2220–2260	medium
C=O	aldehydes, ketones, carboxylic acids, esters	1680–1750	strong
C=C	alkenes	1620–1680	
C–O	alcohols, ethers, esters	1000–1300	strong
C–C		750–1100	
C–Cl		700–800	strong
C–Br		500–600	strong
C–I		About 500	strong

Table 3 *Some characteristic infrared absorptions of various bonds*

The infrared spectrum of ethanol is shown in Figure 10.

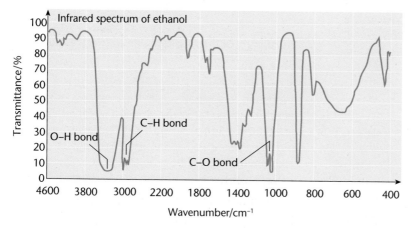

Figure 10 *The infrared spectrum of ethanol*

The infrared spectrum

Ethanol has the molecular formula C_2H_5OH. Its displayed formula is:

$$
\begin{array}{cccccc}
& \text{H} & & \text{H} & & \\
& | & & | & & \\
\text{H} & - \text{C} & - & \text{C} & - \text{O} & - \text{H} \\
& | & & | & & \\
& \text{H} & & \text{H} & &
\end{array}
$$

The infrared spectrum of ethanol is shown in Figure 10.

The spectrum records percentage transmittance (the amount of radiation that passes through the sample) against wavenumber, cm^{-1} (its energy). If no radiation is absorbed by the sample, the transmittance is 100%. If radiation is absorbed to increase the bond vibration, the transmittance is less than 100%. The spectrum may appear to be upside down, but remember that it is transmittance (the amount of radiation passing through) that is measured.

The infrared spectrum of ethanol shows a broad strong absorption around $3400\ cm^{-1}$. This is characteristic of the OH group in alcohols. The absorption at $3000\ cm^{-1}$ is characteristic of the C–H bond. The strong absorption around $1000\ cm^{-1}$ results from the C–O bond.

Ethanoic acid has the molecular formula CH_3COOH. Its displayed formula is:

$$
\begin{array}{ccc}
\text{H} & & \\
| & & \diagup \text{O} \\
| & & \diagup\diagup \\
\text{H} - \text{C} & - \text{C} & \\
| & & \diagdown \\
\text{H} & & \text{O} - \text{H}
\end{array}
$$

The infrared spectrum of ethanoic acid is shown in Figure 11.

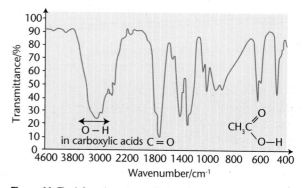

Figure 11 *The infrared spectrum of ethanoic acid*

The characteristic absorptions are:

A medium-strong broad absorption at $2500–3000\ cm^{-1}$, characteristic of the OH group in carboxylic acids.

A strong absorption at $1700\ cm^{-1}$ characteristic of the C=O group in carboxylic acids.

The fingerprint region

The region of the infrared spectrum between 500 and $1500\ cm^{-1}$ is known as the **fingerprint region**. In addition to the wavenumbers absorbed by the C–C and C–O bonds, the wavenumbers in this range are also absorbed by vibrations of the whole molecule. As a result, this is usually the complicated part of the spectrum with many absorptions. The fingerprint region is used to identify a molecule by comparing it with infrared spectra of known samples that you can find in databases and reference books.

ASSIGNMENT 2: CHECKING PURITY

(MS 3.1; PS 1.1, 1.2, 2.3, 4.1)

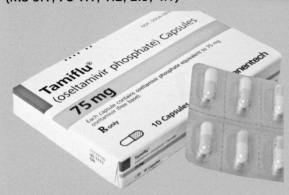

Figure A1 *A batch of medicine can be identified by its batch number, so that if there is a problem, the batch can be recalled.*

The pharmaceutical industry manufactures most medicines in batch processes rather than continuous processes. An advantage of producing medicines this way is that the purity of each batch can be tested and checked. Mass spectroscopy and infrared spectroscopy are often used to confirm the identity of the medicine and to check for any impurities.

Here is an example. 2-hydroxybenzoic acid is a compound that relieves pain. However, it does not taste nice and it can cause stomach irritation. So instead the pharmaceutical industry converts it into a compound called 2-ethanoylhydroxybenzoic acid. You probably know it as aspirin. It is the active ingredient in aspirin tablets. In the body, 2-ethanoylhydroxybenzoic acid hydrolyses to produce 2-hydroxybenzoic acid.

The structural formula of 2-hydroxybenzoic acid is:

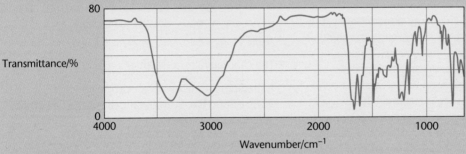

Figure A2 *The infrared spectrum of 2-hydroxybenzoic acid*

Questions

A1. Ethanol is a common solvent in the preparation of herbal medicines. Liquid Echinacea, taken to improve the immune system, contains ethanol as a solvent.

An ethanol sample is suspected of being contaminated by ethanoic acid. Figure A3 shows its infrared spectrum.

a. Identify the absorptions produced by the ethanol molecule.

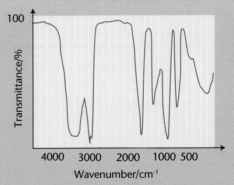

Figure A3 *Is the ethanol contaminated?*

b. Identify any absorptions that would indicate the presence of ethanoic acid. Note that because it is difficult to distinguish between O–H bonds in carboxylic acids and O–H bonds in alcohols, the absorption at 1700 cm^{-1} by the C=O group is used to identify the acid.

c. Is the ethanol contaminated?

Stretch and challenge

A2. When batches of aspirin tablets are produced from salicylic acid, the purity of the salicylic acid can be checked using its mass spectrum and its infrared spectrum (see Figure A4).

a. Identify the groups responsible for the main absorptions.

b. Explain what is happening inside the aspirin molecule when infrared radiation is being absorbed.

c. How would you use the IR spectrum of 2-hydroxybenzoic acid to check the purity of a sample?

A3. The molecular formula of 2-hydroxybenzoic acid is $C_7H_6O_3$. Its presence can be confirmed using mass spectroscopy.

a. What is the formula of the molecular ion?

b. What is the relative mass of the molecular ion peak on a low-resolution mass spectrometer?

c. Using the isotopic masses of carbon-12, hydrogen-1 and oxygen-16 to six significant figures, calculate the relative mass of the molecular ion peak you would expect from a high-resolution mass spectrometer.

d. A high-resolution mass spectrometer would provide a more accurate confirmation of the presence of 2-hydroxybenzoic acid than a low-resolution mass spectrometer. Explain why.

ASSIGNMENT 3: DRUG ANALYSIS

(MS 3.1; PS 1.1, 1.2, 2.3, 4.1)

Infrared spectroscopy is used extensively as an identification tool in forensic science. If a motorist fails a roadside breathalyser test, he must take a further breath test in a police station, or provide a blood or urine sample for analysis. Infrared spectroscopy is used to analyse the amount of ethanol in the sample by measuring the intensity of the absorption of the O–H bond at 3000 cm^{-1}.

IR spectroscopy is also used to analyse suspected drug samples. The IR spectrum of sample A in Figure A1 is of a confiscated substance suspected of being heroin. The IR spectrum of heroin is shown in Figure A2.

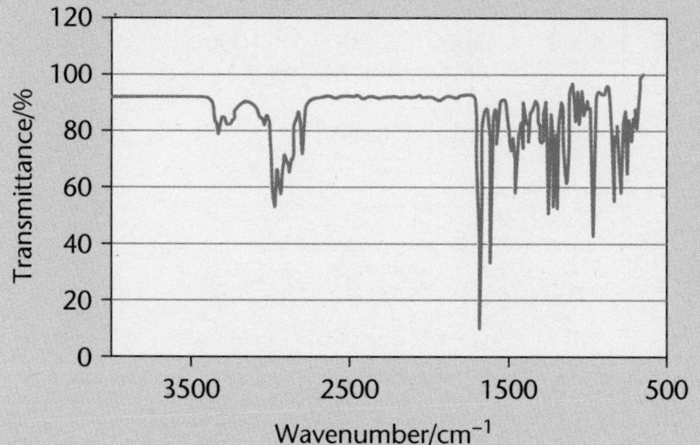

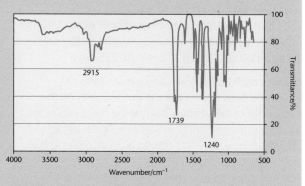

Figure A1 Infrared spectrum of Sample A

Figure A2 Infrared spectrum of heroin

Questions

A1. Suggest possible identities for these absorptions on the heroin spectrum:

   a. 2915 cm^{-1}

   b. 1739 cm^{-1}

   c. 1240 cm^{-1}

A2. Which of these absorptions also occur on the IR spectrum of the suspect sample?

A3. a. What is the region between 1500 and 500 cm^{-1} called?

   b. What is it used for?

A4. Is the suspect sample heroin? Give reasons for your answer.

KEY IDEAS

> Infrared radiation is absorbed by molecules, causing the molecules to vibrate more rapidly.

> An infrared spectrum can be obtained by measuring the transmittance of infrared radiation through a sample of material.

> Different functional groups absorb infrared radiation of characteristic energy (wavenumber, cm^{-1}).

> The infrared spectrum of a compound may be used to identify that compound.

16.4 THE GLOBAL WARMING LINK

The electromagnetic radiation transmitted to Earth from the Sun is mainly UV, visible and the shorter wavelengths of infrared radiation. Earth absorbs the radiation and is warmed by some of it. The radiation emitted from Earth's surface is mostly infrared. Without greenhouse gases in the atmosphere, most of this would be lost into space and it is estimated the temperature on Earth would drop by 35 °C in 50 years.

But greenhouse gases absorb infrared radiation. The major greenhouse gases are carbon dioxide, methane and water vapour, and steady concentrations of these in the atmosphere have kept Earth's climate fairly stable for thousands of years. Their infrared spectra are shown in Figure 12.

The O–H bonds in water molecules strongly absorb infrared radiation at 3000–3800 cm^{-1}, the C–H bonds in methane at about 3000 cm^{-1} (a characteristic of all alkanes) and the C=O bonds in carbon dioxide absorb strongly at 2350 cm^{-1}.

Infrared radiation absorbed by greenhouse gas molecules does not escape the atmosphere. As the concentration of greenhouse gases in the atmosphere increases, the amount of infrared radiation being absorbed increases. This results in an increase in the kinetic energy of the molecules in the atmosphere, which is detected as a temperature increase and referred to as global warming.

Carbon dioxide infrared spectrum

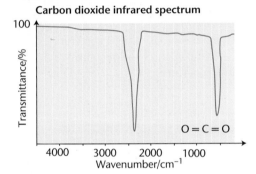

Methane infrared spectrum

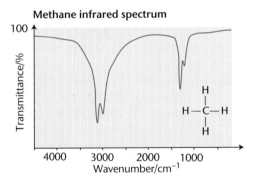

Water infrared spectrum

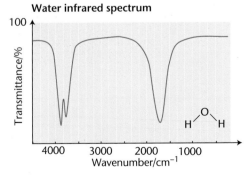

Figure 12 *The infrared spectra of carbon dioxide, methane and water*

328

QUESTIONS

Stretch and challenge

13. Oxygen gas and nitrogen gas are not greenhouse gases. What would you expect their infrared spectra to look like?

14. The C–F bond absorbs infrared radiation between 1000 and 1400 cm^{-1}. The C–Cl bond absorbs it between 700 and 800 cm^{-1}. Sketch the infrared spectrum you would expect from dichlorodifluorocarbon.

PRACTICE QUESTIONS

1. Glucose is an organic molecule. Glucose can exist in different forms in aqueous solution.

a. In aqueous solution, some glucose molecules have the following structure.

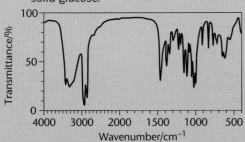

i. Deduce the empirical formula of glucose.

ii. Consider the infrared spectrum of solid glucose.

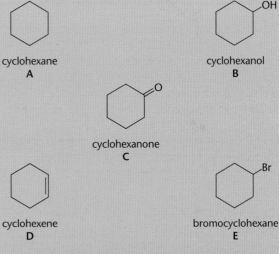

b. State why it is possible to suggest that in the solid state very few molecules have the structure shown.

AQA Jan 2013 Unit 2 Question 5a

2. Consider the five cyclic compounds, A, B, C, D and E.

cyclohexane
A

cyclohexanol
B

cyclohexanone
C

cyclohexene
D

bromocyclohexane
E

The infrared spectra of compounds A, B, C and D are shown here:

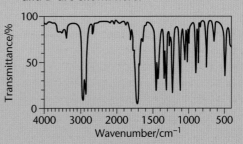

(Continued)

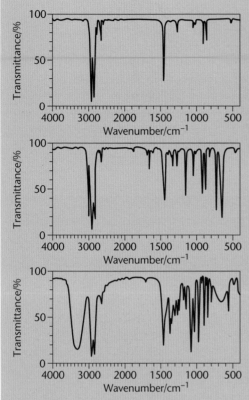

Bond	Wavenumber/cm^{-1}
C–H	2850–3300
C–C	750–1100
C=C	1620–1680
C=O	1680–1750
C–O	1000–1300
O–H (alcohols)	3230–3550
O–H (acids)	2500–3000

Table Q1

He measured the infrared spectra of the two compounds and obtained the following spectra:

A

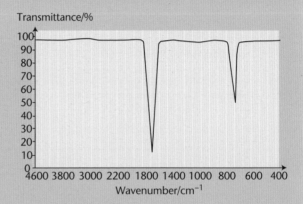

B

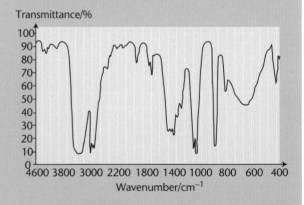

a. Write the correct letter, A, B, C or D, next to each spectrum.

b. A simple chemical test can be used to distinguish between cyclohexane (A) and cyclohexene (D). Give a reagent for this test and state what you would observe with each compound.

c. Cyclohexanol (B) can be converted into cyclohexanone (C). Give a reagent or combination of reagents that can be used for this reaction and state the type of reaction. State the class of alcohols to which cyclohexanol belongs.

AQA Jan 2012 Unit 2 Question 10 a, b, c

3. A chemist employed in the wine-making industry separated two different organic compounds from a wine suspected of being contaminated with ethanoic acid.

a. Give the displayed formulae for
   i. ethanol
   ii. ethanoic acid

(Continued)

b. In spectrum A, state which bonds are responsible for the absorptions at

   i. 3400 cm^{-1}

   ii. 3000 cm^{-1}

c. In spectrum B state which bonds are responsible for the absorption at 1700 cm^{-1}.

d. Which spectrum is ethanol?

e. The chemist demonstrated the reactions responsible for the conversion of ethanol to ethanoic acid. He added acidified potassium dichromate(VI) to a sample of ethanol and refluxed the reactants for several minutes.

   i. Name the apparatus needed to reflux ethanol and acidified potassium dichromate(VI) solution and describe how you would carry out the experiment.

   ii. What type of chemical reaction is this?

4. Mass spectra can be used to identify ethanol, ethanal and ethanoic acid.

a. Copy and complete Table Q2 to show the formula of the molecular ion formed by each compound and its mass.

Compound	Formula of molecular ion	Mass of molecular ion
ethanol		
ethanal		
ethanoic acid		

Table Q2

b. Describe how low-resolution mass spectroscopy can be used to distinguish between them.

c. High-resolution mass spectroscopy provides more accurate information. These are the relative isotopic masses for carbon, hydrogen and oxygen:

$^{12}C = 12.0000;\ ^{1}H = 1.0078;\ ^{16}O = 15.9949$

Use these values to calculate the relative molecular mass of ethanol.

5. Figure Q1 shows the infrared spectrum of carbon dioxide.

a. Name two other gases present in the atmosphere that strongly absorb infrared radiation.

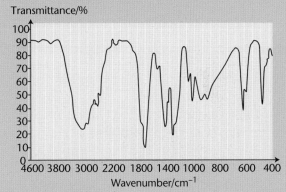

Figure Q1

b. What effect does absorbing this infrared radiation have on the temperature of the atmosphere?

c. The infrared spectrum of CO_2 shows an absorption at 2350 cm^{-1}. Suggest which group is responsible.

d. A sample of carbon dioxide gas is injected into a low-resolution Time of Flight mass spectrometer.

   i. What happens in the ionisation area?

   ii. Write the formula for the ion formed when one electron is removed from a molecule.

   iii. What is the mass of the molecular ion peak in this spectrum?

(Continued)

6. Compound X is known to be be methanol, methanal or methanoic acid. The infrared spectrum of compound X is shown in Figure Q2.

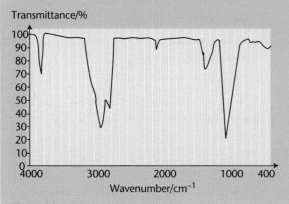

Transmittance/%

Wavenumber/cm^{-1}

Figure Q2

a. Give the displayed formulae for

   i. methanol

   ii. methanal

   iii. methanoic acid

b. i. Use the infrared absorptions on the data sheet and name the compound.

   ii. Explain which two absorption peaks enabled the compound to be identified.

c. Write the formula for the molecular ions formed by

   i. methanol

   ii. methanal

   iii. methanoic acid

d. A high-resolution mass spectrometer can be used to confirm the identity of X. The relative isotopic masses are: ^{12}C = 12.0000; ^{1}H = 1.0078; ^{16}O = 15.9949.

Calculate the relative mass of each molecular ion formed.

Practical skills question

7. Students are identifying four unknown colourless liquids, labelled as W, X, Y and Z. They are told that they are propanoic acid, butan-1-ol, butan-2-ol or cyclohexene.

They carry out the qualitative tests in the table and record their results.

a. Explain the difference between a qualitative test and a quantitative test.

b. Explain why it is important to use fresh samples for tests 1, 2, 3 and 4.

	Test	W	X	Y	Z
1	Adding bromine water	colour change from orange to colourless	no change	no change	no change
2	Testing with indicator paper	pH = 7	pH = 7	pH = 3	pH = 7
3	Adding sodium hydrogen carbonate solution	no change	no change	gas given off that turns limewater milky	no change
4	Adding acidified potassium dichromate(VI) solution and warming	no change	colour change from orange to green	no change	colour change from orange to green
5	Adding Fehling's solution to the products of test 4 and warming	no change	a brick red precipitate forms	no change	no change

Table Q3

(Continued)

c. Identify samples W, X, Y and Z.

d. Explain why the products from test 4 can be identified using Fehling's solution.

e. Give one advantage and one disadvantage of using test tubes to carry out these tests instead of spotting tiles.

f. Give two advantages of using a water bath to warm the reaction mixtures instead of a Bunsen burner.

Stretch and challenge

8. Propan-2-ol is extensively used in perfumes as a solvent. Aromatic compounds dissolve in it and propan-2-ol quickly evaporates on the skin, leaving the aromatic substances on the skin. Ethanol is also used in perfumes for the same reason.

a. State whether propan-2-ol and ethanol are primary, secondary or tertiary alcohols.

b. Describe test-tube reactions that could be used to determine whether the solvent in a perfume was propan-2-ol or ethanol.

c. Describe how low-resolution mass spectroscopy could be used to distinguish between propan-2-ol and ethanol.

d. Students suggested that IR spectra of propan-2-ol and ethanol could be used to differentiate between them. Give two reasons why they could not.

Multiple choice

9. Which compounds will give a positive test with bromine water?

A. 3-methylhexane

B. but-2-en-1-ol

C. butanoic acid

D. butanal

10. A halogenoalkane is warmed with sodium hydroxide solution and a few drops of acidified silver nitrate solution add. A cream precipitate forms that is soluble in concentrated ammonia solution. Name the type of halogenoalkane.

A. Fluoroalkane

B. Chloroalkane

C. Bromoalkane

D. Iodoalkane

11. Which is the correct definition of a molecular ion peak on a mass spectrum?

A. An ion with a positive charge

B. The peak produced by molecules of the sample on a mass spectrum

C. The peak produced by molecular ions of the sample on a mass spectrum

D. The peak produced by negative ions of the sample on a mass spectrum

12. Which of these labels are usually used on infrared spectra?

A. m/z

B. Wavelength

C. Frequency

D. Wavenumber

ANSWERS TO IN-TEXT QUESTIONS

1 ATOMIC STRUCTURE

1. Most substances could be classified using these four elements. The theory was compatible with religious belief.

2. **a.** J J Thomson's experiment with cathode ray tubes.

   b. Rutherford's gold foil experiment.

   c. Moseley and Rutherford's work with X-ray spectra of elements.

   d. Chadwick's experiment using alpha particles.

3. The neutron has no charge.

4. The idea of matter consisting of particles called atoms developed into an understanding of the existence of the subatomic particles – electron, proton and neutron – within the atom. Evidence for their existence was gathered.

5. **a.** 9 protons, 10 neutrons, 9 electrons

   b. 85 protons, 125 neutrons, 85 electrons

   c. 11 protons, 12 neutrons, 10 electrons

   d. 15 protons, 16 neutrons, 18 electrons

   e. 24 protons, 28 neutrons, 21 electrons

6. **a.** 2

   b. 121, 123

   c. 57.25%, 42.75%

   d. 121.86

7. Mass of 100 atoms = $(0.006 \times 234) + (0.72 \times 235) + (99.2 \times 238) = 23780.204$;

   relative atomic mass = $\dfrac{23780.204}{100} = 237.80$

8. 107.97

9. Elements have spectra made up of discrete lines, not a continuous spectrum. The wavelength or frequency of each line can be explained as corresponding to the difference in energy between the electron shells.

10.

Excited electron may also be shown falling back directly from $n = 3$ to $n = 1$.

1. **a.** $[Ne]3s^2\ 3p^4$

b. $[Ne]3s^2\ 3p^1$

c. $[Ar]4s^2$

d. $[Ar]3d^1\ 4s^2$

e. $[Ne]3s^2\ 3p^2$

f. $[Ar]3d^6\ 4s^2$

g. $[Kr]5s^2$

h. $[Ar]\ 3d^{10}\ 4s^1$

2. 2 is the period of the element and the number of the main shell, p is the type of orbital (sub-division of the main shell), 6 is the number of electrons in the 2p orbital.

3. **a.** $1s^2\ 2s^2\ 2p^6\ 3s^2\ 3p^6$

b. $1s^2\ 2s^2\ 2p^6\ 3s^2\ 3p^6$

c. $1s^2\ 2s^2\ 2p^6$

d. $1s^2\ 2s^2\ 2p^6\ 3s^2\ 3p^6\ 3d^{10}\ 4s^2\ 4p^6$

e. $1s^2\ 2s^2\ 2p^6$

14. $M^{2+}(g) \rightarrow M^{3+}(g) + e^-$, $M^{3+}(g) \rightarrow M^{4+}(g) + e^-$

15.

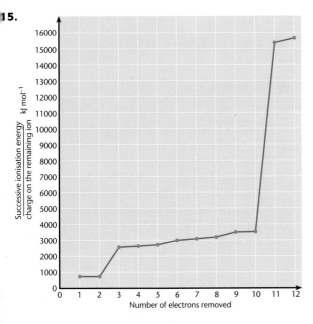

16. There is a large increase in first ionisation energies where the next electron is removed from an inner shell. These have less shielding from the nuclear charge and are more difficult to remove. The pattern suggests two electrons in the first shell, eight in the second and four in the third.

17. X in Group 3(13), Y in Group 1, Z in Group 2

2 AMOUNT OF SUBSTANCE

1. **a.** 64

b. 30

c. 46

d. 208.5

e. 180

2. **a.** 184

b. 74

c. 213,

d. 342,

e. 158

Answers are given here in standard form

3. **a.** 1.807×10^{24}

b. 6.023×10^{21}

c. 1.656×10^{23}

d. 1.054×10^{24}

e. 4.216×10^{21}

4. **a.** 3.016×10^{24}

b. 4.367×10^{23}

c. 1.205×10^{25}

d. 1.506×10^{22}

e. 6.023×10^{20}

5. **a.** 0.25 mol

b. 0.6 mol

c. 3.0 mol

d. 3 mol

e. 0.25 mol

f. 2 mol

g. 4 mol

h. 6 mol

i. 1000 mol

j. 0.1 mol

6. **a.** 26 g

b. 40 g

c. 2070 g

d. 13.6 g

e. 28 g

f. 0.46 g

g. 0.098 g

h. 70 g

i. 140 g

j. 1 g

7. **a.** 6.023×10^{21}

    b. 6.023×10^{22}

    c. 1.205×10^{21}

    d. 1.833×10^{23}

    e. 6.023×10^{20}

8. **a.** 7.529×10^{22}

    b. 6.023×10^{22}

    c. 6.023×10^{20}

    d. 2.842×10^{21}

    e. 6.692×10^{25}

9. **a. i.** 298 K

    ii. 523 K

    iii. 195 K

    b. 123.819 cm^3 or 1.24×10^5 cm^3

    c. 0.007553 m^3 or 7553 cm^3

    d. 0.04 mol

10. **a.** MgO

    b. Fe_2O_3

    c. $FeSO_{11}H_{14}$

    d. C_2H_5

    e. empirical formula CH_2, molecular formula C_2H_4

11. **a.** $2C_2H_5OH(l) + 7O_2(g) \rightarrow CO_2(g) + 6H_2O(g)$

    b. $2Al(s) + 6NaOH(aq) \rightarrow 2Na_3AlO_3(aq) + 3H_2(g)$

    c. $6CO_2(g) + 6H_2O(l) \rightarrow C_6H_{12}O_6(aq) + 6O_2(g)$

    d. $C_3H_8(g) + 5O_2(g) \rightarrow 3CO_2(g) + 4H_2O(l)$

    e. $2Al(s) + 6H_2SO_4(l) \rightarrow Al_2(SO_4)_3(aq) + 3SO_2(g) + 6H_2O(l)$

12. **a.** $FeCl_3(aq) + 3NH_4OH(aq) \rightarrow Fe(OH)_3(s) + 3NH_4Cl(aq)$

    b. $CuCO_3(s) + 2HCl(aq) \rightarrow CuCl_2(aq) + CO_2(g) + H_2O(l)$

    c. $3NaOH(aq) + H_3PO_4(aq) \rightarrow Na_3PO_4(aq) + 3H_2O(l)$

    d. $2Fe(s) + 3Cl_2(g) \rightarrow 2FeCl_3(s)$

    e. $CuO(s) + H_2SO_4(aq) \rightarrow CuSO_4(aq) + H_2O(l)$

13. **a.** 224 tonnes

    b. 89.3%

    c. 45.9%

    d. 680 tonnes

    e. 80.9%

    f. 68.7%

    g. 49.9%

14. **a.** 16.9%

    b. 100%

15. 36.72 g

16. **a.** 0.01 mol dm^{-3}

    b. 0.025 mol dm^{-3}

    c. 0.0125 mol dm^{-3}

    d. 0.25 mol dm^{-3}

    e. 2.00 mol dm^{-3}

17. **a.** 0.25 mol

    b. 0.0025 mol

    c. 0.025 mol

    d. 0.02 mol

    e. 0.02 mol

18. **a.** 2.65 g

    b. 0.25 mol dm^{-3}

    c. 1.06 g

3 BONDING

1. Mg^{2+}: 2,8, O^{2-}: 2,8, Na^+: 2,8, Ca^{2+}: 2,8,8, F^-: 2,8

2. **a. i.** KCl

    ii. CaS

    iii. $MgCl_2$

    v. Li_2O

    v. Al_2O_3

    vi. Mg_3N_2

    b.

    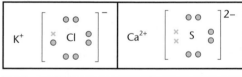

    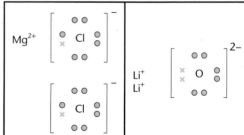

3. **a.** -3

    b. -2

    c. -3

d. +2

e. +3

f. +3

g. +1

h. does not form a simple ion

4. **a.** $SrBr_2$

b. $Al(OH)_3$

c. $Mg(HCO_3)2$

d. NH_4HCO_3

e. $(NH_4)_2CO_3$

f. K_2S

g. $Ba(NO_3)_2$

h. Na_3P

i. Rb_2S

j. $Al_2(CO_3)_3$

5.

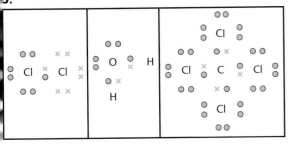

6.

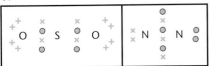

7.

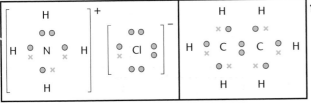

8.

9.

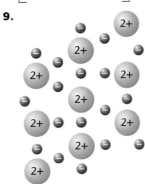

and same again for Al^{3+}, but with 3 + in each circle and an extra grey circle with a negative sign inside it for every two shown in the Mg^{2+} diagram

10. **a.** 2 mol

b. 3 mol

11. **a.**

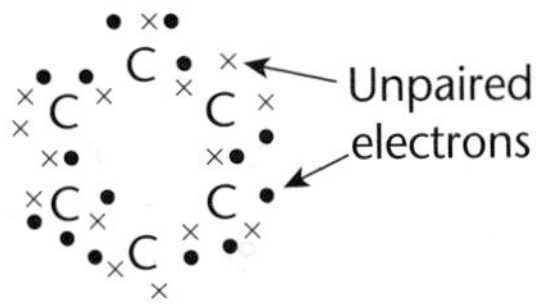

Unpaired electrons

b. Labels need to show graphene's unpaired electrons and these are the reason that graphene reacts easily with other substances. In graphite ther unpaired electrons become delocalised and bond the layers together, so they are not available to form new bonds.

12. The ions in an ionic lattice are held in a fixed position by the strong electrostatic forces and are not free to move. Ionic solids are not malleable. In a metal, the metal ions can slide over each other within the sea of delocalised electrons and metals are malleable.

13. a. The values are compared to aluminium being 1.

b.

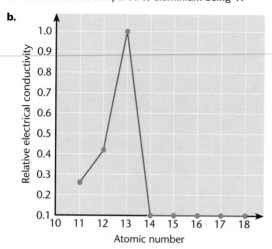

c. From sodium to aluminium the number of electrons in each outer shell is increasing so the number of delocalised electrons is increasing. Conductivity increases with the number of delocalised electrons.

d. Elements from phosphorus to chlorine form covalent bonds to make molecules. These do not conduct electricity. silicon is a semi-conductor. Argon is an inert gas and gases cannot conduct electricity.

14. a. ionic crystal

b. molecular crystal

c. macromolecular (giant covalent) or metallic crystal

d. ionic crystal

15. a. ionic bonds

b. metallic bonds

c. covalent bonds

d. intermolecular forces

16. Ionic: M_r; metallic: A_r; giant covalent: A_r; molecular: M_r

17. a. Three bonding pairs, one lone pair.

b. Two bonding pairs, two lone pairs.

18. a. Noble gases rarely form compounds

b. Electronegativity increases across a period from left to right and decreases down a group.

19. As you go down group 7(17), an extra shell of electrons is added. This shields the attractive force of the nucleus and the element has a lower electronegativity value.

20. a. C and H have similar electronegativities and there are no polar bonds

$$H_{2.1}-C_{2.5}-H_{2.1}$$

with $H_{2.1}$ above and below the central C.

No permanent dipole-similar electronegativities

b. Electronegativities: F = 4.0, C = 2.5. The molecule will have a permanent dipole with the F atom having a δ^- charge and the rest of the molecule a δ^+ charge

$$H_3C^{\delta+}_{2.5}-F^{\delta-}_{4.0}$$

Permanent dipole

c. Electronegativities: F = 4.0, C = 2.5. The molecule will have a permanent dipole with the F atoms having a δ^- charge and the rest of the molecule a δ^+ charge

$$H_2C^{\delta+}_{2.5}-F^{\delta-}_{4.0}$$

with $F^{\delta-}_{4.0}$ below.

Permanent dipole

d. The molecule will not have a permanent dipole because the δ^- charge is spread over the surface of the molecule.

$$F^{\delta-}_{4.0}-C^{\delta+}_{2.5}-F^{\delta-}_{4.0}$$

with $F^{\delta-}_{4.0}$ above and below the central C.

Permanent dipole

Bonds have permanent dipoles, but the molecule is non-polar

21. a. Water about 210–220 K

b. Hydogen fluoride about 180–190 K

22. M_r for ethanoic acid is 60. Hydrogen bonding bonds two ethanoic acid molecules together, giving an M_r of 120

23.

4 THE PERIODIC TABLE

1. **a.** $1s^2 2s^2 2p^5$; p block

   b. $1s^2 2s^2 2p^6 3s^2 3p^6 4s^2$; s block

   c. $1s^2 2s^2 2p^6 3s^2 3p^6 3d^{10} 4s^2$; d block

2. **a.** Period 4

   b. p block

   c. Group 6

   d. Six

3. **a.** Strontium $1s^2 2s^2 2p^6 3s^2 3p^6 3d^{10} 4s^2 4p^6 5s^2$; s block

   b. Fluorine $1s^2 2s^2 2p^2$; p block

   c. Gold $1s^2 2s^2 2p^6 3s^2 3p^6 3d^{10} 4s^2 4p^6 4d^{10} 5s^2 5p^6 5d^{10} 6s^1$; d block

   d. Aluminium $1s^2 2s^2 2p^6 3s^2 3p^1$; p block

   e. Iron $1s^2 2s^2 2p^6 3s^2 3p^6 3d^6 4s^2$; d block

   f. Germanium $1s^2 2s^2 2p^6 3s^2 3p^6 3d^{10} 4s^2 4p^2$; p block

4. The d-block elements of Period 4, scandium to zinc, successively add one electron to the 3d sub-shell, which has a maximum of ten electrons. The p-block elements from gallium to krypton add to the p sub-shell, which has a maximum of six electrons. There are five orbitals in the d sub-level and three in the p sub-level.

5. Names starting with 'unnil' or 'unun' are elements yet to be confirmed. Other names are elements that have been confirmed – they are named after the discoverer or place of discovery.

6. Argon is unreactive and does not form compounds. You cannot measure its covalent radius

7. **a.** Atomic radius increases as you go down Group 1

   b. Atomic radius increases down a group because another shell of electrons is added

   c. Metallic radius is measured

   d. S^{2-} has an extra proton and, therefore, a greater nuclear charge w hich has the effect of drawing the electron shells closer together.

8. **a.** There are strong covalent bonds between the atoms in P_4 molecules and in S_8 molecules. Both have weak intermolecular forces/ van der Waals forces between the molecules

   b. P_4 and S_8 molecules have relatively low melting points because the weak intermolecular forces are easily broken. The S_8 melting point is higher than the P_4 melting point because of the stronger van der Waals forces between the larger S_8 molecules.

5 INTRODUCTION TO ORGANIC CHEMISTRY

1. **a.**

   b.

   c.

2. **a.**

   b.

3. N–N and P–P bonds are weak bonds (they have low mean bond enthalpies) and are easily broken

4. Transferring ('losing') four electrons requires too much energy because as more electrons are transferred, the remaining electrons are attracted more strongly.

5. **a.** $C=C$ is shorter than C–C and has a higher bond enthalpy; C≡C is shorter than $C=C$ and has a higher bond enthalpy.

   b. C–C has a σ bond and bond enthalpy of 347 $kJ\ mol^{-1}$. If $C=C$ consisted of two σ bonds, you could expect the bond enthalpy to be 2×347 $kJ\ mol^{-1}$ or 694 $kJ\ mol^{-1}$. Since it is only 612 $kJ\ mol^{-1}$ the π bond must be weaker than the σ bond.

6. **a.** C_5H_{11}

   b. $C_{10}H_{22}$

   c. $CH_3CH_2CH_2CH_2CH_2CH_2CH_2CH_2CH_2CH_3$

7. **a.** CH_2

   b. C_3H_6

   c. CH_3CHCH_2

8. **a.** C_5H_{12}

   b. CH3CH2CH2CH2CH3

   c.

   d.

9. Formula unit for compounds with giant structures (ionic or covalent) and molecular formula for molecules (though displayed formula are also used)

339

10. Organic compounds can have different structures (isomers) and functional groups with the same molecular formulae, so using structural formulae prevents confusion.

11. a. −OH, alcohol

b. C−H, alkane

c. C=C, alkene

d. −COOH, carboxylic acid

e. −COOR, ester

f. −C−Br, haloalkane

g. C=C, alkene

h. −OH, alcohol

12. Draw the displayed formulae for the compounds in question 8 that are not hydrocarbons.

−OH, alcohol
```
   H  H  H  H
   |  |  |  |
H—C—C—C—C—O—H
   |  |  |  |
   H  H  H  H
```

−COOH, carboxylic acid
```
   H  H  H  H
   |  |  |  |    OH
H—C—C—C—C—C
   |  |  |  |    ‖
   H  H  H  H    O
```

−COOR, ester
```
           H  H  H         H  H
           |  |  |         |  |
H—C—C—C—C       O—C—C—H
   |  |  |   \         H  H
   H  H  H    O
```

−C−Br, haloalkane
```
   H  Br H  H
   |  |  |  |
H—C—C—C—C—H
   |  |  |  |
   H  H  H  H
```

−OH, alcohol
```
   H  OH H  H  H  H
   |  |  |  |  |  |
H—C—C—C—C—C—C—H
   |  |  |  |  |  |
   H  H  H  H  H  H
```

13. a.
```
   H  H  H  H  H  H  H  H  H
   |  |  |  |  |  |  |  |  |
H—C—C—C—C—C—C—C—C—C—H
   |  |  |  |  |  |  |  |  |
   H  H  H  H  H  H  H  H  H
```

b.
```
        H
        |
     H—C—H
   H  |  H  H  H  H  H
   |  |  |  |  |  |  |
H—C—C—C—C—C—C—C—H
   |  |  |  |  |  |  |
   H  H  H  H  H  H  H
```

c.
```
        H
        |
     H—C—H
     H  |   H  H  H
     |  |   |  |  |
H—C—C—C—C—C—H
     |  |   |  |  |
     H  |   H  H  H
     H—C—H
        |
        H
```

d.
```
           H
           |
        H—C—H
   H  H  |   H  H  H
   |  |  |   |  |  |
H—C—C—C—C—C—C—H
   |  |  |   |  |  |
   H  H  |   H  H  H
     H—C—H
        |
        H
```

e.
```
        H
        |
     H—C—H
     H—C—H
   H  H  |   H  H  H  H
   |  |  |   |  |  |  |
H—C—C—C—C—C—C—C—H
   |  |  |   |  |  |  |
   H  H  H   H  H  H  H
```

14. a.
```
H        H
 \      /
  C=C
 /      \
H        H
```

b.
```
   H              H
   |              |
H—C—C=C—C—H
   |  |  |  |
   H  H  H  H
```

c.
```
H          H  H  H
 \         |  |  |
  C=C—C—C—C—H
 /         |  |  |
H          H  H  H
```

d.
```
   H  H              H  H
   |  |              |  |
H—C—C—C=C—C—C—H
   |  |  |  |  |  |
   H  H  H  H  H  H
```

15. a. Propene

b. Pent-2-ene

c. Hex-1-ene

d. 3-methylpent-1-ene

e. 4,4-dimethylpent-2-ene

16. a. chloromethane

b. 1,1,2-tribromopropane

c. 1,1,1-trichloro-2,2-difluorethane

d. 1,1,1-tribromo-2,2,2-trifluoroethane

e. 1-bromo-2,2-dichlorobutane

17. a.

```
    Br  Cl
    |   |
H — C — C — H
    |   |
    Br  Cl
```

b.

```
    Cl  Br
    |   |
H — C — C — H
    |   |
    Br  Cl
```

c.

```
    Cl  F   H
    |   |   |
H — C — C — C — H
    |   |   |
    Br  F   H
```

d.

```
    I   H
    |   |
I — C — C — H
    |   |
    I   H
```

e.

```
    F   Cl
    |   |
F — C — C — Cl
    |   |
    F   Cl
```

f.

```
    H   H   H
    |   |   |
H — C — C — C — H
    |   |   |
    H   Cl  H
```

g.

```
       H
       |
Cl — C — Cl
       |
       Cl
```

h.

```
    H   H   H   Cl
    |   |   |   |
H — C — C — C — C — Cl
    |   |   |
    H   Br  Cl
```

18. e. is 1,1,1-trifluoro-2,2,2-trichloroethane, f. is 2-chloropropane, g. is trichloromethane, h. is bromo-1,1,1,2-tetrachlorobutane

19. a. Propan-1-ol

b. Propanal

c. Propanoic acid

d. Butanone

e. Butan-2-ol

f. Ethanal

g. Butanoic acid

h. Pentan-2-one

20. a. The carbon atom can only form one other bond

b. The carbon atom must be bonded to two other carbon atoms

c. The oxygen atom of the OH group uses one of carbon's four bonds, leaving three other bonds.

21.

```
      H
      |           δ+    δ-
H — N:  +  H — Cl
      |
      H
```

22.

```
         ..  δ-   δ+
H+  +  :O — H
         ..
```

23. a.

```
        CH₃
        |
H₃C — C — Br
        |
        CH₃
```

b.

```
        CH₃                              CH₃
        |              slow              |
CH₃ — C — Br    ⇌           CH₃ — C⁺  +  :Br⁻
        |                               |
        CH₃                             CH₃
```

c.

```
        CH₃                              CH₃
        |            ..  -                |   ..
H₃C — C⁺      ( :O — H )    →    H₃C — C — O — H
        |            ..                   |   ..
        CH₃                              CH₃
```

24. $CH_3CH_2CH_2CH_2CH_2CH_3$, hexane; $CH_3CH(CH_3)$ $CH_2CH_2CH_2CH_3$, 2-methylpentane; $CH_3CH_2CH(CH_3)$ CH_2CH_3, 3-methylpentane; $CH_3C(CH_3)_2CHCH_3$, 2,2-dimethylbutane; $CH_3CH(CH_3)CH(CH_3)CH_3$, 2,3-dimethylbutane

25. Butane has an unbranched chain structure and methylpropane has a branched chain structure. The van der Waals forces bewteen butane molecules will be stronger than those in methylpropane because the butane molecules can lie closer together. More energy is needed to break the van der Waals forces in butane and the boiling point is higher.

26. Pent-1-ene;   Pent-2-en

```
H         H  H  H                 H           H  H
 \        |  |  |                 |           |  |
  C=C — C — C — C — H     H — C — C=C — C — C — H
 /        |  |  |                 |           |  |
H         H  H  H                 H           H  H
```

27. Propan-1-ol

```
      H  H  H
      |  |  |
H — C — C — C — OH
      |  |  |
      H  H  H
```

Propan-2-ol

```
      H  H  H
      |  |  |
H — C — C — C — H
      |  |  |
      H  OH H
```

28. They both have the same molecular formula, but propanal is an aldehyde with a CHO functional group. Propanone is a ketone with a C = O functional group.

29. a.

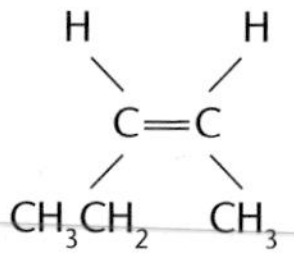

b.

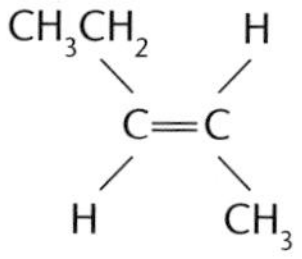

c.

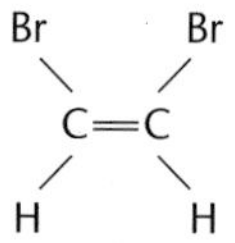

d.

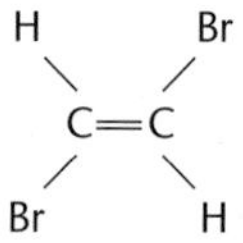

e.

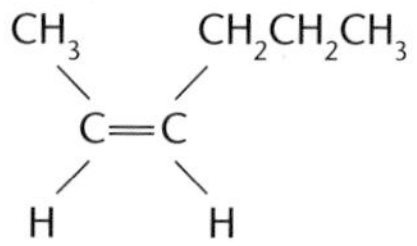

f.

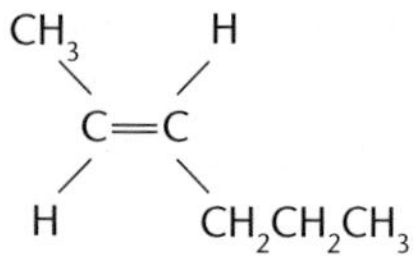

30. a. $C_2H_2Cl_2$

b.

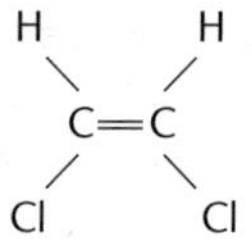

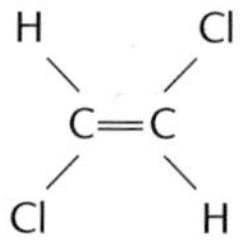

Z-1, 2-dichloroethene   E-1, 2-dichloroethene

31. The same molecular formula

32. (a), (c) and (d) are isomers – positional and chain isomers; (b) and (e) are isomers – chain and positional isomers.

6 THE ALKANES

1. a. LPG

b. petrol or naptha

c. petrol or naptha

2. For example: $CH_3CH(CH_3)(CH_2)_3CH_3$ (2-methylhexane) and $CH_3CH_2CH(CH_3)(CH_2)_2CH_3$ (3-methylhexane),

3. a. $CH_3CH_2CH_2CH_2CH_2CH_2CH_2CH_2CH_2CH_2CH_2CH_3 \rightarrow$ $CH_3CHCH_2 + CH_3CH_2CH_2CH(CH_3)$

b. $CH_2CH_2CH_2CH_3$, (b) $CH_3(CH_2)_2CH(CH_3)(CH_2)_3CH_3$

4. Vacuum distillation requires lower temperature and, therefore, cracking of the longer carbon chains at higher temperatures is prevented.

5. a. $2C_8H_{18}(l) + 25O_2(g) \rightarrow 16CO_2(g) + 18H_2O(g)$

b. $2C_8H_{18}(l) + 17O_2(g) \rightarrow 16CO(g) + 18H_2O(g)$

6. Shorter-chained alkanes have lower boiling points because fewer van der Waals forces operate along the length of the chains. They evaporate more easily.

7. Road dust contains valuable platinum dust from catalytic converters.

8. It must be unreactive and detectable quantitatively

9. To establish if any sulfur dioxide is in the atmosphere from sources other than power station emissions, for example, volcanic activity

10. a. Sulfur dioxide oxidised to sulfuric acid

b. Oxides of nitrogen oxidised to nitric acid

11. Equation 1 is propagation, Equation 2 is propagation

12. a. Initiation, $Cl_2 \rightarrow Cl\bullet + Cl\bullet$

b. Propagation,

$CH_2Cl_2 + Cl\bullet \rightarrow CHCl_2\bullet + HCl$

$CHCl_2\bullet + Cl_2 \rightarrow CHCl_3 + Cl\bullet$

c. Termination,

$CHCl_2\bullet + Cl\bullet \rightarrow CHCl_3$

13. a. Initiation, $Cl_2 \rightarrow Cl\bullet + Cl\bullet$

b. Propagation,

$C_2H_6 + Cl\bullet \rightarrow C_2H_5\bullet + HCl$

$C_2H_5\bullet + Cl_2 \rightarrow C_2H_5Cl + Cl\bullet$

c. Termination,

$C_2H_5\bullet + C_2H_5\bullet \rightarrow C_4H_{10}$

7 ENERGETICS

1.

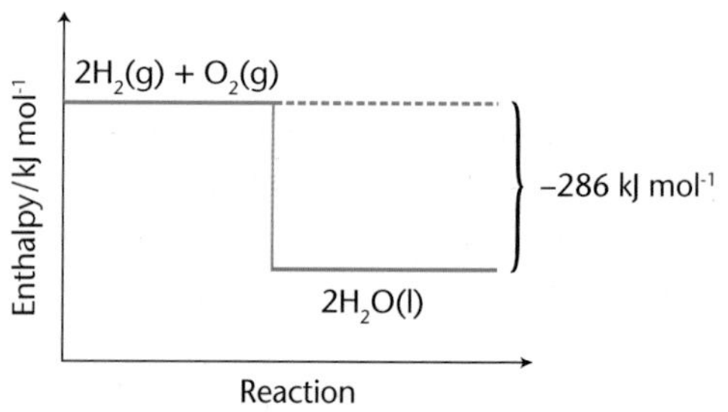

2. **a.** $H_2(g) + \frac{1}{2}O_2(g) \rightarrow H_2O(l)$

b. $4C(s) + 6H_2(g) + O_2(g) \rightarrow C_2H_5OH(l)$

c. $C_2H_5OH(l) + 3O_2(g) \rightarrow 2CO_2(g) + 3H_2O(l)$

d. $5C(s) + 6H_2(g) \rightarrow C_5H_{12}(l)$

e. $C_5H_{12}(l) + 8O_2(g) \rightarrow 5CO_2(g) + 6H_2O(l)$

f. $6C(s) + 6H_2(g) + 3O_2(g) \rightarrow C_6H_{12}O_6(s)$

g. $C_6H_{12}O_6(s) + 6O_2(g) \rightarrow 6CO_2(g) + 6H_2O(l)$

3. Percentage error in 0.76 g is $\pm 13\%$ (error on balance = ± 0.05 g and there are two balance readings); Percentage error in 15.5 °C = $\pm 3.2\%$ (error on thermometer = ± 0.5 °C); Percentage error in 200 cm^3 = $\pm 0.25\%$ (error on 200 cm^3 measuring cylinder = ± 0.5 cm^3)

4. Procedural errors due to heat loss.

5. Major error was energy loss due to heating the surroundings.

6. **a.** $+227$ kJ mol^{-1}

b. -273 kJ mol^{-1}

c. -84 kJ mol^{-1}

d. -110 kJ mol^{-1}

7. **a.** 86

b. i. 4841 kJ

ii. 48410 kJ

8. **a.** -1192 kJ mol^{-1}

b. -175.5 kJ mol^{-1}

c. -726 kJ mol^{-1}

9. -79 kJ mol^{-1}

10. **a.** 464.5 kJ mol^{-1}

b. Data book values include averages for all O−H bonds − they have different environments

11. **a.** -1060 kJ mol^{-1}

b. -1499 kJ mol^{-1}

c. -339 kJ mol^{-1}

d. -1987 kJ mol^{-1}

12. 3251 kJ mol^{-1}

13. $\Delta_c H^\theta$ values are calculated with all components in their standard state at the stated temperature. Bond enthalpies apply to gaseous molecules and vaporisation and condensation also involve energy changes.

8 KINETICS

1. Molecules are constantly colliding and energy is exchanged, causing variations in the kinetic energy of the molecules and a proportion end up with low kinetic energy.

2. Reaction with chlorine should be faster because it has a lower activation energy.

3. Low temperature means molecules in the food have less kinetic energy. Few have the minimum kinetic energy to overcome the E_A barrier and the rates of the decomposition reactions are low.

4. Molecules have more kinetic energy at room temperature, so a higher percentage of collisions have the necessary kinetic energy to react.

5. As the concentration of the reactants decreases, the concentration of the products increases.

6. **a.** $N_2(g) + 3H_2(g) \rightarrow 2NH_3(g)$

b. Molecules are closer together and so more concentrated. There are more effective collisions (ones where the collision energy is equal to or greater than the activation energy).

7. $n/V = concentration$. Rearranging the Ideal gas equation gives: $p = n/V \times RT$. R is the gas constant and if the temperature is constant, RT is a constant. So pressure, p, is proportional to the concentration.

8. Lead compounds bond strongly to the catalyst surface and cannot be easily removed. The catalyst is coated and no longer functions.

9. Catalysts do not lower the activation energy, they provide an alternative route that has a lower activation energy.

9 EQUILIBRIA

1. $CH_3COCH_3(g) \rightleftharpoons CH_3COCH_3(l)$

2. **a.** The rate of reaction slows to the point where it is hardly detectable

b. Two moles

c. The initial rate is the fastest for amounts of products and reactants, then the rate slows, and the reaction reaches equilibrium when the graph is horizontal.

3. Heterogeneous equilibrium

4. (a) and (c)

5. a. $K_c = \dfrac{[HI]^2}{[H_2][I_2]}$

b. $K_c = \dfrac{[NO_2]^2}{[NO]^2[O_2]}$

6. a. $K_c = 12.6$

b. $K_c = 5.10$

7. a. Equilibrium shifts to the forward direction, K_c increases

b. Equilibrium shifts to the reverse direction, K_c decreases

c. Equilibrium shifts to the reverse direction, K_c decreases

8. a. Equilibrium shifts to the left

b. No effect

c. Equilibrium shifts to the right

d. Equilibrium shifts to the left

e. Equilibrium shifts to the left.

9. a. i. a high temperature moves the position of the equilibrium towards the products and increases the rate of reaction

ii. a high pressure moves the position of the equilibrium towards the reactants as there are fewer moles (2 mol on left of equation, 4 mol on right); therefore, yield is higher at low pressures, but the rate is slower.

b. to increase the rate of reaction, which allows lower pressure to be used and, therefore, higher yield.

10 REDOX REACTIONS

1. a. i. $Na(s) + \frac{1}{2}Br_2(g) \rightarrow NaBr(s)$

ii. $Na \rightarrow Na^+ + e^-$
$\frac{1}{2}Br_2 + e^- \rightarrow Br^-$

iii. and **iv.** Na is oxidised, so is the reducing agent; Br is reduced, so is the oxidising agent.

b. i. $2Na(s) + \frac{1}{2}O_2(g) \rightarrow Na_2O(s)$

ii. $2Na \rightarrow 2Na^+ + 2e^-$ or $Na \rightarrow Na^+ + e^-$
$\frac{1}{2}O_2 + 2e^- \rightarrow O^{2-}$

iii. and **iv.** Na is oxidised, so is the reducing agent; O is reduced, so is the oxidising agent.

c. i. $Ca(s) + I_2(g) \rightarrow CaI_2(s)$

ii. $Ca \rightarrow Ca^{2+} + 2e^-$
$I_2 + 2e^- \rightarrow 2I^-$

iii. and **iv.** Ca is oxidised, so is the reducing agent; I is reduced, so is the oxidising agent.

d. i. $3Mg(s) + N_2(g) \rightarrow Mg_3N_2(s)$

ii. $3Mg \rightarrow 3Mg^{2+} + 6e^-$ or $Mg \rightarrow Mg^{2+} + 2e^-$
$N_2 + 6e^- \rightarrow 2N^{3-}$

iii. and **iv.** Mg is oxidised, so is the reducing agent; N is reduced, so is the oxidising agent.

2. a. $2K(s) + \frac{1}{2}O_2(g) \rightarrow K_2O(s)$

b. $2K \rightarrow 2K^+ + 2e^-$ and $\frac{1}{2}O_2 + 2e^- \rightarrow O^{2-}$

c. It donates electrons readily.

3. a. $Ca(s) + 2H_2O(l) \rightarrow Ca(OH)_2(aq) + H_2(g)$

b. $Ca \rightarrow Ca^{2+} + 2e^-$ and $2H_2O(l) + 2e^- \rightarrow 2OH^-(aq) + H_2(g)$

c. It donates electrons readily.

4. a. $+1$

b. $N = +5, O = -2$

c. $P = +5, Cl = -1$

d. $Al = +3, O = -2$

e. $S = +6, O = -2$

f. $P = +5, O = -2$

g. $Na = +1, Cl = -1$

h. $H = +1, N = +5, O = -2$

i. $Mn = +7, O = -2$

j. $K = +1, S = +6, O = -2$

k. $H = +1, P = +5, O = -2$

l. $Cr = +6, O = -2$

5. a. $+2$

b. $+3$

c. $+4$

6. a. 0

b. $+1$

c. -1

d. $+1$

e. $+3$

f. $+5$

g. $+7$

7. $+5$

8. $NO = +2, NO_2 = +4, N_2O_4 = +4$

9. a. Lithium is oxidised (0 to $+1$), chlorine is reduced (0 to -1)

b. Nitrogen is oxidised (0 to $+2$), oxygen is reduced (0 to -2)

c. Nitrogen is oxidised ($+2$ to $+4$), oxygen is reduced (0 to -2)

10. (c) is a redox reaction

11. One atom of chlorine is oxidised in $NaClO$ and the other atom of chlorine is reduced in $NaCl$.

12. Zinc loses electrons to form Zn^{2+} ions. Cu^{2+} ions gain electrons to form copper metal.

13. a. $Cu(s) + 2Ag^+(aq) + 2NO_3^-(aq) \rightarrow Cu^{2+}(aq) + 2NO_3^-(aq) + 2Ag(s)$, redox equation is $Cu(s) + 2Ag^+(aq) \rightarrow Cu^{2+}(aq) + 2Ag(s)$

  b. $Mg(s) + 2H^+(aq) + 2Cl^-(aq) \rightarrow Mg^{2+}(aq) + 2Cl^-(aq) + H_2(g)$, redox equation is $Mg(s) + 2H^+(aq) \rightarrow Mg^{2+}(aq) + H_2(g)$

  c. $Mg(s) + 2Ag^+(aq) + 2NO_3^-(aq) \rightarrow Mg^{2+} + 2NO_3^-(aq) + 2Ag(s)$, redox equation is $Mg(s) + 2Ag^+(aq) \rightarrow Mg^{2+} + 2Ag(s)$

14. K^+ and SO_4^-

15. $6Fe^{2+}(aq) + Cr_2O_7^{2-}(aq) + 14H^+(aq) \rightarrow 6Fe^{3+}(aq) + 2Cr^{3+}(aq) + 7H_2O(l)$

11 GROUP 2, THE ALKALINE EARTH METALS

1. a. $1s^2\ 2s^2\ 2p^6\ 3s^2\ 3p^6\ 4s^2$

  b. $1s^2\ 2s^2\ 2p^6\ 3s^2\ 3p^6\ 3d^{10}\ 4s^2\ 4p^6\ 5s^2$

2. a. The Mg atom has a larger nuclear charge and all the outer electrons are being added to the same shell

  b. Both have the same electron configuration, but Mg^{2+} has a larger nuclear charge.

3. Ca^{2+} has no electrons in the outer shell, so the remaining electrons are attracted strongly by the nuclear charge.

4. a. 20.2%

  b. 40.4%

5. a. Batch processes: advantage – products can be tracked; disadvantages – are more expensive to operate because they are not productive continually, lower rate of production

  b. Continuous processes: advantage – cheaper to operate, higher production rate; disadvantage – products cannot be tracked

6. $+2$

7. a. $Sr(s) + 2H_2O(l) \rightarrow Sr(OH)_2(aq) + H_2(g)$

  b. Oxidation: $Sr \rightarrow Sr^{2+} + 2e^-$,
    Reduction: $2H_2O + 2e^- \rightarrow 2OH^- + H_2$

8. $Mg(OH)_2$ is a solid, and the rest are in solution.

9. $Ba(OH)_2(s) + aq \rightarrow Ba^{2+}(aq) + 2OH^-(aq)$

10. a. Chloride ions and sodium ions

  b. The ions remain in solution

  c. Sulfuric acid contains sulfate ions.

12 GROUP 7(17), THE HALOGENS

1. a. below 2.5

  b. above 184 °C

2. a. $Cl_2(aq) + 2I^- \rightarrow I_2(aq) + 2Cl^-(aq)$

  b. $Cl_2 + 2e^- \rightarrow 2Cl^-$; $2I^- \rightarrow I_2 + 2e^-(aq)$

3. a. Because the nuclear charge increases down Group 7(17) as the number of inner electron shells increases, so the attraction is less; distance of nucleus to outer electrons increases, so attraction is less.

  b. Electron affinity can be quantified from experimental data. It is an objective value with units. Electronegativity is an arbitrary value with no units.

4. Oxidation state of halide = -1 in both reactant and product; Oxidation state of sulfur = $+6$ in both reactant and product.

5. a. Br^- is oxidised

  b. S is reduced

  c. Br^- is the reducing agent

  d. S is the oxidising agent

6. $2Br^- \rightarrow Br_2 + 2e^-$; $4H^+ + SO_4^{2-} + 2e^- \rightarrow SO_2 + 2H_2O$

7. $8I^- \rightarrow 4I_2 + 8e^-$; $10H^+ + SO_4^{2-} + 8e^- \rightarrow H_2S + 4H_2O$

8. a. 2,8

  b. 2,8,8

9. neon and argon, respectively

10. a. sodium fluoride

  b. sodium bromide

  c. sodium chloride

  d. sodium iodide

11. Carbonate ions – silver carbonate is almost insoluble. $CO_3^{2-}(aq) + 2H^+(aq) \rightarrow CO_2(g) + H_2O(l)$

12. Litmus turns red with HCl, and bleaches with HOCl.

13. In the reaction with cold dilute sodium hydroxide, one Cl in Cl_2 changes from 0 to $+1$ and the other from 0 to -1. In the reaction with concentrated sodium hydroxide solution, one Cl changes from 0 to $+5$ and the other from 0 to -1

14. $Br_2(g) + 2OH^- \rightarrow BrO^- + Br^- + H_2O$

13 HALOGENOALKANES

1. **a.** bromoethane

 b. 1,2-dibromoethane

 c. 1,1,1-trichloropropane

 d. 1,1,2-tribromoethane

 e. 1,1-dibromo-2-chloropropane

 f. pentafluoroethane

2. **a.** I, C, H

 b. F, C, H

 c. Cl, F, C, H

3. C and I have the same electronegativity.

4. The molecules are polar because they have one polar bond. However, CX_4 molecules have four polar bonds arranged symmetrically, the net effect being that the molecule itself is non polar.

5. It decreases.

6. **a.** **i.** 18

 ii. 62

 b. iodomethane

7. Since permanent dipoles get weaker down Group 7(17) for halogenoalkanes, boiling points could be expected to decrease. Since they increase, van der Waals forces must become strong enough to compensate for the weakening dipoles.

8. **a.** 1-bromobutane: $CH_3CH_2CH_2CH_2Br + 2NH_3 \rightarrow CH_3CH_2CH_2CH_2NH_2 + NH_4Br$

 b. bromoethane: $CH_3CH_2Br + 2NH_3 \rightarrow CH_3CH_2NH_2 + NH_4Br$

9. **a.** $CH_3Br + CN^- \rightarrow CH_3CN + Br^-$

 b. The nucleophile CN^- is attracted to the positive charge of the carbon of CH_3Br. CN^- donates a pair of electrons to form a covalent bond with carbon. The C–Br bond breaks and Br^- forms.

10. **a.** 3

 b. 4

11. **a.** nucleophilic substition;

 propan-1-ol + chloride ions

 b. elimination; propene + water + chloride ions

12. **a.** Initiation, $Br_2 \rightarrow Br\bullet + Br\bullet$; Propagation, $O_3 + Br\bullet \rightarrow O_2 + BrO\bullet$; $BrO\bullet + O_3 \rightarrow 2O_2 + Br\bullet$

 b. The reaction catalysed by $Br\bullet$ has a lower activation energy.

13. **a.**

$$H-\underset{\underset{F}{|}}{\overset{\overset{Cl}{|}}{C}}-F \qquad CHClF_2$$

 b.

$$H-\underset{\underset{F}{|}}{\overset{\overset{F}{|}}{C}}-F \qquad CHF_3$$

14. **a.**

$$F-\underset{\underset{F}{|}}{\overset{\overset{F}{|}}{C}}-\underset{\underset{H}{|}}{\overset{\overset{F}{|}}{C}}-H$$

 b. CF_3CH_2F

15. **a.** $H-\ddot{\overset{\bullet}{O}}$

 b. It has an unpaired electron.

16. Chlorine gas from swimming pools is very reactive and reacts in the troposphere. CFCs are unreactive and can diffuse to the stratosphere. CFCs require UV radiation from the Sun to form reactive radicals. This radiation reaches the stratosphere, but not the lower troposphere.

17. Antarctica is closest to the Sun in spring and the amount of UV radiation is higher.

14 ALKENES

1. C_nH_{2n-2}

2. $C = C$ consists of π and σ bonds and σ bonds are stronger than π bonds.

3. The carbon atoms with the double bond.

4. The overlapping p orbitals shorten the bond length.

5. Electrons from the π bond

6. The H–Cl bond is stronger than the H–Br bond and more energy is needed to break it.

7. **a.** $HI(aq) + CH_2CH_2(l) \rightarrow CH_3CH_2I$; iodoethane

 b. Electrophilic addition. The H ion from HI has a positive charge. H^+ acts as an electrophile, and the electron density of the double covalent bond in ethene attracts the positive H^+.

8. A nucleophile has a lone pair of electrons, an area of negative charge, and is attracted by areas of positive charge. An electrophile has a positive charge and is attracted to areas with a negative charge.

9. **a.** $C_2H_5OSO_3H + H_2O \rightarrow C_2H_5OH + H_2SO_4$

   b. Acts as a catalyst

   c. React with concentrated sulfuric acid followed by reaction with dilute sulfuric acid (add water).

10. Z-but-2-ene and 2,3-dichloro-Z-but-2-ene are symmetrical; 2-chloropropene and 2-methyl-Z-pent-2-ene are unsymmetrical.

11. **a.** 2-chloropropane

   b. 2-bromopentane

   c. 2-bromobutane

12.

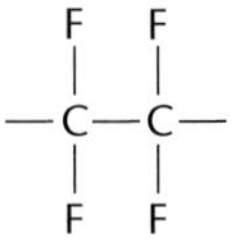

13. **a.** Branched chains of LDPE cannot lie close together, so fewer van der Waals forces form. Less energy is needed to overcome the intermolecular forces so LDPE softens at a lower temperature.

   b. Melting point of alkanes depends on the length of the carbon chain. Poly(ethene) contains a range of long-chained alkanes with a range of melting points.

15 ALCOHOLS

1. **a.** Propan-1-ol

   b. Pentan-2-ol

   c. Butan-2-ol

   d. Methanol

2. **a.** Primary

   b. Secondary

   c. Primary

   d. Primary

   e. Secondary

3. **a.**

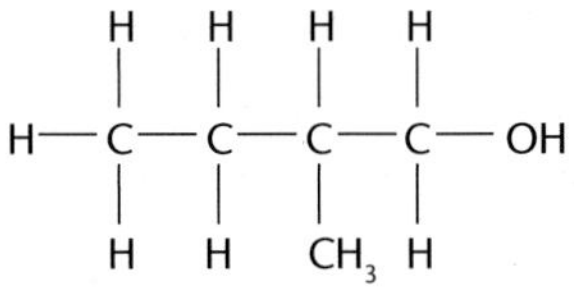

   b.

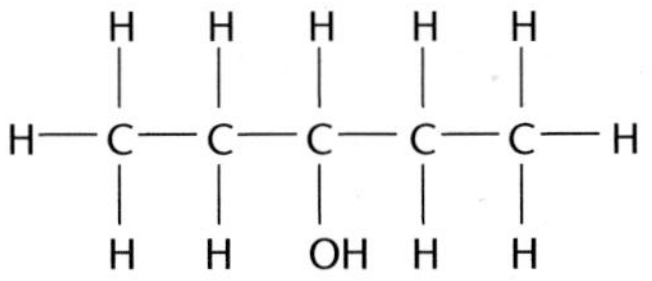

   c.

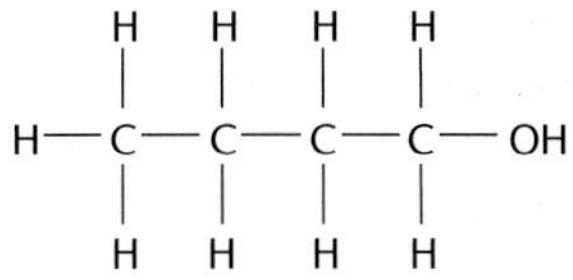

   d.

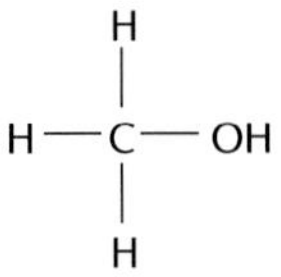

   e.

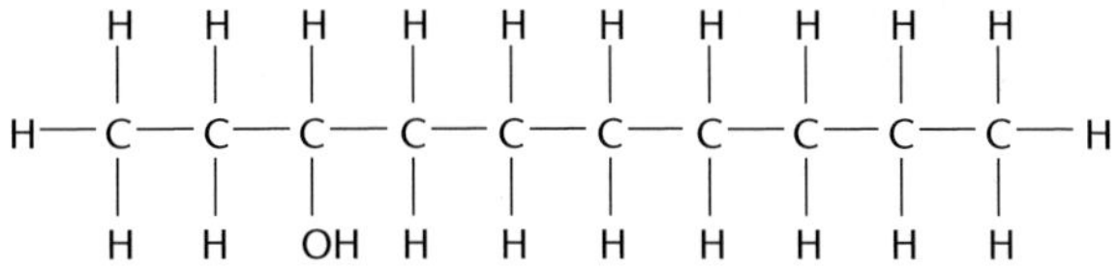

4. Primary alcohol is:

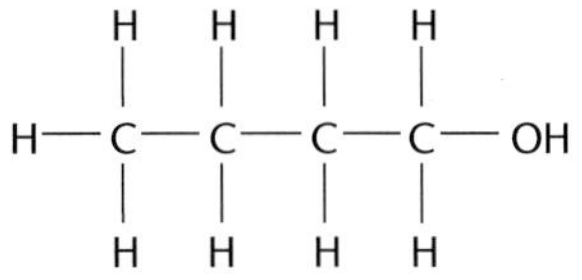

Secondary alcohol is:

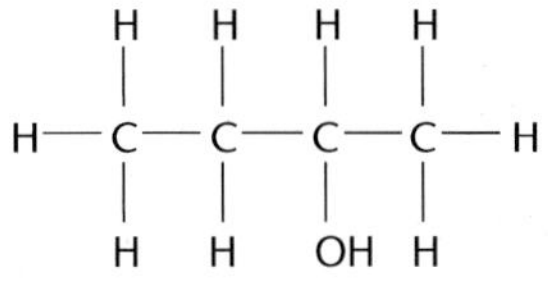

Tertiary alcohol is:

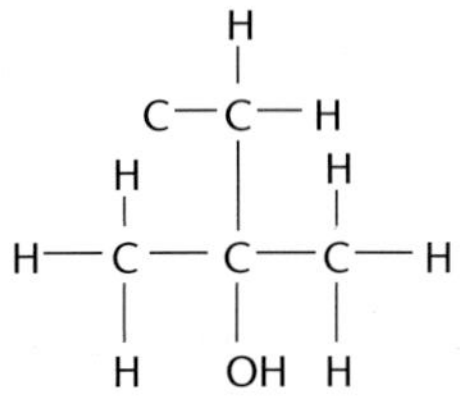

5. **a.** 1:1

   b. More steam is needed because phosphoric(V) acid will absorb some as it is soluble in water

6. **a.** A lower temperature moves the position of the equilibrium in the forward direction, but also decreases the rate of reaction, so a compromise temeprature is used.

   b. Moves the position of equilibrium in the forward direction since two moles of reactants form one mole of products.

   c. Lower pressures have to be used to prevent formation of poly(ethene) in the reaction vessels.

   d. Increases the rate of reaction so equilibrium reached quicker; has no effect on the position of equilibrium.

rew impurities in ethanol produced from ethene and steam are toxic; other substances present in fermentation mixtures (for example, from grapes) provide flavour.

b. It doesn't make sense to manufacture ethanol from ethene since ethene is produced from crude oil and you might as well use petrol. Producing fuels from plants is politically and environmentally acceptable at the moment.

8. The percentage will increase.

9. Land will not be used for biofuel crops that is needed for food production.

10. Carbon dioxide is removed from the atmosphere during photosynthesis and replaced when the fuel is burnt. The only greenhouse gas emissions come from the use of fossils fuels to provide the energy needed for its production.

11. Enzyme digestion is needed to convert the starch in wheat to simple sugars. The enzyme catalysed reaction takes place at a lower temperature and so requires less energy to be transferred to it.

12. a. butan-1-ol primary, butan-2-ol secondary

b. $CH_3CH_2CH_2CH_2OH + [O] \rightarrow$
$CH_3CH_2CH_2CHO + [O] \rightarrow CH_3CH_2CH_2COOH$

c. $CH_3CH_2CH(OH)CH_3 + [O] \rightarrow CH_3CH_2COCH_3$

13. The primary alcohol oxidises to an aldehyde, which in turn is oxidised by Fehling's solution to produce a carboxylic acid and a brick-red precipitate. The secondary alcohol oxidises to a ketone which has no reaction with Fehling's solution.

14. a. enzymes are used to convert the starch to sugar

b. fermentation

c. elimination

d. polymerisation

15. a. $CH_3CH_2CH_2OH \rightarrow CH_3CHCH_2 + H_2O$

b. Propene

c.

16 ANALYTICAL TECHNIQUES

1. The positively charged end of a bromine molecule with an induced dipole.

2. a. 1,2-dibromohexane

b. 1,2-dibromobutane

3. $RX + OH^- \rightarrow ROH + X^-$

4. a. R–O–H

b.

c.

d.

5. a. secondary alcohol

b. aldehyde

c. carboxylic acid

d. tertiary alcohol

e. ketone

6. a. pH < 7

b. Acidified potassium dichromate(VI) solution: green; Fehling's test: blue

c. Warm with NaOH, add acidified silver nitrate solution – cream precipitate

d. As b. but Fehling's test positive

e. Add bromine water – turns colourless

f. Add acidified potassium dichromate(VI) solution – green,

7. a. Add acidified potassium dichromate(VI) solution

b. Warm with NaOH, add acidified silver nitrate solution

c. Add bromine water

d. Warm with Fehling's solution or Tollen's reagent

8. a. $[CH_3CHO]^+$, 44

b. $[CH_3COOH]^+$, 60

9. a. 72

b. 72

10. a.

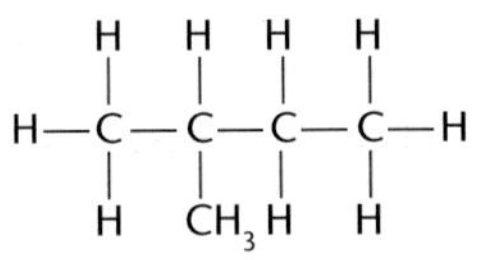

$$H—C—C—C—C—H$$

(with H, H, H, H on top; H, CH_3, H, H on bottom)

b. i. $m/z = 72$, $[CH_3CHCH_3CH_2CH_3]^+$

ii. the molecular mass

c. Molecular ions containing isotopes of carbon and/or hydrogen will give additional molecular ion peaks due to the presence of carbon-13 and hydrogen-2. Similarly, the fragmentation pattern will be similar, but with many more peaks due to the ions containing carbon-13 and hydrogen-2. However, the relative abundance of such peaks will be very low.

11. a. 16

b. 16.0312

12. NO at 29.9980, HCHO at 30.0095

13. There are no absorptions at all in either spectra.

14.

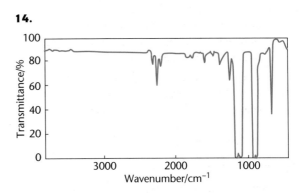

Graph should show absorptions between 1000 and 1400 cm^{-1} for the C–F bond and between 600 and 800 cm^{-1} for the C–Cl bond. Other peaks not needed.

Absolute zero The temperature $-273\,°C$ or $0\,K$, at which the volume of a gas is theoretically zero and has no energy.

Actinides The 14 elements, Z 90 to 103, which follow actinium, Z 89.

Activation energy The minimum energy required by particles in collision to bring about a chemical reaction.

Addition polymers Unreactive polymers formed from alkenes and substituted alkenes.

Adsorption The binding of atoms, ions or molecules to a surface.

Alkane A hydrocarbon with the general formula C_nH_{2n+2}. The first three alkanes are CH_4, C_2H_6 and C_3H_8.

Alkene A hydrocarbon with the general formula C_nH_{2n}. Alkene molecules contain double bonds. The first two alkenes are C_2H_4 and C_3H_6.

Alkyl group A group of atoms with the general formula C_nH_{2n+1} forming part of a molecule.

Amine An organic compound containing the functional group $-NH_2$.

Amorphous Describing the structure of a solid where the particles are not arranged in a fixed pattern.

Atomic number, symbol Z The number of protons (and therefore of electrons) in an atom.

Atoms Single units of an element.

Bond enthalpy The amount of energy absorbed when a bond is broken. It equals the amount of energy released when the same bond is made. It is measured as the energy in kilojoules per mole of bonds.

Bonding pair A shared pair of electrons in a covalent bond.

Bromine number A measurement of the degree of unsaturation of an oil.

Cahn–Ingold–Prelog The rules that decide the priorities when naming compounds.

Calorimeter The apparatus used to measure temperature changes when energy is transferred during a chemical reaction (calorimetry).

Calorimetry This is a method used to measure enthalpy changes in chemical reactions.

Carbocation A group in which a carbon atom bonded to three other atoms has an unpaired electron at the position of the fourth bond that gives the group a positive charge.

Catalyst A substance that usually speeds up the rate of a chemical reaction. It changes the nature of the intermediate compounds formed between reactants and products and so reduces the activation energy. A catalyst remains chemically unchanged at the end of the reaction.

Catalytic converters Devices attached to cars that reduce the emissions of carbon monoxide and oxides of nitrogen.

Catalytic cracking The process which uses high temperature and a catalyst to break up long-chain hydrocarbon molecules from crude oil into smaller, more useful molecules such as ethene.

Chemical equation A chemical reaction expressed in chemical symbols showing reactants and products. The equation shows the number and type of molecules or atoms that react together and are formed.

Chemical formula This shows how many atoms of each element combine together to make a substance.

Closed system A system from which reactants and products cannot escape, and to which substances cannot be added. A chemical equilibrium is possible only in a closed system.

Co-ordinate bond A covalent bond in which the shared electron pair originates from the same atom. It can be written as $X{\rightarrow}Y$, showing that the shared electron pair originated from X.

Collision energy The combined kinetic energy of two colliding molecules.

Collision theory A theory to explain how chemical reactions occur, which states that for a reaction to happen, particles have to collide with a minimum amount of energy.

Complete combustion The exothermic reaction between a fuel and excess oxygen. The oxidation products contain the maximum amount of oxygen (highest oxidation state).

Compromise pressure The pressure used in the Haber process, between 5000 and 20000 kPa.

Compromise temperature Refers to the actual temperature used in an equilibrium reaction to give the best economical conditions between the rate and the equilibrium yield.

Covalent bond A bond in which two atoms share one or more pairs of electrons. A hydrogen molecule, $H–H$, has a single covalent bond. In a double covalent bond there are two shared pairs of electrons, such as in the oxygen molecule, $O{=}O$.

Covalent radius Half the shortest distance from one nucleus to the next in a covalently bonded molecule.

Cracking The process of breaking up the long-chain hydrocarbons in crude oil into shorter-chain hydrocarbons that can be used in the chemical industry and for fuels.

Crystalline Describing the structure of a solid where the different types of particle are arranged in a fixed repeating pattern.

D block The elements in the Periodic Table between Groups 2 and 3 in Periods 4, 5 and 6, known as the transition elements. In each period, a d orbital within an outer s orbital is being filled.

Dative covalent bond Another name for a co-ordinate bond.

Dehydration An elimination reaction in which an alcohol loses a molecule of water to form an alkene.

Delocalised electrons Electrons that are not located at one particular atom, but are free to move between all atoms in the structure.

Displacement reaction A chemical reaction in which an atom in a compound is replaced by another atom, e.g. chlorine displaces bromine in potassium bromide.

Disproportionation A reaction in which a substance is simultaneously oxidised and reduced.

Dissociation The process in which molecules or ionic compounds split into smaller particles.

Dot-and-cross diagram A drawing representing the electrons in a molecule. Dots and crosses are used to indicate their atom of origin.

Double covalent bond A bond in which four electrons are shared (two bonding pairs).

Dynamic equilibrium A stage in a reaction where rate of the forward reaction equals the rate of the backward reaction so that there is no net change in the concentration of the substances involved in the reaction.

***E-Z* stereoisomers** Isomers resulting from different spatial arrangements of functional groups relative to a double bond.

Electron A negatively charged particle. Electrons orbit the atomic nucleus in energy levels. Atoms of different elements have different numbers of electrons. In a neutral atom the number of electrons is always equal to the atomic number of the element.

Electron configuration The arrangement of electrons in an atom, written in number and letter symbols.

Electron density A measure of the probability of an electron occurring in a specific location in an atom or molecule.

Electron pair Bonding electron pairs occur in a covalent bond between two atoms, and include one electron from each atom, except in the case of a coordinate bond where they come from one atom only. Non-bonding electron pairs, or lone pairs, take no part in bonding.

Electron shell A group of electrons with similar energy levels surrounding the nucleus of an atom.

Electronegativity The tendency of an atom to gain or retain electrons. Elements whose atoms gain electrons easily are the most electronegative. Fluorine is the most electronegative element. Elements whose atoms lose outer electrons easily are described as being electropositive. Caesium is the most electropositive element.

Electrophiles Molecules and groups attracted by regions of negative charging (the name means 'electron-loving').

Electrophilic addition Addition reactions involving and electrophile.

Electrostatic attraction The attraction between particles of opposite charge.

Emission spectra A characteristic set of frequencies of radiation given out by excited electrons as they fall back to the ground state. Each element has its own unique and characteristic emission spectrum.

Empirical formula The simplest formula of a compound, showing the ratio of the numbers of atoms in the molecule, e.g. CH_3 is the empirical formula of ethane, C_2H_6.

Energy profile A diagram showing the energy changes in a chemical reaction.

Enthalpy change An amount of energy that is transferred (absorbed or released) during a chemical reaction at a constant pressure.

Enthalpy Energy associated with a chemical reaction.

Equilibrium constant The ration of concentrations when equilibrium is reached in a reversible reaction.

Equilibrium The state reached in a reversible reaction at which the rates of the two opposing reactions are equal, when the system has no further tendency to change. This is a dynamic equilibrium, as reactants and products are both still being formed, but at equal rates.

Ethene The main alkene produced by the process of thermal cracking.

Exothermic reaction A chemical reaction in which energy is released.

F block The elements of the Periodic Table based on the lanthanide and actinide series, in which electrons are filling an f orbital within outer shells.

Fingerprint region The region of the infrared spectrum between 500 and 1500 cm^{-1}.

First ionisation energy The energy required to move the first electron from an atom in its gaseous state.

Flue gas desulfurisation The process of treating waste gases with calcium oxide or calcium carbonate in order to remove sulphur dioxide.

Formula unit A way of showing the ratio of each type of atom in the lattice structure of a substance.

Free-radical substitution A reaction mechanism involving an initiation stage, in which free radicals are formed, a propagation stage and a termination stage.

Functional group A reactive atom or group of atoms in an organic molecule that largely determines the properties of the molecule.

Half-equations Equations representing the oxidation or reduction reaction in a chemical reaction.

Halide ions Halogen atoms that have a noble gas electron configuration and a negative charge.

Halides Compounds containing a halogen atom. They can be ionically bonded, as in metal halides, e.g. NaCl, or covalently bonded, as in non-metal halides, e.g. HCl or PCl_5.

Halogenoalkane An alkane molecule in which one or more of the hydrogen atoms are replaced by a halogen atom.

Hess's law If a chemical change can occur by more than one route, then the overall enthalpy change for each route must be the same, provided that the starting and finishing conditions are the same.

Heterogeneous equilibrium An equilibrium in which the reactants and products are in different states.

Heterolytic bond fission The breaking of a single covalent bond (a bonding pair of electrons) so that the two electrons remain on one atom. Ions are formed: the atom taking the two electrons is negatively

...he atom with no
... from the bond is positively
...ged.

Homogeneous equilibrium An equilibrium in which all the reactants are in the same state.

Homologous series A series of organic compounds with the same general formula, each successive member of the series having one more carbon atom in its molecule, e.g. the alkanes, C_nH_{2n+2}.

Homolytic bond fission The breaking of a single covalent bond (a bonding pair of electrons) so that one electron remains with each atom. The species formed are called radicals and each has an unpaired electron.

Hydration The addition of water. In organic chemistry examples include the conversion of alkenes into alcohols, and in inorganic chemistry the formation of aqua ions by the addition of water ligands to metal ions.

Hydrogen bonding The intermolecular bonding between dipoles in adjacent molecules in which hydrogen is bonded to a strongly electronegative element, e.g. intermolecular hydrogen bonding in H_2O, NH_3 and HF.

Hydrogen bonds A permanent dipole–permanent dipole attraction – the strongest intermolecular force.

Ideal gas A gas made up of particles of zero size, with no forces acting between them. Ideal gases exist only in theory, not in practice, but nitrogen, oxygen, hydrogen and the inert gases behave like ideal gases at high temperature and low pressure. Under these conditions they obey the ideal gas equation.

Ideal gas equation A mathematical description of the relationship between volume, temperature and pressure for an ideal gas: $pV = nRT$ where pressure p is measured in N m^{-2} (Pa), volume V is measured in m^3, n is the number of moles of gas, R is the molar gas constant with the value of 8.314 J K^{-1} mol^{-1} and temperature T is measured in K.

Incomplete combustion Combustion reactions in which there is insufficient oxygen for full oxidation, e.g. incomplete combustion of

hydrocarbons produces carbon monoxide instead of carbon dioxide.

Induced dipole An uneven charge distribution in a particle (molecule or atom), occurring when a charge in an adjacent particle causes movement of electrons within the particle. One region of the particle becomes negative and another part positive: the particle is said to be polar.

Initiation The first stage in a free radical substitution mechanism in which radicals are formed.

Inorganic chemistry The study of compounds that do not contain carbon. (Carbon dioxide and carbon monoxide are the usual exceptions.)

Inorganic compounds Compounds that that are made from atoms other than carbon.

Instantaneous dipole A dipole (molecule or ion with a positive charge in one part and a negative charge in another) that lasts for only an instant.

Intermolecular forces Forces that are able to exist between molecules.

Ionic bond A chemical bond between two ions of opposite charge.

Ionic crystals Crystals consisting of positively and negatively charged ions held together in a regular arrangement by their electrostatic charges.

Ionic equations A concise method of writing down the important changes to the ions in a chemical reaction.

Ionisation energy The energy required to remove one mole of electrons from one mole of atoms of an element to a distance at which the electrons are no longer influenced by the positive charge of the nuclei.

Ionisation region Part of the mass spectrometer where gaseous atoms of the sample are bombarded by electrons from an electron gun and positively charged ions are formed.

Irreversible reaction A reaction where the equilibrium lies to the right, i.e. the reactants react to make the products, but the products do not react to make the reactants.

Isomers Compounds that have the same molecular formula but different structural formulae.

Isotope signature The ratio of isotopes of an element in a sample that is characteristic for a particular set of conditions. For example, organically produced vegetables have different ratios of nitrogen isotopes to non-organically produced vegetables.

Isotopes Atoms that have the same atomic number but different mass numbers. They are atoms of the same element, with the same numbers of protons and electrons but different numbers of neutrons, e.g. isotopes of chlorine are written as chlorine-35 and chlorine-37 (or ^{35}Cl and ^{37}Cl), where 35 and 37 are mass numbers.

Kinetic theory Also called the kinetic molecular model, it describes the movement due to kinetic energy of particles in solids, liquids and gases.

Lanthanides The 14 elements, Z 58 to 71, which follow lanthanum, Z 57.

Mass spectrometer A scientific instrument used to produce a mass spectrum.

Maxwell–Boltzmann distribution For a sample of gas at a particular temperature, it represents the number of gas molecules at each energy over the whole range of energies present, and can be represented by a curved graph.

Mean bond enthalpy The average enthalpy change when one mole of bonds of the same type are broken in gaseous molecules under standard conditions.

Metallic bonding The bonding found in metals. The metal ions are attracted to the delocalised electrons and this attraction holds the structure together.

Metallic radius Half the distance between the nuclei of two metal ions in a solid metal.

Molar enthalpy of combustion The amount of energy released by the complete combustion of one mole of a substance.

Mole An amount of substance that contains 6.022×10^{23} particles. The particles may be atoms, ions, molecules or electrons.

Molecular formula A formula showing the number and types of atoms present in a molecule, e.g. the molecular formula for calcium carbonate is $CaCO_3$.

Molecular ion A molecule that has lost one or more electrons.

Molecular ion peak The peak on a mass spectrum which represents the molecular mass of the sample. It is the heaviest ion produced in the mass spectrometer.

Molecule A particle containing two or more atoms bonded together.

Monomers Molecules that can react with many other similar molecules to build up a large molecule, a polymer, e.g. the monomer ethene gives rise to the polymer poly(ethene).

Neutron A neutral (uncharged) particle in the atomic nucleus. Its mass is approximately 1 atomic unit.

Nitrile An organic compound containing a CN group.

Nuclear charge The positive charge on the nucleus.

Nucleon A particle in the nucleus of an atom.

Nucleophiles Atoms or groups of atoms that are attracted to a positive charge, e.g. NH_3, $-OH$ and H_2O.

Nucleophilic substitution A chemical reaction in which one nucleophile replaces another in a molecule. For example, in the reaction of bromoethane with alkali, the nucleophilic hydroxide ion replaces the bromide ion in the bromoethane molecule: $C_2H_5Br + OH^- \rightarrow C_2H_5OH + Br^-$.

Nucleus The central part of an atom occupying a tiny fraction of its volume, with electrons orbiting round it. The nucleus consists of tightly packed positively charged protons and neutral neutrons.

Orbital The region in an atom where an electron is most likely to be found.

Organic compounds Compounds made from carbon atoms, plus other atoms such as hydrogen, oxygen and the halogens.

Oxidation A process in which a species loses one or more electrons; also defined as an increase in oxidation state of an element. Oxidation and reduction occur together in a redox reaction.

Oxidising agent An element or compound that gains electrons from a reducing agent which itself loses electrons. The oxidising agent is reduced, and the reducing agent is oxidised.

Oxyanion An anion made from atoms of two elements, one of which is oxygen.

P block Elements of the Periodic Table in Groups 3 to 0.

Periodicity The repeating pattern of trend in properties of the elements across a period.

Permanent dipole A molecule in which one part always has a slightly positive charge, and another always has a slightly negative charge.

Photochemical reaction A chemical reaction triggered by the absorption of light energy.

Polar bond A covalent bond between two atoms in which electrons are not shared equally, giving one atom a slight positive charge and one a slight negative charge.

Polymerisation The reaction of monomers to form polymers.

Position of equilibrium A way of expressing the proportion of products to reactants in a reaction, referring to the ratio of quantities indicated in a chemical equation.

Potential energy The energy in a particle derived form its position rather than its motion.

Precipitate An insoluble (solid) product formed from a reaction in solution.

Primary distillation The first distillation of crude oil to obtain fractions containing a mixture of hydrocarbons.

Propagation The second stage in a free radical substitution mechanism in which a radical is included in the reactants and the products.

Proton A positively charged particle in the atomic nucleus. Its mass is approximately 1 atomic unit.

Quantised Having fixed energy values (describing energy levels in an atom).

Quantum theory A theory of matter and energy which states that energy exists as small units called quanta.

Rate curves The curves of a graph showing rate of reaction as the concentration of a product or reactant against time. The gradient on the curve at a particular point on the curve will give the rate of reaction at that time.

Redox reactions Reactions in which oxidation and reduction both occur. One species is oxidised, while another is reduced. The two processes can be shown as half-equations.

Reduction A process in which a species gains one or more electrons. It can also be defined as a decrease in oxidation state for an element. Reduction and oxidation occur together in a redox reaction.

Relative atomic mass, symbol A_r The mass of one atom of an element compared with one-twelfth of the mass of one atom of carbon-12.

Relative isotopic abundance The amount of each isotope of an element as it occurs naturally, expressed as a percentage or proportion (ratio).

Relative mass A way of describing the mass of atomic particles: the relative mass of a proton is 1, a neutron is 1 and an electron is 5.45×10^{-4}.

Relative molecular mass, symbol M_r The mass of one molecule of an element or compound compared with one-twelfth of the mass of one atom of carbon-12.

Repeat unit The section of a polymer chain that is added when one molecule of the monomer is added.

Residence time The amount of time the reactants experience the reacting conditions in the reaction vessel.

Reversible change A chemical reaction which can take place in both directions and so does not go to completion. A mixture of reactants and products is obtained when the reaction reaches equilibrium. The composition of the equilibrium mixture is the same whether the reaction starts from the substances on the left or the right of the reaction equation.

S block Elements of the Periodic Table in Groups 1 and 2.

Saturated Describing an organic compound that contains the maximum number of hydrogen atoms possible. It contains only single bonds between the carbon atoms.

Saturated hydrocarbons Alkanes – compounds containing only single bonds.

Secondary distillation Distillation of the fractions obtained from primary distillation of crude oil. This stage can separate mixtures into pure substances.

Solute A substance that dissolves in another, usually a liquid.

Solution A solute dissolved homogeneously in a solvent.

Solvent The liquid component of a solution, such as water or alcohol.

Specific heat capacity The amount of energy in joules required to raise the temperature of 1 g of a substance by 1 K.

Spectator ions Ions present in a solution that do not take part in the reaction.

Spin diagram A diagram which shows the direction of spin of electrons in an atom.

Square planar structure The shape of a molecule with four groups around the central atom. All angles are 90° and the structure has a flat shape.

Stable isotopes Isotopes that are not radioactive.

Standard enthalpy of combustion, symbol $\Delta H_c°$ – The energy transferred when one mole of a substance burns completely in oxygen under standard conditions.

Standard enthalpy of formation, symbol $\Delta H_f°$ – The energy absorbed when one mole of a substance is formed from its elements in their standard states.

Standard form A way of expressing large numbers as a number between one and 10.

State The form that a substance takes: solid, liquid, gaseous or aqueous.

Stem The main part of the name of an organic compound.

Stereoisomers Isomers that have the same molecular formula and structural formula but different arrangements of their atoms in space. Optical and *E-Z* geometric isomers are stereoisomers.

Structural isomers Isomers having the same molecular formula, but different structural formulae. Butane and 2-methylpropane are structural isomers of C_4H_{10}.

Successive ionisation energies The energy values for removing the second and subsequent electrons from an atom in its gaseous state.

Surroundings In chemistry, the term is used to mean anything outside a reaction mixture.

System In chemistry, the term is used to mean the reaction mixture.

Termination The third and final stage in the free radical substitution mechanism where two radicals combine to produce a molecule and the reaction is terminated.

Thermal cracking Using heat to break C–C and C–H bonds in the cracking of crude oil.

Time-of-flight mass spectrometer A mass spectrometer that determines an ion's mass-to-charge ration using a time measurement.

Titration An experimental procedure to measure a precise amount of one substance of a known concentration which reacts exactly with a measured amount of another.

Titre The volume of reactant added from the burette during a titration.

Transition elements Elements whose atoms form ions with an incomplete d-sub-shell. Most d block metals are transition metals.

Transition state In a chemical reaction when bonds are being broken and new bonds are being formed, the transition state is an intermediate form between reactants and products.

Unbranched alkanes Alkanes made up of straight chains of carbon.

Unsaturated hydrocarbons Compounds that contain one or more double bonds between the carbon atoms.

Vacuum distillation Secondary distillation of the fraction of crude oil that boils above 350 °C, carried out under reduced pressure to lower the boiling points.

Volumetric analysis A titration method used to determine the unknown concentration of a reactant by measuring the exact volume of each reactant needed for a reaction. The concentration of one reactant must be known.

INDEX

ACKNOWLEDGEMENTS

The Publishers gratefully acknowledge the permissions granted to reproduce copyright material in this book. Every effort has been made to contact the holders of copyright material, but if any have been inadvertently overlooked, the Publisher will be pleased to make the necessary arrangements at the first opportunity.

Practical work in chemistry

p1 left: bikeriderlondon/Shutterstock, right: Lisa S./Shutterstock; p2 top: bogdanhoda/Shutterstock, bottom: phloxii/Shutterstock; p3 top: Decha Thapanya/Shutterstock, bottom: Yenyu Shih/Shutterstock

Chapter 1

p4: background NASA/JPL/Caltech/Public Domain; p10: Bobyramone/Shutterstock; p13: NASA/JPL-Caltech/Univ. Of Arizona/Public Domain; p14 left: Peter Bernik/Shutterstock, right: Jens Goepfert/Shutterstock; p16: Science Photo Library/Alamy

Chapter 2

p30 background: NattapolStudiO/Shutterstock; p34: Andrew Lambert Photography/Science Photo Library; p35: Graham J. Hills/Science Photo Library; p37: GL Archive/Alamy; p38: Carlos Caetano/Shutterstock; p43: BASF Antwerp; p46: KilnBZ; p47: Africa Studio/Shutterstock; p51 top: NFPA Design, middle: JayTecGlass, bottom: sciencephotos/Alamy; p53 left: sciencephotos/Alamy, right: SciLabware; p54: Dorling Kindersley/UIG/Science Photo Library; p55: JayTecGlass; p56: Hurst Photo/Shutterstock;

Chapter 3

p61 background: zhu difeng/Shutterstock; p62: Pi-Lens/Shutterstock; p67 top: Andrey_Popov/Shutterstock, bottom: Fpdress/Shutterstock; p68: CSIRO; p69 top: 3Dalia/Shutterstock; bottom: Fablok/Shutterstock; p70 left: Andrew Lambert Photography/Science Photo Library, right: Lafoto/

Shutterstock; p76: Martyn F. Chillmaid/Science Photo Library; p79 top left: Foto76/Shutterstock, Right: Africa Studio/Shutterstock, bottom left: Paul Prescott/Shutterstock; p83: Andresr/Shutterstock; p84 Top: Peter Gudella/Shutterstock, bottom: Nfoto/Shutterstock

Chapter 4

p89 background: CERN; p97: 4science

Chapter 5

p102 background: haru/Shutterstock; p104: Tomacco/Shutterstock; p110: Brocken Inaglory; P111 right: Elena Vasilchenko/Shutterstock, left: Monticello/Shutterstock; p122: Dr P. Marazzi/Science Photo Library; p123: Lucky Business/Shutterstock

Chapter 6

p128 background Tom Grundy/Shutterstock, In-Text: CHAIYA/Shutterstock; p129 top left: Mangpor2004/Shutterstock, middle: Underverse/Shutterstock, bottom left: Olga Mark/Shutterstock, right: Anaken2012/Shutterstock; p130 left: Alexander Maksimov/Shutterstock, right: Paul Rapson/Science Photo Library; p133: Vanderwolf Images/Shutterstock; p134: Niloo/Shutterstock; p133 Top: Gopixa/Shutterstock, bottom: Fairy_Tale/Shutterstock; p137: Arthur Jan Fijałkowski; p138 left: Paolo Bona/Shutterstock; right: Trekandshoot/Shutterstock; p142: LauraD/Shutterstock

Chapter 7

p147 background Hank Morgan/Science Photo Library, in-text: Justin Kase Zninez/Alamy; p148: Oleg Golovnev/Shutterstock; p154: Payless Images, Inc./Alamy; p155 Left: Dikiiy/Shutterstock; Right: Ruhrfisch/CC BY-SA 3.0; p156: Ilya D. Gridnev/Shutterstock; p157: Brian A Jackson/Shutterstock; p159: Lucie Lang/Shutterstock; p161 top: Ggw1962/Shutterstock, bottom; John A Davis/Shutterstock;

p165 left: Hodag Media/Shutterstock, right: Carolina K. Smith MD/Shutterstock

Chapter 8

p171 background: Alexandra Lande/Shutterstock; in-text: Deymos.HR/Shutterstock; p172: Jack Cobben/Shutterstock; p174 Top: Lightpoet/Shutterstock, bottom: Luke Schmidt; p181 top: Public Domain; bottom: David Orcea/Shutterstock; p182: Ruzanna/Shutterstock; p183 top: Loraks/Shutterstock; bottom: Adam J/Shutterstock; p185: Adam J/Shutterstock

Chapter 9

p191 background Vladimir Nenezic/Shutterstock, In-Text: Dmitry Naumov/Shutterstock; p193: www.biophotonicsworld.org; p197: wk1003mike/Shutterstock; p198: Charles D. Winters/Science Photo Library; p200: chemicals-technology.com; p201 top: NASA/Public Domain, bottom left: KPG Payless/Shutterstock, bottom right: Andrew Lambert Photography/Science Photo Library; p203: Memorisz/Shutterstock; p204: NASA/Public Domain

Chapter 10

p209 background: Shaiith/Shutterstock, in-text left: Diana Taliun/Shutterstock, in-text right: Wutthichai/Shutterstock; p210: Pavel Vakhrushev; p214 top: Steffen Kristensen/Public Domain, bottom: Africa Studio/Shutterstock; p215: Sigur/Shutterstock; p216: Mountainpix/Shutterstock; p217: Andrew Lambert Photography/Science Photo Library; p221 top: Avarand/Shutterstock, bottom: Fusebulb/Shutterstock

Chapter 11

p225 background: thailoei92/Shutterstock, in-text: Marcelclemens/Shutterstock; p226 top left: Science Photo Library, top middle: CSIRO, top right: Matthias Zepper/Public Domain, bottom left: Alchemist-Hp, bottom middle: Matthias Zepper/Public Domain, bottom right: CTK/Alamy, far bottom right: Lmfoto/Shutterstock; p227 top: Sciencephotos/Alamy, bottom: 4science; p229 left: Mighty Chiwawa/Shutterstock, right: Robert Red/Shutterstock, bottom left: Public Domain; p231 right: Sciencephotos/Alamy, top left: Andrew Lambert Photography/ Science Photo Library, bottom left: E.R. Degginger/Alamy; p233 top: Filip Ristevski/Shutterstock, bottom: Public Domain, Agricultural Research Service; p234: Tony Hertz/Alamy; p235: Hrustov/CC By SA 3.0; p236 left: Hong Xia/ Shutterstock, right: Charles D. Winters/Science Photo Library

Chapter 12

p242 background: Andrey Armyagov/Shutterstock, in-text: Avatar_023/Shutterstock; p245: Andrew Lambert Photography/ Science Photo Library; p246: Cole Goldberg/Creative Commons; p250: Andrew Lambert Photography/Science Photo Library; p249: Cychr/Creative Commons; p251 left: Pavel Zelentsov, middle: Lilly M/CC By SA 3.0, right: Scilabware Ltd/ reproduced with kind permission;

p252 top left: Masa44/ Shutterstock, top middle: Paket/ Shutterstock, top right: Chepko Danil Vitalevich/Shutterstock, bottom: Photodiem/Shutterstock; p254: Tyler Olson/Shutterstock; p255: Dartmouth College Electron Microscope Facility/Public Domain; p256 top: Final Gamer/ CC-by-SA 3.0; bottom: Emilio100/ Shutterstock; p25 left: Aleksandr Markin Shutterstock; right: Marmaduke St. John/Alamy

Chapter 13

p263 background: Curioso/ Shutterstock, in-text: German Aerospace Center/CC-By-SA 3.0; p268: Cychr/Creative Commons; p273: Wellphoto/Shutterstock; p275 left: Vizual Studio/ Shutterstock, right: Emka74/ Shutterstock

Chapter 14

p280 background: Elnur/ Shutterstock; p283: PhilippN/ Public Domain; p288 Top: Peter38/Shutterstock, bottom left: Digitalreflections/Shutterstock, bottom right: Knot Nattapon/ Shutterstock; p291 top right: Bonzodog/Shutterstock, left:

Joefotoss/Shutterstock, bottom right: Neil Langan/Shutterstock

Chapter 15

p295 background: Izf/Shutterstock, in-text: Mario Roberto Duran Ortiz/ CC-By-SA 3.0; p297: Auremar/ Shutterstock; p298: Aleksandr Kurganov/Shutterstock; p299: Nancy Dressel/Shutterstock; p300: Vivergo Fuels; p303 left: Ivaylo Ivanov/Shutterstock, right: Andrew Lambert Photography/ Science Photo Library; p304 top: Baminnick/CC BY-SA 3.0, bottom far left, left, right and far right: Scilabware Ltd.; p305: top left, top middle, top right: Scilabware Ltd., bottom left and bottom right: Phasmatisnox/CC-By-SA 3.0; p308: Image Point Fr/Shutterstock; p309: Thum Chia Chieh/Shutterstock

Chapter 16

p314 background: mezzotint/ Shutterstock, in-text: Yenyu Shih/ Shutterstock; p318 left: Vallefrias/ Shutterstock, right: Coprid/ Shutterstock; p321: Nattanan726/ Shutterstock; p322: US Navy Photo/Public Domain; p326: Stuart Monk/Shutterstock